ARCHAEOASTRONOMY IN THE AMERICAS

ARCHAEOASTRONOMY IN THE AMERICAS

Edited by

Ray A. Williamson

A BALLENA PRESS/CENTER FOR ARCHAEOASTRONOMY
COOPERATIVE PUBLICATION

Ballena Press
381 First Street
Los Altos, California 94022

The Center for Archaeoastronomy
University of Maryland
College Park, Maryland 20742

Library of Congress Cataloging in Publication Data

Main entry under title:

Archaeoastronomy in the Americas.

(Ballena Press anthropological papers ; no. 22)
Selected papers presented at a conference held at Santa Fe, N.M. , 1979.
Bibliography: p.
Includes index.
1. Indians- -Astronomy- -Congresses. 2. Astronomy, Prehistoric- -Congresses. 3. Indians- -Antiquities- -Congresses. 4. America- -Antiquities- -Congresses.
I. Williamson, Ray A., 1938- . II. Series.
E59.A8A7155 520'.973 81-19147
ISBN 0-87919-094-9 (pbk.) AACR2

Printed in the United States of America.
1st Printing.

CONTENTS

TABLES

FIGURES

FOREWORD

by Travis Hudson

Ideally, every cultural reconstruction would reflect the totality of past human behavior, taking into account all of man's interactions with his natural and social environments, through technology, social organization, and ideology. As scholars concerned with understanding human behavior by studying man's material traces--architecture, utensils, tools, ornaments, art objects, etc.--archaeologists have readily accepted that the ideal is unobtainable. The problem arises from the incompleteness of the archaeological record upon which every cultural reconstruction is based. The agencies of time and destruction, and of incomplete data recovery (surveys, excavations, analyses) and reporting, have all contributed (along with other factors) to ensuring the deficiencies of cultural reconstruction. The problem is only slightly ameliorated when a culture becomes historically known and written records are more profuse and meaningful, or the culture is described ethnographically. Even with these data sources, significant gaps in understanding remain.

Realistically, the material remains in the archaeological record foster cultural reconstructions which emphasize human technology and its relationship to the natural environment. A typical reconstruction of a prehistoric society, for example, reflects the relationship of an archaeological assemblage (artifacts, animal and plant remains, pollen, context, etc.) to paleoecological parameters. This results in a better understanding of such things as what individuals ate or how the food was procured and prepared, but little understanding of their social structure or ideological beliefs. Unfortunately, it is far more difficult to reconstruct a language, kinship system, or a shamanic postulate from a potsherd or projectile point than it is to reconstruct (within time-space parameters) the process of a tool's construction and relate its physical function to other physical remains in the assemblage.

Cultural reconstructions based upon technology and the natural environment have thus had a major place in our efforts to understand the past, and they will continue to assume a major role. But does the importance and dominance of technology in the archaeological literature mean that social and ideological aspects of life were less important in human achievement? We have come to *assume* somehow that man is the product of his technology, but in actuality technology is the product of man, for it must have its origins in the human mind before it can receive creative expression, use, or meaning. The achievements of civilization, with its science, mathematics, writing, monumental architecture, art, and so on, are not the physical expressions of human technology alone, but the *ideas* from which they were derived as well.

It is in the ideological realm of past human behavior that the archaeological record has been most silent; silent, that is, until the recent advent of a new and fledging field, archaeoastronomy. The archaeological record which most concerns scholars in this field is that which has been traditionally ignored in cultural reconstructions (e.g., rock art, alignments of structures or cairns, placement of windows and doors in buildings, etc.). These material remains reflect man's articulation with his natural, celestial environment. Archaeoastronomy's concern is with understanding the importance of the sky for human ideas and achievements. Not surprisingly, initial results are beginning to demonstrate not only our ability to deal with an ideological system within the archaeological record, but also the relationship between that system and other prehistoric achievements in science, mathematics, metrology, and so forth. Of course, those of a materialistic persuasion argue that all of these intellectual achievements, and particularly those achievements related closely to the emergence of civilization, were again a product of technology. The argument (or better, assumption) is that as food production was achieved, astronomy and calendrics were invented as a response to the new need for regulating and controlling crops or flocks. But given the many gaps in the archaeological record concerning the ideological dimension of man, especially hunters/gatherers (the non-food-producers of the ancient world), can we as scientists continue to accept an assumption based not upon evidence, but upon a better-known hypothesis as to the course of man's technological achievements?

Recent ethno- and archaeoastronomical studies of California's hunters/gatherers indicate that the time has come to discard the theory that the production of food is a basic requirement for the development of complex astronomies and calendrical systems. Not only were Native Californians keen observers of the sky who used their information to regulate socio-political and ritual behavior, they were also time-keepers in a manner which reflects significantly upon the very origins of writing and mathematics. In much of northern California, for example, a knotted cord was used to keep track of days, lunations, and solstices, a cord whose obvious counterpart is the *quipu* of Inca Peru. What perhaps stands out even more is that Pomo time-keepers, using bundles of sticks to keep track of calendrical data up to 512 years old, identified the bundles with increasing numerical values, a sort of positional notation which had its counterpart in the Maya long count. While the California data do not directly relate to the Maya or Inca, it should be evident that the accomplishments of these hunters/gatherers could well have been preceded by similar ones in the Late Archaic period of Mexico and Peru, accomplishments which were part of the basis of the great civilizations which followed.

Just how old astronomy and calendrics in the Americas may prove to be is a question for future research. What is important at this point is that man, regardless of his technological achievements, has continually devoted time to thinking and wondering about the universe, and has drawn inspiration from his observations that have served to shape his concepts of gods and religion, of life and afterlife, and of supernatural power. The importance of these ideological constructs is undeniable for a richer and more complete understanding of a past way of life. But until it is recognized more fully that the spade is less likely to contribute to this understanding than the theodolite, cultural reconstructions will doubtless continue to emphasize the importance of technology. The various papers in *Archaeoastronomy in the Americas* reflect a positive endeavor by the many contributors--archaeologists, astronomers, and humanistic scholars--to advance our recognition of this fact; hopefully it is not far off.

PREFACE

by Ray Williamson

In reflecting on archaeoastronomy and its place in the intellectual world, it now seems obligatory to remind ourselves at least once of Sir Eric Thompson's remark that "Maya astronomy is too important to be left to astronomers." It is well that we do so, for it is hard for us to accept the fact that western science is as much a product of western culture as Bach's *Art of the Fugue* or Durer's *Melancholia*. Astronomy, even scientific astronomy, exists in the milieu of politics, fads, and fallacies; i.e., in a cultural context. The data astronomers gather may perhaps be timeless, but their interpretation and use within society are not. As Remington reminds us in her review of Mesoamerican astronomy, we have to reach well beyond our own cultural bonds and deep into an alien worldview to even begin to understand the role of astronomy in traditional New World cultures.

Judging from the variety of disciplines represented at the Santa Fe conference, the participants have heard Thompson's plea and responded. One wonders, though, whether they might actually have heard a variation on Thompson's message: "New World astronomy is too important to be left to just a small coterie of archaeoastronomers." Approximately 150 persons attended the Santa Fe conference, which was sponsored by St. John's College, Santa Fe, and the National Endowment for the Humanities. Participants from a variety of disciplines--anthropology, archaeology, art history, astronomy, geology, history, and mathematics, to name a few--presented a total of 40 papers detailing their archaeoastronomical studies in North America, Mesoamerica, and South America.

In the present volume, 25 authors contribute the fruits of their recent research. In addition, three authors summarize the status of archaeoastronomy in the major geographical divisions. The volume's first section, however, is composed of a group of position papers on archaeoastronomy as a field. As New World archaeoastronomy has developed and matured, questions pertaining to archaeoastronomy's stature as a research topic have surfaced from time to time. Are we doing the right things? Is archaeoastronomy a fad? Should archaeoastronomy be institutionalized? Several of us had already begun to address these questions formally, most notably at a day-long symposium at the Smithsonian Institution (Williamson 1978) and at an evening seminar in Mexico City (Young 1979a). It therefore seemed particularly appropriate, at a time when many of the scholars who had studied New World ethno- or archaeoastronomy for a number of years were to be together in Santa Fe, to organize a small symposium to think through some of these questions. In order to provide a context for discussion, I asked each symposium participant to address a specific question; this is reflected in his position paper. Aveni agreed to chair the discussion, Young served as rapporteur (Young 1979b), and Krupp presented his impressions of the symposium and of the conference on the last day of the conference. His remarks (which are included here) ably captured much of the flavor of the pre-conference Symposium and of the conference as a whole.

In listening to the conference papers, and even more in the editing, I was struck by the multiplicity of ways the cultures of the Americas have chosen to reflect their interest in the sky. I have emphasized this in my review paper on North American astronomy, but it is a valid statement for the other two geographical regions as well. Even within the same latitudinal bounds, cultural expressions of this interest vary considerably. There are, however, certain important similarities. For example, consider the following statement:

> August 3
> The people are trying to have the Snake Chief watch the sun very closely, and so to call out at the right time. Sometimes, if the Snake Chief does not watch the sun right, they dance early, then it freezes early too. This is the reason why they must try and watch the sun very closely. Also they have to watch the crops [Crow Wing 1921].

To me, this statement from the notebook of the Hopi Indian Crow Wing epitomizes Native American astronomy--astronomy, ritual, and agriculture are all intimately related. Lacking an accurate calendar and effective ceremonies, the crops will fail. But as McCluskey warns us, native calendrics must not be thought of as unvarying or static. His study of the changing Hopi calendar provides an object lesson for ethno- and archaeoastronomer alike--the calendar is subject to change according to the needs of society. In another paper, Farrer and Second provide an exciting insight into the manner in which astronomical notions may be integrated into a culture's art, ritual, and indeed its entire world view. For the Mescalero Apache, the star motif assumes a central ceremonial role. Their Pueblo neighbors to the north strongly emphasize the sun and solar ritual. But what about the stars? In the third of this trio of papers on Southwestern historic practices (i.e., ethnoastronomy), Young and Williamson argue that Pueblo interest in the stars may be stronger than we had thought; the Zuni recognize a constellation, The Chief of the Night, which extends across the entire sky--the chief never sets!

In an examination of an earlier period in the Southwest, the Autrys and Evans and Hillman demonstrate how broadly conceived the architectural manifestation of calendrics can be within a narrow geographical area. Like the Medicine Wheels a thousand miles northeast, Zodiac Ridge reminds us that an arrangement of unmortared stones make a perfectly effective calendar device. Both it and Casa Grande, that remarkable adobe building north of Tucson, show evidence of alignments to the extreme lunar stand-stills. Should these initial data prove persuasive upon further examination, they will be the strongest evidence we yet have that the Indians north of Mexico tracked the horizon motions of the moon in addition to watching its phases.

The far West also supported a wealth of native astronomical lore and practice, much of which has only recently been discovered (or re-discovered). Hedges reports on two California sunwatcher's sites in Kumeyaay territory, which apparently integrate both natural horizon-features and man-made stone features to mark the winter solstice position.

Five thousand kilometers east, in New England, carefully wrought stone chambers, numerous standing stones, and other artifacts suggest to some the presence of an earlier culture. Considerable controversy surrounds these remains. Dix and Mavor present the results of their meticulous survey of one such assemblage in central Vermont and demonstrate that alignments do exist; whether they represent the observational intentions of a pre-sixteenth century culture will require further archaeological study of a more traditional sort, however.

Archaeoastronomy has also reached the middle of this continent, with predictable results--the Indians of central Ohio also seem to have tracked the horizon motions of the sun in order to organize their ritual calendar. Heilman and Hoefer suggest that portions of the "Incinerator Site" in Dayton were deliberately laid out in accordance with the local calendar.

After all of our apparent success in proving the hypothesis that most Native American groups regularly watched the sun, it seems appropriate to sound a warning note: we may run roughshod over good science in our enthusiasm. In her comprehensive study of the southeastern mounds, Daniel-Hartung warns us of this pitfall and urges us to use more rigorous methods. She also raises the question of Mesoamerican contacts with North America. In both the Southeast and the Southwest, there is ample suspicion of extensive cultural contact between north and south, but little proof; the astronomy doesn't resolve the issue, nor may it ever provide proof. After all, the celestial phenomena are there for everyone to see and interpret. Given the apparent diversity of interpretations of the data, north and south, it may ultimately prove impossible to separate indigenous and imported constructions. Nevertheless, archaeoastronomers in all geographical regions will want to keep informed about each other's work; that is part of the reason for assembling papers concerned with a variety of geographic contexts in one volume.

The Mesoamerican section leads off with Remington's enlightening personal view of Middle American archaeoastronomy. She reminds us of a fact that many of us are apt to forget--that archaeoastronomy really began there nearly 100 years ago in the first attempts to understand the Mayan calendar. As Milbrath demonstrates, calendric problems are still important; she presents us with an art historian's view of several pages of the Madrid Codex and illustrates just how complicated the arguments can be. As we so often hear, the Mayan calendar was highly accurate, but it was also subservient to another aspect of Mayan intellectual interest--numerology. Carlson ably summarizes Mayan numerological interests and shows us how they might have affected the Mayan calendar. Collea's analysis of some Mayan hieroglyphs demonstrate that they convey astronomical data in a variety of ways; not only do they contain specific data about astronomical dates and cycles, they incorporate a group of specific elements called celestial bands about which little is known. Collea has collected a number of examples of celestial bands and has analyzed them for their iconographic role in the hieroglyphs. Similarly, Lamb looks at Maya celestial lore contained in western writing. He has compiled a rich list of references to the celestial objects and to the astronomers who tracked them from colonial and later Maya dictionaries.

It is striking that the Mesoamerican section contains only a single paper that discusses building alignments. This seems to be less a reflection on the lack of importance of building orientation than it is on the state of Mesoamerican archaeoastronomy. It is time to reflect on the meaning of the massive body of data which has been already collected. How do these findings correlate with other cultural knowledge about the region? Malmström, in looking at the group of alignments known as the 17° family of alignments, has attempted to understand the genesis of these puzzling alignments as a reflection of the diffusion of sacred astronomical lore from Izapa to other areas in Mesoamerica. However acceptable his or other competing theories will prove to be in the course of time, it is clear that an impressive foundation for such testing has been laid by the work of the last decade in Mesoamerica.

The fact that this volume contains a paper reviewing our understanding of South American native astronomy truly illustrates just how far archaeoastronomy as a whole has progressed since the Colgate conference in 1975. How different the sky appears near the equator! As ethnologist Urton explains, this special view has engendered a cosmological vision of the world which is quite different from the worldview of temperate climes. And in companion papers, astronomer Aveni and anthropologist Zuidema team up in print to report on their recent joint studies on the *ceque* system of the Incas. They show that though they represented it quite differently from their neighbors to the north, the Incas also had a sophisticated astronomy.

Two final papers close this volume. In the penultimate one, Hicks presents his work on the Irish henges and argues that it may be possible in Ireland to use Irish folklore to explore the possible astronomical associations of the henges; thus, in a way, New World archaeoastronomy serves as a model for Old World archaeoastronomy. The last paper in the volume discusses the role of archaeoastronomy in teaching. This is a particularly fitting closure, for it is ultimately in the classroom that archaeoastronomy will retain its vitality and strength as a way of understanding both native peoples and ourselves. Smith suggests new ways to use the results of archaeoastronomical studies, and asks several penetrating questions of archaeoastronomy. It is worth noting that students have a great deal of interest in the subject; six of the twenty authors in this volume were students at the time of the conference. Rather than slavishly following in the footsteps of older scholars, they are forging new ground, applying their own special interests in new ways to the task of understanding traditional New World astronomy.

In looking back through the contents of this volume, I am impressed with the breadth and depth of these contributions. There are still many omissions, in both geographical coverage and subject matter; perhaps the most striking such omission is Chamberlain's recent study of Pawnee starlore (Chamberlain n.d.), which proved much too long for inclusion here. It will be published by Ballena Press as a separate monograph. Many people contributed their help and hard work to the production of this volume: first, the authors whose carefully articulated research makes this a highly readable volume; and second, the anonymous reviewers who read and commented on the articles, both those which were accepted for publication and those which were not. Jean Monroe, whose superb organizational skills and editorial eye made the work of editing this volume manageable, is due special thanks, as is Anthony F. Aveni, without whose encouragement and support I might have given up too soon. William Dunham, retired Vice President of the St. John's College, Annapolis campus, offered much help and advice in the early planning of the conference. Jane Young ably prepared the index. Finally, I wish to thank the National Endowment for the Humanities for their financial support of the Santa Fe conference and the preparation of this volume, and Sarah Williamson for her help in preparing the bibliography.

1 Some Thoughts on Archaeoastronomy Today

John A. Eddy

It is heartening to see so much interest in the area of archaeoastronomy today. There is a promising side to this, and a somewhat frightening one as well. The heartening fact to me is the entrance into the field of more scholars and students and practitioners of anthropology. I suspect the biggest revolution in our hyphenated field in the last few years has been this shift of emphasis in who is active in the area. Traditionally archaeoastronomy has been the province of astronomers and engineers. Except perhaps in Mesoamerica, archaeologists have played the role of what might be called curious or cautious hospitality. That is the nature of archaeology and we should be glad it is so conservative a field. Now we see the open interest of archaeologists in looking at the astronomical side of things; there are conferences scheduled by them on the subject, graduate students working in the area, provisons in salvage archaeological projects, and so on. This is surely the real hope for the field: a genuine two-way interaction between astronomers and archaeologists in studying certain things, like the collaboration of Aveni with Zuidema or of Williamson with Ellis. This meeting in Santa Fe is, I think, a living example of this mutual interest and collaboration.

I look upon this stage, however, not as a final goal, but as the second of three steps toward it (the first being the sometimes flamboyant knocking at the door by astronomers, holding up this or that to get the archaeologists' attention; it is in this first category, by the way, that I would put my own work on the Wyoming Medicine Wheel or Hawkins' work on Stonehenge or the far earlier studies of Norman Lockyer and others).

In the second step, we will develop criteria for believability, and I am sure (as Aveni has for so long urged) we astronomers will come to the realization that to do anything useful in this area we must first and foremost understand the cultures of the people whose astronomy we surmise, and we will find out that it will take more than the superficial interest of two diverse scientists to make real progress.

Some will use this second phase, as they are now, to try to enforce their own particular philosophy of science or how to do science, and that is all right. I think we have always been a little overbearing in this regard in astronomy. I am amused at present, however, by the efforts of some who would like to teach us again what was known so many years ago as the "scientific method": the necessity to construct a hypothesis and so on. Whenever I hear this I am reminded of the *New Yorker* cartoon in which a helmeted astronaut steps out on the moon. His landing there has interrupted two alien beings, a man and a woman, who sit beneath a tree in which there is a snake and a single apple for which the man is reaching. "Stop, stop!" are the words of the landing astronaut, who is, of course, familiar with the scene and its inevitable consequences. It's not quite the same, of course, and I don't want to say "Stop, stop" to anyone. I would say "Wait, wait," for I think it is an unnecessary formalization of the process of discovery--a thing so fragile and elusive that it can hardly be defined anyway. Does anyone really want to put it in the straight-jacket of a set of parliamentary rules, or to saddle it with a pedantic load of step-by-step instructions? I don't think Thomas Edison would have found anything of use at all had he been strapped and bound to a cradle-board of "scientific method" terminology and rules at an early age. I should like to know what Einstein would have said if he were asked what he thought about the school-taught "scientific method." I will wager that his eyes would have twinkled, and if he were really frank, he would have admitted that like a lot of other things we teach at school it's all right in the school room but it has no business in the real world; this ends my tirade.

I would like to list the three steps to perfection which I would like to see archaeoastronomy follow. The first is already history, and the second is where we are now, with mutual collaboration. But the third and final step, I would hope, would be the complete assimilation of archaeoastronomy into the fabric of anthropology; not as a separate, recognizable discipline requiring its own journals or formal organizations, but as one of the tools of anthropology whose strengths, and weaknesses, were well understood by most of the regular practitioners. That is why I like to spend my time in archaeoastronomy teaching anthropology students; I'm sure many of us feel the same way. And it is surely obvious that the final assessment or verdict of whether this or that site is astronomically aligned, or whether it is important, or whether this or that document is or is not an ephemeris, must reside with those who eventually will pay for the consequences. And that is surely the anthropologists, not the astronomers.

In the present phase and the one before it, we have played the enviable role of consultants who move in, make declarations, and then move out--leaving behind a sometimes bewildered customer who has to live with our advice, or at least has to make up his own mind about whether it is worth taking or not. It is my view that for the most part archaeoastronomy (if by this we mean the study as limited to pre-literate peoples) can give only restricted service to modern astronomy or astrophysics, and that statement comes from one whose principal research today involves using old records for modern astrophysical knowledge. I have found that the really useful and critical records for the most part come from historical times, which I think is generally regarded as outside the area that we discuss here at Santa Fe. My point is that we, as astronomers, are not so much partners in archaeoastronomy as we are salesmen, knocking at the door trying to sell our product, albeit altruistically--a product that we understand and, therefore, know will be useful to the customer. But our anthropology partners are also the customers and if they have bought the product, they're the ones who have to live with it and to make the payments after we leave. I have never ever blamed them for reluctance, even when I knew that my own stuff was above reproach!

That is why I would like to see the day come when there are no archaeoastronomers--at least none operating outside the general field of anthropology. There should be some, indeed, but they should be in the middle of the field in which they work and through which they can make useful contributions, and by which they can be criticized and helped and tested and evaluated. I would hope that we won't even need the name "archaeo-astronomer," any more than we now need names for archaeobasketologists, or archaeobone-fleshing-tool-ologists, and so on. If there were in the future archaeologists trained narrowly in astronomical techniques, I would not trust them either, for one always wants to sell his own expertise.

This is the goal toward which I would push archaeoastronomy at present. I think we do need this sort of organization now and we do need a newsletter of the sort that Carlson and Williamson have done so much with. I do not think that we should ever have our own journal in the real sense of a journal; I have opposed this in other splinter fields as well. The reasons must surely be clear to all of us. I am not a strong believer in the "scientific method," but I have the greatest faith in the "method of science" as it has been traditionally practiced, subject to checks and balances and the often painful and sometimes unfair slings and arrows of referees and all of the other things that make conventional, broad science move so slowly and conservatively. A specialized journal, be it in sun-weather relationships, or archaeoastronomy, or UFOs, intentionally pulled away from the mainstream of the fields it serves can only serve to isolate its practitioners and to make them ingrown, and to a great extent hide their product from real criticism. I want to see papers in archaeoastronomy kept in wide-field journals and in the traditional journals of anthropology; and if they are not accepted there, or if we are told they are too speculative or too long or too mathematical, etc., we should look into ways of making our case more convincing--but not start up our own splinter journal that will only fragment and split apart what now looks like a promising if delicate marriage.

The point that frightens me about the current state of archaeo-astronomy is one which concerns others too: that it has become a current fad, a subject that newspapers and magazines and book publishers are always hungry for. We should all feel blessed that archaeology and astronomy have both enjoyed such a wide and popular appeal among laymen; they surely draw more popular readers than does, say, nuclear physics or chemistry or structural engineering. And as I have pointed out before, their confluence in archaeoastronomy is a sure winner, as all the book publishers and magazine editors know. And that, of course, is the danger: how do you tread a path of honesty in an area so diffuse and with so few clues, when you know the hungry clamoring of the public for something exciting? I think that is more of a strain than most scientists are used to working under. I don't think any of us intentionally steps over that line, but I am sure it has an effect at a subliminal level. I am sure that when faced with the kind of imponderables and the maze of statistical arguments that characterize archaeoastronomy, we are always left with a leap of faith or some kind of judgment call. I am one who has yet to be convinced that man can do really objective science; I'm not sure we have proven that point yet. The history of science is certainly filled with a record of mistakes, and some of them came from rushing into print to beat someone else to a point, or in modern times to get funding for one thing or another, or to secure a job, or to make or to sustain a reputation. We like to think that we are objective creatures, but are we really in any way? I have proposed that science might be much better done were all papers published anonymously, but that could be defeated were one paid (as would then seem necessary) by the word, or by the number of articles turned in. It just might be that science is best done by ascetics, or by the impossible combination of very young scholars, fresh and bright and honest, who also have the security of tenure; or maybe by computers.

It is this sort of philosophy that leads me to take an ever more conservative approach toward archaeoastronomy. We do deal with extremely faint and threadbare facts. We lean when necessary on statistical arguments, and we all know how fickle they can be. We have traditionally allowed ourselves a wider latitude of conjecture than we would apply to our own more restricted fields. Somehow, once we have put that hyphen in between the words, we feel that it is all right to speak as an expert in either field. And then there is the real but terrible temptation of the book publishers, the television producers, and the stream of offers to speak or travel here or there, that tempts us while we are working with these tiny shreds of evidence. What is remarkable is that we have reached phase two of progress in this field. We should all be proud of that. But now that we have got the public's attention and the far more meaningful attention of a segment of the working community of anthropologists, we are more than ever obliged to clean up our act and to keep as fastidious and conservative and careful as we can possibly be. We need much more criticism from within and from without. We need a lot less of patting each other on the back and of publishing, in one way or another, everything that bears the word archaeoastronomy. We need competing theories from within our own field to explain certain things. It hurt to hear Ovenden's (1979) criticism of my work with Kehoe on the Moose Mountain Medicine Wheel, and to read a competing theory; then I realized how wrong this was and how different from what I thought my reactions were in other fields of my endeavor. I realized that I had become "territorial," and that (though common in science) is a thing we should all avoid.

What Carlson and Williamson have done at Maryland and with the *Archaeoastronomy Bulletin* and with meetings like this one in Santa Fe has been to make possible the kind of testing of ideas, and criticism, and the open exchange of ideas that the field needs. Perhaps that is what the new Historical Astronomy section of the American Astronomical Society would also accomplish. What we don't want to do, however, is to institutionalize or fragment the field, or start a new advanced degree, Doctor of Archaeoastronomy, or make ourselves too much into a kind of fan club, or into a band of the disinherited who think of themselves as persecuted or deprived (as has happened, for example, with much of the UFO work).

2 Archaeoastronomy Today

Anthony F. Aveni

Listen to some of the volleys we've hurled at one another during the turbulent decade and a half since the resurrection of archaeoastronomy:

> Confidence of this order is a rare and precious possession, beyond the grasp of most scientists [Atkinson 1966:212, to Hawkins].

> Some of the dates he used had, unfortunately, been wrongly read. I was able to demonstrate that the incorrectly read dates produced a higher percentage of astronomical phenomena than those correctly read [Thompson 1974:96, to Ludendorff].

> His result here is wrong all through. The probability quoted is wrong; the method of testing the hypothesis is wrong, and the restriction of the possible sight-lines to 50 is wholly inadmissible [Atkinson 1966: 213, to Hawkins].

> In the light of such deductions, which one can only designate as highly intemperate--one is inclined to say that Maya astronomy is too important to be left to the astronomers [Thompson 1974:97, to astronomers].

> I did not at that time dream that our greatest prehistoric monument would actually be taken over by astronomers and recognized as a celestial observatory and calculating machine [Hawkes 1967:174, to Hawkins].

> She is a silly little fool--perfectly inexcusable. She ought to have her pants taken down, be turned over and spanked for not doing her homework [Movius in Trotter 1972:126, to Rosenfeld].

> Three of Kelley's tables, though attractive as art, fail to convey information to this astronomer [Mulholland 1978:1197, to Kelley].

> Had astronomers interested in Maya astronomy such a background (i.e., Maya culture and history), we would be spared assertions that the Maya *calendar* was in full swing before 3500 B.C., when in fact, agriculture in the New World had hardly got under way . . . [Thompson 1974:97, to astronomers].

> I object on general principle to the notion of giving full professors well paid vacations to persue [sic] arcane interests . . . [Anonymous reviewer (1976) of one of my NSF proposals].

Such statements, though amusing to listen to, reflect the misunderstanding and unwillingness to engage in cross-disciplinary cooperation which characterized the early development of our field. Our own traditional disciplines, after all, were separated by what C. P. Snow (1959) characterized as a "gulf of mutual incomprehensibility." Too often we were suspicious of how a foreign disciplinarian approached our subject matter, a domain we regarded as our own private turf. Our wars became territorial and petty with no element of reason and understanding entering the deliberations. Precious time and energy were wasted.

The establishment of the paradigm of world archaeoastronomy finally occurred with the general acceptance of the relative sophistication of British Megalithic astronomy in the journals, and especially as embodied in the recent works of Wood (1978) and MacKie (1977), who attempted to tie astronomical and geometrical findings to our knowledge of the society and historical chronology of Western Europe in the Bronze Age. Simultaneously, there occurred an outpouring of archaeoastronomical research papers from the Americas. The wave of evidence from the written and unwritten record which had built up for more than a decade had finally broken. Accordingly, the Time-Life yearbook for 1977 featured a lengthy presentation on Native American (particularly North American) archaeoastronomy by George Alexander, Science Editor of the *Los Angeles Times*. A *New York Times* (1978) science editorial called our field a "growth discipline." The number of lectures delivered by archaeoastronomical scholars to educated audiences at respectable institutions increased by an order of magnitude. Even the American Astronomical Society saw fit to sponsor a section on archaeoastronomy at its Mexico City meeting in January 1979.

Today we find ourselves in the undertow of a new wave of interest in the interdisciplinary study of the astronomies of ancient civilizations which promises to attract an even larger cadre of scholars into the academic melting pot. In my view, recent developments portend both positive and negative scenarios for the future of our work. Accordingly, I would like to deliberate some of the issues openly and candidly with you at this conference.

FINDING A HOME

It has been suggested that archaeoastronomy is a discipline which will attain respectability once it finds a home (i.e., once its methods, procedures, and results can be couched in the framework of an established discipline such as cultural anthropology, archaeology, history of science, or astronomy). But is the establishment of the Center for Archaeoastronomy at the University of Maryland a statement from one scholarly faction that the discipline ought to be left to develop on its own as it did in Great Britain for so long? On the other hand, should we consider further the idea of establishing an Archaeoastronomy or History of Astronomy section within the American Astronomical Society? Our own studies in Mesoamerican and Andean archaeoastronomy have begun to take root and grow quite naturally in the field of cultural anthropology, where great interest and excellent channels for feedback and evaluation of ideas prevail. The meetings of the American Anthropological Association and International Congress of Americanists have provided receptive audiences for the evaluation and dissemination of new ideas. When we speak of the Maya "cosmovision," their attitudes about the heavens and their methodology for observing and recording time intervals, we are dealing with people and their cultural history, not simply the natural world. Given the subject matter we study, it seems only appropriate that while most of us still define *ourselves* as astronomers and scientists, we should build our sturdiest bridges across the gulf separating us from the interested community of cultural anthropologists.

HYPOTHESIS FORMING

Should archaeoastronomers simply gather data, leaving them for a better informed audience to interpret, or must they formulate a problem and define its scope before undertaking an investigation? This question was debated in an exchange of letters to the editor of *Science* by Reyman and Eddy (1975). Re-reading the literature of that debate, I am led to make two salient points, one on behalf of each of the proponents. (1) I agree with Eddy that archaeoastronomy can never be raised to what Mulholland (1978:1198) calls the "critical mass needed for significant progress" unless organized data-gathering occurs on a widespread basis; i.e., unless a methodology is defined and practiced universally. As Eddy has suggested, the early history of our discipline was, indeed, characterized by an overindulgence in speculation, and not a lack of it. In the past five years we have done much to correct this situation. (2) On the other hand, Reyman and Thompson (see above) are correct to suggest that when individuals who are totally misinformed begin to dabble in another discipline and extend their ideas prematurely to the printed page, often they can create falsehoods which propagate through the literature with interminably long decay times. To be honest, I think we have too many part-time archaeoastronomers and not enough scholars working full-time with unabated commitment to these very serious studies. While many of us have demanded that anthropologists and archaeologists learn astronomy (a short-term proposition), we have not advocated strongly enough the more difficult maxim that astronomers ought to become anthropologically oriented. I believe the former involves learning to use a *tool*, while the latter requires the development of an *attitude*. We must become more conscious about the relationships ancient astronomers sought between the heavens and agricultural cycles, meterology, and environmental affairs in general. My own opinion on this issue may be extreme and perhaps a bit Boasian, but I believe our discipline is secure enough for me to suggest that rather than seeking a piecemeal investigation of a problem, it is better to do no investigation at all. We must seek out explanations no matter what the cost in time. I would prefer to see one thorough ten-page paper on archaeoastronomy appear in a journal than ten one-page papers.

CONFLICT WITH THE HISTORY OF SCIENCE

While I have sought to link archaeoastronomy more strongly with cultural anthropology, I also consider as one of my goals the funneling of the results of our studies into the mainstream of the History of Science, a field which until now has dealt only with the Old World and the written astronomical record. I believe this is an effort shared by such traditional historians of science as Gingerich and Hoskin. The effort has already been undertaken by Hoskin, editor of the *Journal for the History of Astronomy*, to bring out a special issue on Native American Astronomy in the near future (*Archaeoastronomy*, Supplement to the *JHA*, No. 1, 1979). On the other hand, if we wish to make such a move, we have a lot of educating to do, for that community of scholars often attempts not only to interpret the "science" of foreign cultures, but also to evaluate and judge it by western standards. Many western scholars do not realize that the Maya tendency to avoid the lunar nodes, sidereal periods, and the true right-angle or true arch in their architecture does not indicate that they only climbed to a lower position on the intellectual ladder. Still, if our common goal is to understand human as opposed to western intellectual achievement, at least some of the knowledge we archaeoastronomers disseminate belongs in the realm of the History of Science literature.

THE EDUCATION OF FUTURE ARCHAEOASTRONOMERS

In the past few years I have seen many excellent undergraduates matriculate through my own and other quality institutions who possess a deep interest in pursuing a career in archaeoastronomy. I advise them to enroll at a graduate institution in one of the established fields mentioned earlier, provided the graduate institution exhibits a sympathetic attitude for programs which may be at least partially self-designed. In practically no cases do I find total satisfaction on the part of the students with the existing programs. While I firmly believe that the edifice of one's graduate education in archaeoastronomy must be firmly implanted within a traditional discipline, I still suggest that the graduate institutions have been too rigid and inflexible about how they treat these incoming students, many of whom I regard as scholars of considerable ability. Nevertheless, what we are dealing with here is an issue in the philosophy of education, a field which we find moving through a rather conservative period, after the chaotic revolution of the 1960s. Unfortunately for these students, our own revolution in archaeoastronomy has come at the wrong time. One of the roads to the future we must carefully pave is that of communication with the graduate schools. We should make them aware of the advantage of acquiring the special breed of student who already has done some thinking about his future in a growing interdisciplinary field and wishes to exercise a measure of control over that future. I have the feeling that if we de-emphasize the fact that these students wish to become "archaeoastronomers" in our letters of recommendation and, instead, stress that they desire to apply the methodology and attitudes of, say, anthropology to the study of ancient astronomy, our students will receive a more sympathic attitude. In the past few years I have had the opportunity to read a number of quality papers by very able graduate students whose work is often neither appreciated nor encouraged by their academic advisors because it is too far out of their advisors' realm. I believe that any academic advisor ought to be concerned enough about the direction of his student's research to see to it that the work is evaluated. We must encourage the established graduate departments to permit advisors from outside institutions who specialize in archaeoastronomy to sit on thesis committees relating to our field.

I realize in this brief position paper that I have only raised a few issues and have surely resolved none of them. But, hopefully, I have provided at least a platform for a serious dialogue in which we might engage these few days. My comments about where we stand today have been made solely out of a deep interest which I assume we all share--that of exploring and disseminating information about the ancient past in the field of cosmology as viewed through the eyes of the cultures of Native America. We have come very far in a relatively short period of time. Let us hope that when the afterglow of the limelight fades away, we can remain worthy of the credit which is so abundantly being heaped upon us.

3 Anthropology and Archaeoastronomy

R. T. Zuidema

My first attraction in Andean culture was to kinship and its concomitant systems of social organization. Astronomy and kinship clearly have something in common. All cultures have to deal in kinship with the same limited set of elements: father, mother, and children. Other, wider, kin relationships are built on these primitive ones, like that of grandparent or uncle and aunt. Although we can define what kinship is about and what kin terms are, and what the limits of such a system are, there are probably no two cultures in the world that have exactly the same type of kinship system. We can classify kinship systems and we recognize some major types. Here, more than anywhere else in anthropology, we have advanced in our theoretical understanding, since the rigorous analysis that is possible is similar to that in linguistics. In fact, the first Spaniards who analyzed American Indian cultures recognized this similarity; they often included a chapter on kinship in their grammars of Indian languages. Finally, kinship provides a good first approach to other parts of culture that are far more difficult to analyze theoretically such as, for example, mythology. The frequent use of kin terms allows us to set up a firm primary frame of reference.

The parallel between the study of different cultural systems of astronomy and that of kinship is obvious. All people observe the same primitive elements in the sky: sun, moon, and stars, distinguishing among the latter the planet Venus. Day and night and the seasons are recognized as related or as derived elements. On this primitive set of concepts peoples have built astronomical, cosmological, and calendar systems with as much variation as they have with kinship systems. We should be able to analyze the former systems, therefore, with the same rigorous and theoretical detail as the latter.

On the one hand, it has been said that the origin of science is to be found in the astronomy of the Babylonians and the Greeks. On the other hand, it has been argued by anthropologists that Australian cultures, with their intricate kinship systems, developed an equally abstract interest in models (drawn in the sand or made with sticks and pebbles) in order to explain to one another variations in their kinship systems. I would argue that we could go a step further: an interest in kinship, as well as astronomy and time have induced many peoples in the world to speculate abstractly and theoretically about the systems that they have built about both types of phenomena. Apparently the possibility for abstract speculation is not dependent on the level of material or political development reached by a specific society or culture. Bordieu (1977) and Turton and Ruggles (1978) pay similar attention to kinship and to astronomy or calendars.

My attraction to astronomy was triggered by the system of 41 *ceques* (radial lines) in Cuzco, leading from the Temple of the Sun and continuing in the same direction beyond the immediate horizon. Having the Temple of the Sun as the center would induce us to think of the system as primarily astronomical, involving observations of risings and settings of the sun, moon, and stars. However, as Aveni and I have shown, the astronomical system is in a sense of a derived nature, since horizon points were observed for astronomical purposes from various points in Cuzco, and not solely from the Temple of the Sun.

The Inca valued horizon points and directions and classified them according to subjective criteria, taking into consideration social rank and position. The *ceque* system--the system of lines (*ceques*)--was one of social classification: moieties, clans, etc. But one purpose of allocating people to any of these groups was for an equal distribution of tasks throughout the year, in other words for calendrical purposes. The *ceque* system was a system of total cultural classification in a social system conceived of as a local unit. The chronicler who gave us the description of the *ceque* system, with all the *huacas* or sacred places that were organized by it, also defined most clearly this total cultural purpose:

> Many other men and women are in charge of the fixed guacas (sacred places) of which I have given a special account in the general Map of Cuzco, which is common in all parts of the kingdom and which contains all the places that distinguish themselves in anything around the village or town, up to the mountaintops if it is difficult terrain. They call these places (on the horizon) *apachetas*. Other places (that they worship) are big stones and all the (water) wells and sources of water (of rivers and irrigation canals), or flat places on a slope or specific trees, or where they sow corn for their sacrifices; because all these places are divided by their *ceques* and lines around every village or town as they are put in charge of persons who do there the different sacrifices for various purposes. In some places they sacrifice in order that women will become pregnant; in others they say that out from there comes the cold or hail; from others rain. In this way the Inca king taught them this division of places in all places that he conquered. It is therefore necessary that people paint this map in all the villages or towns, and when they see the example of Cuzco then they will do it [Polo 1571:114].

Our research bears out the contention of Polo de Ondegardo that the system of sightlines classified the skyline according to prominent features related to (1) wind directions (winds that bring different kinds of rain, hail, snow, etc., at different times of the year); (2) roads of access to the valley (especially the highest points of these roads as visible from town, the so-called *apachetas* that also were used for astronomical purposes); and (3) prominent places in rivers and irrigation canals, etc. But Polo also mentioned places where women would become pregnant or where children could be cured. We need a total knowledge of the culture to find the sightlines. The effort, however, is also totally rewarding: the Inca astronomical and calendrical system, like their elaborate system of kinship, allows us to reconstruct their general concepts of cultural classification.

Polo de Ondegardo insinuated that the *ceque* system of Cuzco was but a typical example of what all villages and towns in the Andes had developed. Modern ethnographies reveal that the "*ceque*-ideology" has not been completely lost in modern villages and that customs such as those related to the feast of the third of May and to the cleaning of the irrigation ditches refer to it.

The way that the horizon was used in Inca astronomy might, however, have been spread much wider than just in the Andes; McCluskey (1977) gives an excellent description of a very similar system. He makes the following interesting observation:

> The use of horizon observations as a means of regulating the calendar is best adapted to a sedentary society that has the opportunity to identify suitable local landmarks denoting the seasons of the year or to erect markers for this purpose. Nomadic or cosmopolitan societies, on the other hand, would tend to use an astronomical system that is valid over an extended geographic area, hence the concern of the Greeks and the Egyptians with heliacal risings of stars as a means of regulating the calendar. Since the calendar of an expanding society is almost certain to be portable, it is not likely to be based upon horizon observations [McCluskey 1977:174].

McCluskey then comments on how the native calendars of Northern Europe and America were supplanted by Roman calendars; the latter historical process is well-known. In Peru and Mexico, however, the earlier pre-Roman/Christian use of horizon observations can still be studied. And the Inca case demonstrates that McCluskey's conclusion that a system of horizon observation cannot be used, or elaborated upon, by an expanding society is not necessarily true. It was exactly the Andean ability to extend sightlines over the horizon, connecting far-away towns and temples, that gave the Incas a tool for political integration on the scale of an empire.

This sophisticated use of horizon observation may again have been of far wider importance than just the Andean case. Aristotle, in his Constitution of Athens, describes a situation involving the intimate connection of the body politic, with its subdivisions of twelve groups and thirty-six subgroups, to the calendar, to the directions of the compass, and to land use outside Athens.

We have distinguished systems of horizon observation from those based on heliacal risings (without claiming at all that they should not coexist in a culture). Other general distinctions that we can make are between tropical and non-tropical astronomical systems and between those in the mountains, in flat countries, or on the sea. Maybe we should look also more at systems in higher latitudes where the sun and the moon stay above the horizon for part of the year. Anthropologically speaking, we should not look at these systems from the point of view of the prehistory of western science and astronomy, but from the standpoint of theoretically analyzing the human propensity to classify the social and physical universe. Kinship and astronomy (in their broadest sense) complement each other, and provide equal opportunities, in a non-evolutionary way, of studying human variation in using limited sets of variables for the general purpose of classification.

4

The Role of Architecture and Planning in Archaeoastronomy

Horst Hartung

It is now generally accepted that archaeoastronomy depends strongly on interdisciplinary studies. The intention of this paper is to explore the relationship of architecture and planning to this new field of investigation, as seen by an architect/planner.

Astronomers and architect/planners both deal with space. Astronomers observe the heavenly bodies in celestial space and come to certain conclusions. Meanwhile, planners/urbanists define the exterior terrestrial space on a larger or smaller scale, and architects primarily occupy themselves with the design of the interior spaces of buildings. Astronomical studies proceed on a strictly scientific basis, but the works of an architect include (or at least should include) the functional and spiritual needs of the people for whom they are destined. Both professions require knowing about space, astronomy in terms of large dimensions, architecture in limited ones. Both are creative, but the latter must deal with human factors.

The spatial imagination necessary for an astronomer facilitates the architect's understanding of the physical space which he or the planner defines, but this does not necessarily mean that the astronomer in turn is able to evaluate the resulting architectural space *per se*. In particular, interior space can be delineated only partly by the drawings or models of one trained in astronomy. An appreciation of interior space is possible only by "being in the space," or even better, by walking around in it. An examination of certain sculpture illustrates that delimited space is easier to understand than the space surrounding figures. It may be worthwhile--when treating delimited and surrounding space--to refer to the Maya city of Tikal. Its towering temples dominate the urban space; at the same time these temples contain only reduced interior spaces.[1]

---

[1]Today the invading jungle prevents a full spatial appreciation of the central part of Tikal (Fig. 4.8). Because the main temples rise above the tree line, measurements between these buildings can reveal geometric and astronomical relations (Hartung 1977). The court of the Temples of the Crosses at Palenque illustrates a similar situation on a reduced scale--masses and volumes in space (Fig. 4.9). An evaluation of the exterior space has been published in the proceedings of the Segunda Mesa Redonda de Palenque, 1974 (Hartung 1976). The results of probable astronomical connections present valuable data: (1) The Temple of the Sun faces the winter

HOW AND WHY IS ARCHITECTURE RELATED TO ARCHAEOASTRONOMY?

We all live in architectural spaces. We all appreciate (or complain about) architecture, but not everyone is trained to recognize its complex aspects. The same is true of astronomy; we can all witness the night sky filled with stars and admire the composition of the universe, but only a few of us can give a scientific explanation of the laws that govern the complicated forms we observe.

Our living and working spaces generally are astronomically oriented, usually to the sun. The orientation may be in favor of or in opposition to sunlight, depending on the particular situation. Sometimes houses or certain rooms are located to overlook the sunrise or, much more frequently, the sunset. Some Japanese houses even contain terraces from which to observe the moon (e.g., the well-known example at the Katsura Palace).

In order to see more clearly how architecture is related to archaeoastronomy, I will present stages of astronomical observation from open to closed spaces. (For detailed previous studies, see Hartung 1975, 1977.)

Stage A

Places in the open landscape suitable for observations include the peak of a mountain, the entrance of a cave, or a prominent large rock (Fig. 4.1), from which the rising and/or setting positions of the heavenly bodies may be seen to correspond to features on the horizon. One of the best examples of this in Mesoamerica is Cerro El Chapin, near the archaeological site of Alta Vista in northwestern Mexico. The observation point is on a large boulder, atop an oval-shaped plateau, into which a pecked cross is chiseled (Aveni and Hartung n.d.). The reference is a prominent peak situated on the other side of a wide valley, where the sun rises on the summer solstice. In addition to using a "scientific" approach to trace the aforementioned line, the selectors of the site must have also taken its relation to the surrounding terrain into account, for this was part of the planning, whether conscious or intuitive. In modern days the site selection

---

solstice sunrise (Carlson 1976); (2) "the orientation of the Temple of the Foliated Cross to the setting position of the bright star Capella is interesting if not coincidental" (Aveni and Hartung 1979:173); (3) these two temples form a compositional and symbolic line to the Temple of the Inscriptions' central doorway, or below the entrance to the crypt (Hartung 1971); and (4) the front alignment of the Temple of the Inscriptions, extended to the east, points to the central doorway of the Temple of the Cross. From this spot on the summer solstice, one can watch the sun set over the famous tomb, a fact that should be considered in investigating the great ruler of Palenque who is buried there. It is generally reported in the literature that on the winter solstice the sun sets behind the Temple of the Inscriptions as seen from the Palace. I have had the opportunity to observe both sunsets, and the winter solstice sunset is much less impressive and cannot be referred to a precise or even approximate point.

most probably would have been made in collaboration with a landscape architect.

Stage B

The next step toward architecture would be to define the desired astronomical directions with sculptural elements and/or walls at the observation point or distant foresight points. For example, at the site of Alta Vista there exists a building with a square plan (Hall of Columns), oriented in such a way that its corners point to the cardinal directions; in one of these is the above-mentioned peak, to which astronomical sightings were made from the Cerro El Chapin. Unusual walls to the east of the main building may have served as practical observation lines at the time of the equinoxes (Fig. 4.2).

Fig. 4.1 A pecked cross and other glyphs in western Mexico

At Copan in the Maya zone, in addition to the well-known Stela 12 to Stela 10 relationship (Aveni and Hartung 1976), the positions of several stelae in the ceremonial center seem to have been planned around astronomical considerations. Stelae with inscriptions are particularly suggestive since they may contain astronomical references (Fig. 4.3).

"Mass-elements" such as columns, stelae, altars, boulders, walls, etc., determine open (exterior) spaces, and a particular space is conceivable around them. Generally there exists an interaction of one space with another: the space of the surrounding landscape penetrates the delimited "inner" space. Any sensitive person feels the impressive space of the Great Plaza at Copan, but only a few are able to analyze it. At Copan we have essentially an interrelationship of spaces within the plaza; at Stonehenge the relations are primarily outwards.

Stage C

The third step directly involves architecture, and includes what generally is called urban planning. (Architecture refers mainly to the *interior* spaces of a building and the *immediate exterior* space which it surrounds, forming part of the urban space.)

Astronomical observations in connection with architecture should

Fig. 4.2 The Labyrinth, Alta Vista, Mexico

Fig. 4.3 The Great Plaza at Copan

be from inside a building or from a spot immediately in front of it or along its side. Given the preference for using exterior space in Mesoamerican planning, there seem to have been more observations made from outside than from inside of buildings, or from the point between the outside and inside: i.e., the doorways of buildings. The importance of doorways (especially the central one, if there are several) for relational lines (of astronomical, geometric, and/or design origin) has been proven in the planning of ceremonial and urban centers (Hartung 1971).

The best-known and most impressive example is the Palace of the Governor at Uxmal (Figs. 4.4, 4.5, 4.6). It seems plausible that the astronomical observation was made in this case from the rather obscure second chamber in the center (light enters only through the doorway of the front chamber). Of course, one can not rule out the possibility that the central front doorway was used, or even a point just outside of this doorway, where a glimpse along the front of the building to the north encounters the first landing of the stairway to the Chenes Temple of the Pyramid of the Magician (Fig. 4.7).[2] The transverse axis through the central doorway points eastward to the southernmost rising point of Venus (about A.D. 850); it is marked as a foresight by the pyramid of Nohpat at a distance of more than 6 km. Here the facade, nearly 100 meters in length, faces (at 90°) the Venus rise position at its southern extreme. At the same time the building with its large, mostly artificial platform, confronts the immenese flat landscape, where on the horizon only the overgrown pyramid of Nohpat serves as an accent.

Fig. 4.4 Palace of the Governor at Uxmal

---

[2]As we see later, this relation does not seem to be fortuitous, as both buildings are related to Venus extremes.

Fig. 4.5 View from inner central chamber of Palace of the Governor

Fig. 4.6 The Pyramid at Nohpat

Fig. 4.7 View northwards along facade of Palace of the Governor

In addition to a view along a transverse axis, for which a marker (such as, perhaps, a pair of crossed sticks in the doorway) would be useful in determining the line accurately, the vertical corners of a doorway can also define exact directions, using the outer left and the inner right corners (or vice-versa), as the observer looks from the interior of the chamber. The latter procedure was applied when defining directions in the so-called windows of the Caracol at Chichen Itza (Aveni et al. 1975).

Because of the rather short depth, only well-built doorways and windows can be measured with confidence. Possible restorations, stabilizations, and even smaller alterations should be taken into account; these are often not mentioned in reports. Usually these changes can be detected by careful observations *in situ*, and here collaboration with an archaeologist and/or architect is recommended. Time often equalizes old and new materials, particularly where an ancient building has been disassembled and reconstructed, using partially new elements.

For example, inspection revealed that the west (and only) window of Temple 22 at Copan was relatively free from reconstruction errors, thus permitting astronomical studies to be carried out (Aveni and Hartung 1976; Closs et al. n.d.). However, measurements of the window now existing in Room 44 at Wupatki (Arizona), which aligns "to the solstitial and equinoctial sunrise points," were found to be of little use because Reyman (1978:732) later discovered that the room "originally had no window in its east wall."

WHAT DOES ARCHAEOASTRONOMY GAIN FROM ARCHITECTURE AND URBAN PLANNING?

The architect/planner specializing in the field of Pre-Columbian studies can detect possible sites, places, and buildings which might have had an astronomical significance. Sometimes he will proceed to make the corresponding measurements, and will compare the suspected alignments with astronomical events. The astronomer has to be involved at this stage, although it is recommended that he undertake his measurements in collaboration with the architect at the site, so that the most probable original lines can be determined. For example, in the case of the rather irregular, unrestored monumental stairway of the Collonade at Ake, Yucatan, a line could be determined within five minutes of arc by selecting about six points on the steps which, in the opinion of the architect, seemed to be in their original places.

Astronomical alignments can be suggested by (a) direct observations in the field, and (b) map interpretations. In addition, epigraphic, ethnohistoric, archaeological, and other data may also indicate possible alignments. For example, (a) a perpendicular line from the central doorway of the Palace of the Governor at Uxmal to the pyramid of Nohpat (noted as a remarkable elevation on the horizon) was observed *in situ* in 1964 and 1966; and (b) a "suspicious" line leading from the doorway of the Chenes Temple of the Pyramid of the Magician at Uxmal via the ball court to the main entrance of the West Group, observed by map interpretation in 1968, was later found to be directed to the southern extreme of Venus, as determined by astronomer Anthony Aveni (Hartung 1971; Aveni 1975).

One may ask whether the view from Nohpat to Uxmal might have some astronomical significance. Could the northern setting position of Venus be related to the visible buildings at Uxmal? One can hypothesize that Nohpat was consciously built on the "Venus line" (in relation to the Palace of the Governor), with the exact point on this line selected for special reasons, since there seems to exist no natural elevation nor remarkable feature in the landscape at this place. Nohpat has not been investigated in detail nor excavated to my knowledge. In January 1976, and again in February 1979, we asked our guides about archaeological sites which can be seen from Nohpat. On both occasions they pointed only to Xcosh (or Xcoch), in a northeastern direction. The situation of Xcosh and its pyramids is very similar to that of Nohpat; John L. Stephens visited and described them both in his famous book (1843). The visual connection from Xcosh to Uxmal and Nohpat (or vice-versa) seems to be more than casual and should be investigated, although it might have a geometric or symbolic basis rather than an astronomical one.

WHAT DOES ARCHAEOASTRONOMY GIVE TO ARCHITECTURE AND PLANNING?

There is strong evidence to suggest that the astronomical knowledge of the Maya must have played a part in the planning of their centers and (at least in part) the orientation of their buildings. But archaeoastronomers did not get seriously interested in Mesoamerica until 1970, and even in the early years of the seventies several vague propositions and conclusions still persisted which had been present in earlier works. There was an early insistence upon an adequate comprehensive preparation and actualization of archaeoastronomical studies (Reyman 1975).

Scholars in Pre-Columbian architecture and planning obtained convincing astronomical data which confirmed earlier suggestions and encouraged further studies. The situation of the Palace of the Governor--to stay with this extraordinary building--was first analyzed in terms of architecture and planning, but now its orientation has been given a surprising explanation. The Maya emphasized this direction by constructing the pyramid at Nohpat, on the horizon line as seen from Uxmal, thus forming a "point de vue."

FINAL REMARKS

Interdisciplinary work is extremely necessary in archaeoastronomy, and a mutual understanding of expert opinions from other fields is important. The archaeoastronomer should have a general knowledge of the archaeology of the zone in which he is working, but if a problem involves complex constructions or sometimes even details, he should consult an archaeologist or an architect specializing in Pre-Columbian buildings, depending on the particular case. It is always dangerous to move to another field of specialization, and the archaeoastronomer is severely exposed to this temptation; he should not repeat the many "faux pas" of the archaeologists and anthropologists who falsely felt competent about Mesoamerican architecture.[3]

---

[3]Here we should mention only two examples, one theoretical and

Fig. 4.8 View from doorway of Temple I at Tikal

Fig. 4.9 Group of the Temples of the Crosses at Palenque

---

the other practical, both of which are closely related to recent archaeo-astronomical investigations: (1) aesthetic judgments about architecture like Eric Thompson's about the Caracol, although it is a very witty one; and (2) the many restored and altered Pre-Columbian structures in Meso-america such as Teotihuacan, Monte Alban, Uxmal, and numerous other sites.

5 Archaeoastronomy and Related Problems: Old World Approaches vs. New

Ronald Hicks

Interest in the ancient knowledge of astronomy as reflected in architecture and monumental construction is hardly new, but only in the past decade or so has it at last gained enough strength among archaeologists, astronomers, and people in related fields to become a recognized area for study. Archaeoastronomical studies are being carried out in much of the world today, but the two primary lines of research involve megalithic monuments of the Neolithic and Bronze Age in the British Isles and neighboring parts of Western Europe, and the astronomical knowledge of the Indians of Pre-Columbian America. It may be useful at this stage to compare briefly the directions taken by research in these two geographical areas.

Archaeoastronomers in both Europe and the Americas are faced with interpreting remains from a period before history, but an immediate difference that influences their work is the remoteness of that time. In the Americas, much of the astronomical knowledge was still in use when the first Europeans arrived, making it possible for us to use surviving written records and ethnographic studies among living peoples for clues to the extent of prehistoric knowledge and, perhaps more importantly, for clues to which celestial bodies and events were watched. In Western Europe, by contrast, we are dealing with monuments dating from before 3000 B.C. down to perhaps as recently as 500 B.C. (in Ireland), with the bulk of them apparently belonging to the period from roughly 2500 to 1500 B.C. Even at best this leaves us several centuries removed from the earliest written records for the region. Moreover, some would argue that the earliest historical inhabitants of that part of Europe, the Celts, did not arrive in Britain until 100 B.C., meaning that we have no knowledge at all of the megalithic builders. As has often been emphasized, the druids, a Celtic priesthood, could therefore have had nothing to do with such monuments as Stonehenge. If this assessment is true, we have nothing to work with but the mute stones and earthen monuments.

Not everyone, however, would agree with this assessment. Sir Norman Lockyer, founding editor of *Nature* and a strong advocate of astronomical studies of megalithic sites, felt much information could be gleaned from folklore, particularly that surrounding the seasonal festivals with their clearly pre-Christian roots. Unfortunately, he approached this material very uncritically, with the result that the source was discredited and has been almost completely ignored in recent work, except among the lunatic fringe who use it to bolster their own questionable ideas. If archaeologists look at the folklore at all, they tend to approach it in the way Leslie Grinsell did recently in an otherwise very useful book entitled

The Folklore of Prehistoric Sites in Britain. In his introduction Grinsell says that he has omitted all references to druids on the grounds that "all or nearly all . . . date from 1600-1900" (1976:74), which is to say that they are almost solely the result of the druid revival of the eighteenth and nineteenth centuries (i.e., of the Romantic movement). This movement is the source of the neo-druids who today are allowed to perform their midsummer solstice ceremonies at Stonehenge. The general view of Grinsell and many others is that while the folklore is very interesting, the monuments are from too dim and distant a past for it to have much relevance to their interpretation.

I am not so pessimistic, perhaps because my work has concentrated on Ireland rather than Great Britain. Traditional interpretations of British prehistory see it as involving a series of invasions, from the Beaker people of the Neolithic/Bronze Age transition through the Halstatt and La Tene of the Early Iron Age to the historically documented Romans and Anglo-Saxons, with dustings of Urnfield people, Vikings, and the like. This interpretation, though it is now being reevaluated, has made it difficult to argue for much continuity of tradition from the time of the megaliths. Ireland, on the other hand, is remote enough that most evidence for early invasions appears rather insubstantial, and the Romans and Anglo-Saxons never got there at all. Ireland's society remained, therefore, essentially Iron Age until the coming of the Norman aristocracy in 1169, though the spread of Christianity after the arrival of St. Patrick and his colleagues in the fourth and fifth centuries A.D. caused the slow fading away of the old beliefs and practices. If abundant literature survives from this transitional period, and if a case can be made for continuity from the megalithic period into this Celtic/druidical time, we can have hopes of untangling some of the astronomical threads. Some of the earliest Irish references do in fact indicate that stone circles were still in use (not really surprising since our own religious monuments sometimes continue in use after a millennium or more, and after so substantial a structure as a henge or stone circle is in place there is little need to replace it). Likewise, some of this early Irish literature associates the ancient sacred sites with druids. There are other bits of evidence for Neolithic-Iron Age links: Iron Age pottery from Stonehenge, Roman coins as apparent votive offerings at Newgrange passage grave in Ireland, and some henge-like earthworks that are certainly Iron Age in date. The druids did not believe in writing down their knowledge, but if we are to believe Julius Caesar, the priesthood was not only strongly interested in astronomy but was also introduced into Celtic Gaul from Britain; it may therefore be British rather than Continental Celtic in origin. From the distribution of certain later types of monuments, a case can be made for some sort of missionary movement out of Britain or Ireland in the sixth century B.C. or thereabouts. Consequently, though we can probably gain little from the classical sources, distorted and limited as they are, I feel that the druid question needs to be reexamined and that much more attention should be paid to the ethnographic and folkloric evidence in Britain and Ireland.

A second area that has received considerably more attention on this side of the Atlantic is the interpretation of pictographs, petroglyphs, and effigy figures in astronomical terms (e.g., see Brandt et al. 1975; Britt 1975; and Cowan 1975). Abundant art exists in connection with astronomical monuments in the British Isles, as in the daggers and axes cut into the sarsens of Stonehenge, the elaborate decoration of the stones of passage graves such as Newgrange in Ireland, carved stone objects found in some of the monuments, and in the very shapes of the stones themselves at others such as Brogar in the Orkneys and Avebury a few miles north of Stonehenge. But while there have been discussions of possible symbolism (Crawford 1957; O'Kelly 1973; Brown 1976:292-293), no one has looked at the placement of such art on the monuments from an astronomical point of view, beyond making the so-far unfruitful suggestion that groupings of the circular depressions called cup-and-ring marks may represent constellations. It seems very likely to me that particular symbols or shapes may have been intended to call attention to certain orientations and to represent the deities associated with them. There are also at least two slabs from Irish passage graves decorated with radial, fan-like designs coupled with other elements that could conceivably be horizon-based sky maps or at least mnemonic devices. Some cautious research into these matters could prove extremely fruitful.

As with the availability of ethnographic data, some other differences between megalithic and native American archaeoastronomy are dictated not by approach but rather by conditions and the nature of the evidence. There is no point, for instance, in talking about zenith passages of the sun in Britain, which is well north of the latitudes where such events occur. In addition, while Lockyer, Alexander Thom (1967), and John Barber (1973) have all discussed stellar alignments, their emphasis has been on the sun and moon, at least partly because most of the monuments with which archaeoastronomers in the British Isles are working are not so well dated as those in America and without reasonably accurate dates stellar alignments are difficult if not impossible to identify. Even some of Thom's early work based on radiocarbon dates has been invalidated because of the radiocarbon recalibration. Nonetheless, it seems to me there are some significant omissions in the work done so far. We have ethnographic evidence for the use of certain stars as timekeepers in Western Europe, and we know that the Pleiades were important, but little effort has been made to use this information. More seriously, the planets have been almost completely ignored, aside from a passing reference by Hawkins (1965:105). My own studies in Ireland suggest that one mythological character (Finn MacCumaill, possibly identical with the Celtic god Lugh) may have been associated with Venus. These possibilities need to be looked into.

Not all the research gaps are in megalithic studies as viewed from America. The moon has been much more intensively studied in British archaeoastronomy than in American. Since it is common knowledge that the Indians of the Eastern Woodlands and elsewhere in the New World expressed time in moons, I cannot help feeling that lunar orientations or evidence for knowledge of the Metonic cycle should appear somewhere in the Americas, as some very recent evidence suggests may be the case. On the other hand, I remain highly skeptical of Thom's (1971) arguments that the ancient inhabitants of Britain possessed an extremely accurate knowledge of lunar movements, down to minor variations.

Eclipses provide a topic about which too little is known on both sides of the Atlantic. Much of the controversy over Hawkins' interpretation of Stonehenge has centered on his argument that it could have been used for keeping trace of the likely times for eclipses (Hoyle 1966a, 1966b, 1977;

Newham 1966; Colton and Martin 1967; Sadler 1967). While most agree that it *could* have been, such use seems to involve enough complications that not everyone--and particularly not many archaeologists--are convinced that it actually was, at least not for exact prediction of eclipses. In American archaeoastronomy, on the other hand, most of the discussion of eclipses has centered on the so-called eclipse tables in the Dresden Codex (Owen 1975; Smiley 1975; Gibbs 1977:27-28; Kelley 1977:59ff), which of course make it clear that people at a level of development roughly comparable to that of the Stonehenge planners could indeed have had a knowledge of eclipse cycles. What has been left unanswered is the question of how and when either the megalithic astronomers or the Maya arrived at this knowledge. And do any of the American sites allow eclipse prediction without resort to tables? How difficult is it to develop this much sophistication in astronomy?

In another area America offers documentation where Britain allows only speculation: ancient calendars. We know the Maya and Aztecs had highly accurate calendars, which we understand well despite the difficulties of correlating them with our own dates. In Western Europe once again we have only the stones--aided a bit by more recent traditions. We know that in ancient Britain and Ireland the solstices were carefully noted and that even today they are the occasions for large bonfires. We also know that the old quarter days--at the beginnings of February, May, August, and November, mid-way between the solstices and equinoxes--were considered the beginnings of the seasons and were likewise marked by fires and rituals. If indeed we can assume some continuity from Neolithic into historic times, it is not surprising that we find solstices, equinoxes, and quarter days alike marked by alignments in the monuments. Lockyer (1909), Boyle Somerville (1922), and Thom (1971) all have found what is usually referred to as a May Day alignment to be one of the most common. This May orientation is the same required for sunrise early in August, and given a carefully chosen location, with horizons at a similar elevation on east and west, it can be used as well for sunset on the other two quarter days. It was the discovery that an unlikely percentage of the henges I was studying in Ireland had similar entrance orientations that led me to look into archaeoastronomy in the first place, and that orientation turned out to be the one toward sunrise at the beginning of May, the old Celtic festival of Beltaine, "Bel's fire." Entrances pointed toward the corresponding point in the southwest are nearly as common as the northeasterly ones. So we have evidence for a calendar. From the evidence of folklore and recent practice, it seems most likely that the actual day chosen for seasonal celebrations was linked to a phase of the moon falling near the solar date. This conforms to the evidence provided by the surviving fragment of a Celtic calendar from Coligny, which seems to be based on the Metonic cycle, using lunar months and correlating the solar and lunar calendars over a nineteen-year period with intercalary months inserted as needed (MacNeill 1926-28). Thom, however, pursues the megalithic evidence to what seem to me to be uncomfortable lengths, postulating the division of the year into sixteenths, giving months 22 to 24 days in length (Thom 1967: 107-117).

Other British concerns which have not become the center of such controversy in America, even though studied here, are geometry and metrology. Much of Alexander Thom's work has revolved around the idea that the megalithic sites not only incorporate astronomical sight lines but also are laid out with geometrical precision using a standardized unit of length Thom refers to as the megalithic yard. To anyone working with the stone circles and henges, it quickly becomes apparent that many are not perfect circles. Instead some are elliptical, ovoid, flattened in one or more arcs, or of a shape that can perhaps most charitably be described as "roughly circular." To most archaeologists this has simply indicated that the builders were a bit careless or found it difficult to transform their plans into reality, or perhaps merely that time had not been overly kind to the sites. Thom has argued instead that all such deviations were planned, and that the designs were based on various combinations of Pythagorean triangles to produce three types of flattened circles, two types of egg-shaped rings, ellipses, simple circles, and compound circles (Thom 1967:27-33). Further, these variants were used in order that as many measurements as possible within the monument would be integral multiples of their standardized megalithic yard, i.e., 2.720 ± 0.003 feet. In addition, Thom has postulated various fractions and multiples of this yard, namely the megalithic inch, half yard, fathom, and rod (Thom 1955, 1962, 1968; Thom and Thom 1972).

The idea of such prehistoric units is not unique to Thom. The antiquarian William Stukeley proposed a "druid cubit" early in the eighteenth century, and even such a respected archaeologist as Flinders Petrie once argued for an "Etruscan foot" (Brown 1976:165, 261). In America similar ideas have been put forward by Hugh Harleston, Jr. (the standard Teotihuacan unit, 3.475 feet) and Warren Wittry (the Cahokia yard, 3.425 feet) (Norrish 1978: 7). Similarly, Rene Millon (1967:40) has argued that Teotihuacan was laid out on a grid using units of approximately 187 feet, and James A. Marshall has advocated the view the same grid unit was used for the Middle Woodland enclosures of Ohio (Marshall 1977:31; 1979), which he also feels were based on various right triangles. It is worth noting that many of the people who have made such speculations are neither archaeologists nor astronomers but civil engineers, and thus presumably accomplished surveyors.

Thom is very likely right that the ellipses and egg-shaped rings are deliberate deviations from the circle in megalithic construction; they seem too clearly planned not to be. However, I find his other shapes much less convincing, and as I have pointed out elsewhere (Hicks 1977), I am completely unconvinced by the argument for a megalithic yard. In only two out of 145 cases cited by Thom (1967) did the unit used exactly match the proposed yard. The actual range was from 2.54 to 3.00 feet, and the distribution of lengths was very close to what one would expect if the traditional nose-to-thumb yard were used. This certainly can't be used to prove that a standardized measure was not used in the American sites, but it suggests we should examine any such claim with great care.

One final comparison between Old World and New needs to be discussed, namely the work of Euan MacKie of the Hunterian Museum of the University of Glasgow. His writings provide us with one interesting new approach. In the discussion up to this point, any influence of the archaeoastronomical work in one hemisphere on that of the other has flowed from east to west, from work on megalithic astronomy and related matters to that on native American. MacKie (1977:208ff) comes closest to reversing this pattern by looking to America, specifically to the Classic Maya civilization, for a model of a type of society that could have produced the astronomical constructions of megalithic Europe. Certainly caution is necessary in making comparisons between people following a largely pastoral way of life in the mid-latitude mixed-forest environment of the British Isles and a maize-based economy in the tropical forests of Mesoamerica, but I find the attempt worthwhile. MacKie's discussion voices some of my own suspicions, such as the possibility that some of the large henges served primarily as monastic enclosures for communities of priests associated with the astronomical monuments, as is likely the case with Durrington Walls, just downhill from Woodhenge and only two miles east of Stonehenge. A better model would undoubtedly be available if we knew more about the Middle Woodland society that produced the earthworks in the area from western New York, Pennsylvania, and West Virginia across Ohio into Indiana and northern Kentucky. Here we have a copper-using society very similar in level of development to that of Late Neolithic Britain, living in a similar environment, and building sacred enclosures that would look perfectly at home among the henges of an earlier time in Britain. Here a genuinely useful comparison could be made. However, our knowledge of these people is even scantier than that for the megalithic builders, and so many of their monuments have been destroyed over the past few decades that it may prove impossible even to determine with certainty if these too had astronomical overtones. On the other hand, it is possible that we can again assume some continuity through time and make useful comparisons with the Indians of the Eastern Woodlands as they were when first encountered by the Europeans. Geoffrey Wainwright (1970:38) and Colin Renfrew (1973:233-236) have both used such a comparison with the historic Creek and Cherokee to suggest a tribal gathering-place function for the henges within a society organized as a chiefdom.

To summarize, if work in both hemispheres is reviewed, it is possible to identify a series of some seven general problems that need to be considered. First of course is the one topic that has received equal attention on both sides of the Atlantic, namely architecture--the orientation of structures, ceremonial centers, and cities. But certain questions, it seems to me, have not been given equal attention in the two regions--such as the part played by planets and constellations in Britain, and the possibility of lunar orientations and the use of site structure instead of tables for eclipse prediction in the Americas. A second focus of attention is on the information that can be provided by ethnographic studies of folklore, traditional practices, and early texts concerning astronomy and related topics. This area has been largely ignored in Britain because of the inadequately examined assumption that there is not a sufficient cultural continuity from megalithic to historic times. The information to be gleaned from a close study of art--petroglyphs, pictographs, the shapes of the stones themselves--provides a third problem for study. Here again much more work has been done in the Western Hemisphere, but Western Europe is rich in materials that may be relevant. New World archaeoastronomy is also far ahead in a fourth area, the understanding of ancient calendars. In Britain, Alexander Thom has proposed a megalithic solar calendar that bears little resemblance to the combined lunar/solar calendar that appears to have been in use in early historic times, a difficulty that at present I'm not sure can be resolved. On a fifth question, work on the two continents has taken rather different tacks, dictated at least in part by the differences in the sites being considered. This is the question of geometrical knowledge and the use of standardized measurements. An examination of the British work, however, suggests that this is a subject that needs to be approached with considerable caution. Another problem area that will undoubtedly receive increasing attention, and in which attempts have recently been made to compare the

European and American cases, is that of the social structure within which archaeoastronmy was carried out--tribe, chiefdom, theocracy, or what? And finally, wherever work in archaeoastronomy is being done, a central question must be that of evaluating the reliability of our data and our interpretations.

In this brief and necessarily superficial review I have tried to point out some of the similarities and the differences in the studies of archaeoastronomy and related topics as they are being pursued in Western Europe and in the Americas. By doing so, I hope I have brought into focus some of the gaps in our knowledge of each region that may provoke new questions and provide fruitful areas for future research.

6 Archaeoastronomy and Education

John B. Carlson

> Why study "archaeoastronomy" in the first place?
>
> Who do we want to reach and with what result?
>
> What do we actually expect to accomplish in the classroom? . . . in seminars? . . . with public lectures? . . . in the field? . . . in the planetarium? . . .
>
> Who is qualified to teach "archaeoastronomy"?

When I began to collect ideas on archaeoastronomy and education for my portion of this seminar, the result took an alphabet-soup format. For better or worse, let's look at some of these possibilities using ideas gleaned from the dozen or so people who have taught courses or lectured on "archaeoastronomy." (The use of quotation marks here is intended to convey the broadest and most varied interpretation of the subject.)

Archaeoastronomy has a unique appeal to the public at large. The subject involves both astronomy and archaeology, each usually perceived in a rather romantic light--professionals included. Consider Stonehenge, druids, medicine wheels, Maya astronomer-priests, the "wisdom of the ancients," moon landings, exploding stars, "2001"--and, yes, ancient astronauts, Atlantis, pyramid power, Velikovsky, and astrology. In the popular press, we now have a gallery of ley lines and earth zodiacs seated alongside research that precipitated genuine re-evaluation of the whole of European prehistory. Science, pseudoscience, scientism, and nonsense all contribute to the popular culture and feed current anti-rational trends. I gave questionnaires to University of Maryland students on the first day of an "archaeoastronomy" course which asked what they thought the course content should be--and what they would like to learn. A surprisingly large number expected "Von Daniken" and "Chariots of the Gods" with their horoscopes thrown in for good measure. A few expected a history of western astronomy. Most had heard about *Stonehenge Decoded*. The situation, then, is that there exists a great variety of interests, curiosities, and active misconceptions, which places archaeoastronomy or something like it in a special position. We have the opportunity to teach the methods and results of a new interdisciplinary science, examine the humanistic side of the origins of astronomy in different cultural contexts, and encourage skeptical inquiry rather than "religious" belief in such things as astrology and catastrophism. We would like to help ourselves and our students form and defend their own views and function effectively in the modern *Two Cultures* (Snow 1959) environment. This is a tall order.

This is the place to narrow the discussion of "who we want to teach" down to the general public and to the undergraduate, who may be seeking credits for a science requirement; seminars for professional archaeologists and graduate students in the techniques and results of archaeoastronomy form a separate educational context. What are some of the specific topics and approaches that have been taken for popular undergraduate courses? Let's look at a list:

(1) A "Selected Topics (readings) in Archaeoastronomy" course based in part on student input. Selections of readings are made available and a term paper is required. The aim is to teach the subject as a new interdisciplinary field in its own right.

(2) A "history/philosophy of science" course with examples from nineteenth and twentieth century archaeoastronomy (and history of astronomy) used in a history-of-ideas context. Emphasis is on archaeoastronomy as an emerging discipline.

(3) An approach to the astronomically-related "pseudosciences," such as Von Daniken's *Chariots of the Gods*, Lowell's Mars, Velikovsky's catastrophism, pyramidology, ley lines, and astrology. This course or lecture series takes an open-minded look at the evidence and the history of these ideas. In any case, questions on such subjects do come up naturally in a course on archaeoastronomy.

(4) An appreciation of "Naked Eye Astronomy" in a world-wide context. How do pre-technological peoples perceive the world without the aid of modern instruments. The planetarium is a useful tool in such a course or series of lectures.

(5) Oral traditions, folklore, mythology, and ethnoastronomy form the basis of a literature-oriented course. Ethnographic and ethnopoetic literature is used in conjunction with the results of current research.

(6) Archaeoastronomy may be used as part of an introductory astronomy course to give a different interpretation of the roots of modern astronomy and cosmology and to show a more humanistic side of the subject. This is a good choice for the university general sciences requirement course. The goal is to show how science (and interdisciplinary science) really works.

(7) The use of topics and results of archaeoastronomy specifically to look at the development of non-Western thought and culture. A course in "Native American Astronomy" is an example. This offers a multidisciplinary look at a foreign mental system.

(8) The results and methods of archaeoastronomy may be used in architecture and history of architecture courses. Any discussion of ancient architectures should probably not be taught without the inclusion of the cosmological and cultural background.

(9) Archaeoastronomy and its techniques and results should become a natural part of most traditional anthropology or prehistory courses. Consider anthropology as including archaeology, cultural anthropology (ethnology), physical anthropology, and linguistics.

Reviewing our original questions, "archaeoastronomy" is worth teaching to students and the general public as a subject in its own right. Furthermore, its unique appeal makes it particularly well suited for the special educational tasks outlined above.

We would like to reach students of all ages with the idea that skeptical inquiry may be more enjoyable than uncritical acceptance of new ideas; that interdisciplinary cooperation and collaboration are useful, desirable, and rewarding in an age of deep specialization; that the sciences and humanities need not be taught separately--the organism of human culture does not recognize the dichotomy of the *Two Cultures*; that the religious and mental life of ancient peoples may be examined in conjunction with the "hard" evidence of the "new archaeology" to the advantage of all concerned. The questions that archaeoastronomy raises naturally demand a more integrated approach.

Furthermore, the popular idea has been generated that our ancestors, the ancients who built the megalithic ruins and temple-pyramids of antiquity, were somehow too inept or unsophisticated to do the job by themselves. It has been suggested that they required the advanced technological assistance of visiting extraterrestrial humanoids, the "gods" who created man in their own image. The study of archaeoastronomy and its proper presentation to the public should certainly help with the reassessment of this speculative and unsubstantiated claim.

These discussions are an individual statement, perhaps idealistic, but they should strike familiar chords when we look at what we expect to accomplish whenever we as teachers (or students) try to share our specialized knowledge. We must strive to convey the substantial *and* controversial sides of our subject without succumbing to the sensational--to convey the excitement and enthusiasm without fostering uncritical acceptance of whatever we say.

7 A Glance Into the Smoking Mirror

E. C. Krupp

In the New World, no less than the Old, our ancestors watched and pondered the sky. They found meaning in it, and they expressed that meaning in their traditions, temples, and tombs. In most of ancient America there was no written record of this activity, and in Mexico, where the written record existed, it was almost completely obliterated in the Conquest. But archaeoastronomy and ethnoastronomy are piecing together, from the clues and fragments that remain, an intriguing picture of the significance of the sky to the ancient inhabitants of the New World.

A quickening pace in interest and research has prompted this conference, "Archaeoastronomy in the Americas." As the third major conference devoted to this issue, and the first since 1975, it is an appropriate occasion to reestablish and refine our perspective on these inquiries into our ancestors' world views and the roles of their astronomies.

A world view, usually tempered by the sky, enables a group (or for that matter an individual) to define its place in space and time, and therefore its identity. This is the same thing as our deciding who we are, where we are, and when we are.

Ray Williamson sensed that the community of those active in archaeoastronomy shared this need for the cohesion that a sense of identity and cosmology provides. He arranged therefore an informal seminar--more accurately, a structured conversation--prior to the conference itself. This was intended to permit a little preconference exploration of some of the philosophical questions about which most of us here have been thinking. I realize this borders dangerously close on archaeopolitics, but it is a risk I believe we all have to take.

As chairman of the preconference discussion, Anthony Aveni imposed a useful regimen. It facilitated the emergence of issues that concern us all, the fundamental questions of identity. Considerable anxiety--an identity crisis, in fact--has plagued archaeoastronomy for at least a decade. The heart of the problem seemed to reside within a single mystery: what to call this material. We weren't concerned that we, who are involved with it, couldn't recognize it, but no one was quite sure what to name it. All of the coined names seemed awkward, and none satisfied everyone.

It is the diversity of our professions and viewpoints that have made something so prosaic as a label for this area of inquiry assume the significance of an heroic quest. We are, after all, astronomers, anthropologists, archaeologists, architects, art historians . . . and these are just the *As*; I could continue through the alphabet. With so many different disciplines involved, it is difficult to know where we belong, if indeed anywhere, in the minds of the other professionals with whom we must all interact. A name, however, does imply an identity. And there is in naming always the danger of creating an identity that not only others misperceive, but that we misperceive.

Formalizing our identity is the same thing as institutionalizing archaeoastronomy, and I think finally a consensus on this issue has emerged. We have abandoned the effort to guess Rumplestiltskin's name. And at least for the present there is the sense that formalizing archaeoastronomy as a division or section or similar component in a parent organization or forming an independent society may not only be a bit pretentious but, more importantly, stifling.

There are, perhaps, organizations in which we might naturally orient our activities, but residing in almost any one of them runs the risk of filtering some people from the archaeoastronomical conversation. That would make archaeoastronomy more exclusive and more isolated and would make us all the poorer.

Every one of us clearly senses that it is the interdisciplinary character of this subject that gives it its strength. What propels us forward is the exchange, the conversation, the almost chance matching of data and idea. It's not what you know that's important in this business; it's not even who you know. It's what who you know knows that counts.

To refer glibly to "interdisciplinary study" is insufficient. If mere lip service is paid, "interdisciplinary" becomes synonymous with "superficial." Perhaps we should demand depth and follow-through. If we say interdisciplinary, we must mean it. Access to this inquiry, a study of people's responses to the sky, must be genuinely open to all who have something to contribute. And each of us must expand our vocabularies, skills, and understanding into areas remote from our own training and expertise.

Sir Norman Lockyer, a renowned British astrophysicist and the founder-editor of *Nature*, was among the first to attempt a true interdisciplinary style. He was responsible for the broadest and most formalized approach to archaeoastronomy prior to the 1960s. Although considerable controversy surrounded his work on Stonehenge and other British monuments and his work on Egyptian temples, his two books on those subjects remain important resources. Several of his hypotheses still await adequate testing. Although it is true that even before Lockyer, as early as the eighteenth century, attempts had been made to understand Stonehenge astronomically, Lockyer stands out as the first systematic interpreter. His interdisciplinary approach, while perhaps undisciplined, is evident.

An important stimulus to archaeoastronomical research occurred in 1963, when Gerald S. Hawkins published his first paper on Stonehenge. Despite all the controversy and criticism surrounding his subsequent book, *Stongehenge Decoded*, Hawkins can be credited with creating a climate in which serious interest in archaeoastronomy would grow.

The work of Professor Alexander Thom and his collaborators is one other salient component of the environment from which the present situation evolved. Thom's immense body of reliable measurements, thoughtful analyses, and imaginative hypotheses command attention. His conclusions are controversial; many may be disproved. But his work has set high standards for data acquisition and handling.

Astronomer Anthony Aveni should be recognized, however, as the first investigator to make the "inter" in "interdisciplinary" work. In collaboration with Horst Hartung and Tom Zuidema, Aveni has helped bring archaeoastronomy home to the New World. Hartung is an architect; Zuidema is an anthropologist. The first archaeoastronomy conference, held in Mexico City in June 1973, was cochaired by Aveni and Hartung, and Aveni organized and chaired the second conference, in September 1975, at Colgate University. At both conferences it was evident something interesting was happening. A great broadening of the discussion occurred. Investigators with a variety of professional backgrounds found evidence for observational, institutionalized, and symbolic astronomy in many societies widely separated in space and time. More significantly, the people most interested in these matters began to learn of each other's existence.

"Archaeoastronomy in the Americas" in June 1979, is the largest such conference yet. The quality of the papers is high, and the variety is great. The conference has been truly interdisciplinary and, in fact, it seems to have remedied what had been another nagging (and related) problem.

It had seemed that perhaps the archaeoastronomical conversation had become too ingrown. The sweet but deadly habit of keeping the same company had begun to set in. Too many of the same people were beginning to say too many of the same things to each other. Useful criticism and review were hard to come by. New research was slow in coming. This not only makes for bad science, it means less fun. This subject--whatever we name it--is for us, I think, a refreshment, and exhilaration, and a freedom. The present conference has been exciting and intense because the number of participants has grown. It is our best indication that archaeoastronomy is alive and well.

There are still difficult, crucial issues pending before us. Standards for research are yet to be established. Archaeoastronomy is still seeking its criteria for proof. How is archaeoastronomy to be done? What is the right way? How are its results to be evaluated?

We have seen different attempts, even quantitative attempts, to work this out, but I don't believe--and I'm not sure anyone else believes--we are really going to be able to develop a universal, structured system of evaluation. Clear definition of problem and argument is vital. But overly restrictive definition, for the sake of what we might perceive as scientific certainty, will most certainly screen out valuable, unquantifiable data.

I don't know myself how to define a good line of reasoning and evidence, but I know it when I see it: it is comprehensive; it holds together; it rings true. This is a subjective judgment, of course, but when the circumstantial case we draw--and in archaeoastronomy it is almost always circumstantial--is ample and self-consistent and meaningful, I think we can recognize it. Any attempt to formalize this may prove futile, but let me suggest we adopt, at least provisionally, the principle of converging lines of evidence. A successful case most often is one supported by multiple and independent arguments that compel us to the same conclusion. And any conclusion we draw must fit the broader context to which it applies. When we go wrong, and surely we shall, we will have to depend on the self-correcting process of science and the consensus of all interested parties to keep interpretation in tune with evidence.

We can impose upon ourselves the responsibility of conducting investigations comprehensively. This means sensing and outlining a problem's entire context. We must ask all the questions and face all the answers. The issues we ignore, our sins of omission, will be what compromise the quality of this research.

We face another problem. The future development of archaeoastronomy and ethnoastronomy depends upon the next generation of investigators. These are the students now in, or soon on their way to, graduate school. Can they integrate their interest in this material with the academic structures that now exist? It is not as though they really have a choice. No departments of archaeoastronomy exist, and that is actually a relief. There are academic disciplines within which an archaeoastronomical inquiry might be appropriate--cultural anthropology, for example, or art history. We must make it possible for this to occur. Genuine interdisciplinary exchange and responsible research will be what make the difference. It is important that academic disciplines perceive archaeoastronomy as something within--not independent of--their spheres of activity. By trying to meet other disciplines on their own ground, we may avoid the self-consciousness and pretension a self-imposed isolation creates. There may be archaeoastronomy, but there are no archaeoastronomers.

Finally, the most timely issue we can address is "What should we do next?" It now seems unlikely that a special organization, or even a journal, will be established. A means of communication, something informal and newsy, that tells who's doing what, is vital, however. *Archaeoastronomy*, the bulletin that has served us so well under the tutelage of John Carlson and Ray Williamson, fulfills this function. It is in all of our best interest to support it.

The need for conferences like this will arise spontaneously and naturally. Presumably the agent to conduct them now sleeps, as Arthur in Avalon, but will appear when needed. For now, a more useful activity may be meetings of smaller groups, in workshop format.

I began this commentary by observing that we are concerned with who we are now and what we ought to be. No real mysteries have been revealed. I simply affirm the temperate approach most of us have already taken. We are not an independent discipline. We are not even a subdiscipline of anything else. It is our meeting like this that creates our identity, and when we part, our identity blends, like protective mimicry, into the network of science.

This meeting has included an incredibly diverse group of papers, and yet there is a common urge that brings our disparate interests together. We want to know what people thought, how they felt, how they lived. Whatever names we choose to call this enterprise, it is still the same thing: our quest to understand the interaction between the brain and sky.

Archaeoastronomy is useful. It has a humanizing influence. It changes the way we look at people, at our ancestors. Through that it changes the way we look at ourselves. And attitude is the real bonanza. Whenever you have something that prompts people to see themselves and the world more clearly, more sympathetically, more humanistically, you are talking about enhancing social cohesion and the richness of life. That sounds like a tool for survival to me.

8 North America: A Multiplicity of Astronomies

Ray A. Williamson

In the last ten years, we have witnessed a remarkable growth of interest in traditional cosmologies of the Americas, both prehistoric and historic. Perhaps our Western-oriented scientific viewpoint had previously blinded us to appreciating the fact, but Native American astronomical concepts are enormously complex. Native Americans not only have had a keen interest in the heavens, they have demonstrated a strong desire to integrate astronomical knowledge into their total conceptual framework. Thus their astronomy, in sharp contrast to Western historic astronomy of the past two millennia, has tended to be more down to earth, less abstract, and apparently less the province of a class of specialists. Native American astronomy is indissolubly linked with ritual practice and related to natural, earthbound cycles.

Nowhere in the Americas is this more evident than in North America, where, as we are beginning to understand, there is a rich complexity of astronomical practice incorporated into the ceremonial life of the Native American. As is evident in the papers presented at this conference, great strides have been made in understanding the cosmological views of native North American groups and their use of astronomy to organize calendrical systems. In this review I will point to a few milestones along the way which have appeared since our last major conference at Colgate University in 1975 (Aveni 1977a).

Beginning geographically on the west coast, the work that comes immediately to mind is Travis Hudson and Earnest Underhay's studies which resulted in *Crystals in the Sky* (1978). Relying heavily on John P. Harrington's vast store of ethnographic data as well as their own field studies of Chumash rock art and lore, they have put together an impressive, if sometimes highly speculative, study of Chumash astronomical knowledge.

As those of us who have attempted to understand or interpret rock art are well aware, the potential pitfalls along the path are deep, and clarity about the significance of any particular painting, carving, or panel is extremely difficult to achieve. Hudson and Underhay have tried to relate ethnographic data to interpretations of some Chumash rock art and have shown, I think persuasively, that the Chumash shamans had a tradition of using their astronomical lore and observations to predict the weather, set the calendar, and to heal the sick and troubled.

The ethnographic data they used are a good example of the wealth of material which anthropologists gathered early in this century and which contains numerous references to stellar and solar lore and practice. Fragmentary, incomplete, but tantalizingly rich, the data of Harrington and other ethnographers are enormously helpful in unraveling the prehistoric interest in astronomy as well as in understanding the practice of historic Native American groups who still maintained vital societies less than a hundred years ago.

As Hudson and Blackburn (1978) have further emphasized, the motions of celestial gods and goddesses, personified by sun, moon, north star, and morning and evening star, were carefully observed as a reflection of the cosmic balance. The Chumash and other California shamans made a conscious effort to know, understand, and manipulate changes in the celestial scene for the ultimate good of society. They understood that changes in the natural cycles around them were closely linked to changes above. In particular, they understood the critical nature of the solstices. Maintenance of the proper course by the sun was essential to maintaining the cosmic balance. Among the shaman-priests' duties was that of encouraging the sun to return along its accustomed path, especially at the winter solstice.

Observations of the solstices were wide-spread in California. As Hudson et al. (1979:41-43) have shown (Fig. 8.1), at least 34 out of 49 California tribes observed either the winter solstice or both solstices. None apparently observed merely the summer solstice. Their methods, where known, vary considerably, but they commonly used a horizon calendar for direct observations of the sun. They also probably made use of caves and other natural features, together with rock paintings, for indirect observations and to celebrate the solstice in ritual activity (Hedges 1977; Hudson and Underhay 1978:51-73). Though data are sparse, it appears that in many of the California cultures the individuals in charge of watching the skies for signs of imbalance were also the time keepers (Hudson et al. 1979:53). In the structure of their societies, these astronomer-time keepers were also members of the elite, both by class and by the power their knowledge gave them, for as shamans they were expected to maintain the earthly balance and to manipulate the celestial one.

Largely because of the dual role these astronomer-shamans played in the California tribes, it may be possible today to trace the places their prehistoric predecessors observed from as well as to determine some of their methods and modes. Dorothy Mayer (1975; 1977) has suggested that some California rock art may depict stellar patterns. In another context (Mayer 1976; 1979) she has pointed to the possibility that rock art in California and elsewhere may contain pictorial references to the solstices or other solar events. On the assumption that such depictions may also indicate observatory sites, Hudson et al. (1979:54-58) surveyed the California rock art literature to find motifs suggestive of solar-horizon relationships. They located some 27 potential solstice observation sites by this method (Fig. 8.2). Proof of their hypothesis awaits further testing. Suffice it to say here that they and their colleagues have demonstrated beyond a reasonable doubt that rock art was an important component in solstice observations at some sites, a practice which was (as we shall see) part of Southwest astronomical tradition as well.

Several investigators have also suggested that Native Americans

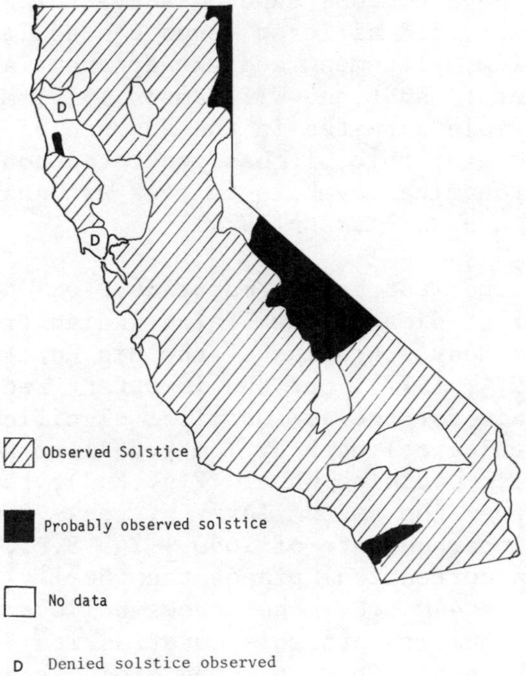

Fig. 8.1 Distribution of solstice observations among California Indian groups (Courtesy Travis Hudson)

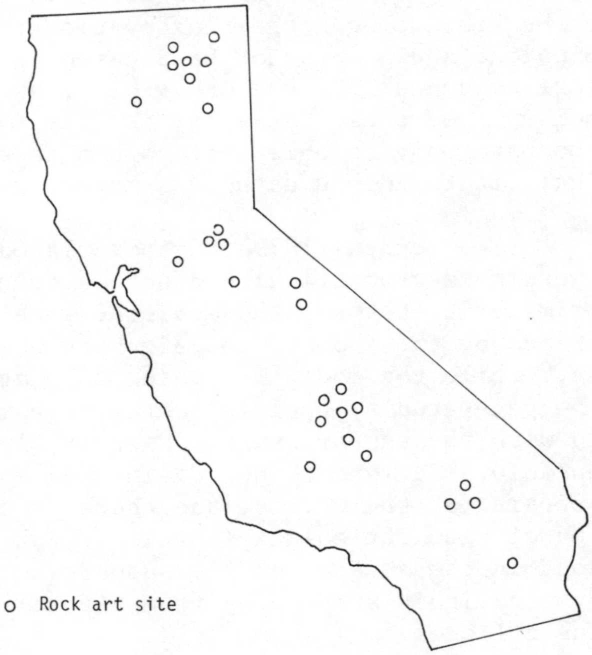

Fig. 8.2 Distribution of California rock art sites that may also be locations of direct solstice observations (Courtesy Travis Hudson)

recorded specific astronomical events. Since William Miller (1955) first suggested that a petroglyph and a pictograph depicting a crescent and circular object together were records of the A.D. 1054 supernova event (the Crab Supernova), investigators have identified as least seventeen additional western examples in which the geographical placement and the apparent age are appropriate for records of the event (Brandt and Williamson 1979; Mayer 1979). Although the hypothesis must remain untested in the absence of more accurate dating of rock art than is now available, it has generated considerable interest in the study of astronomical symbols in rock art and in Native American art as a whole (Brandt and Williamson 1979).

Moving north and east from the west coast, major developments have taken place in archaeoastronomical studies of the Plains Indian tribes. Nearly all of us are familiar with John Eddy's studies of the Big Horn and Moose Mountain medicine wheels (Eddy 1974; 1977), but the important recent finding which lends credence to his claim for the astronomical significance of these wheels is the result of archaeological work by Tom and Alice Kehoe (1979). In excavating the Moose Mountain medicine wheel (Fig. 8.3), they found numerous small bits of charcoal from the bottom layer of rocks composing the central cairn to have a radiocarbon date of 2650 $\pm$ 245 B.P., which (with the appropriate radiocarbon correction) places the wheel's initial construction date between 1080 and 440 B.C.; the Kehoes believe the later date to be the most likely one. Thus its probable construction date is only a few hundred years prior to the best fit Eddy found from his astronomical alignments--A.D.0. In their excavations of Cairn E and the inner ring, they find no evidence to suggest a different construction date for the rest of the wheel. It is highly significant that Eddy's work was completed before the Kehoes began their excavations; their findings strongly support Eddy's conclusions and he can not be accused of forcing his data to fit archaeological evidence. On the other hand, as Michael Ovenden has argued, the near coincidence of these dates may be only that--coincidence. After all, the Kehoes have only a single radiocarbon date, and determining the best astronomical fit to Eddy's data is a matter requiring considerable judgment.

Ovenden (1979) has postulated that the basis for the design of the wheels is geometrical and not astronomical. He further suggests a striking similarity between the medicine wheels and British megalithic structures described by the Thoms. The Big Horn wheel shows similarities to "flattened ovals," while the Moose Mountain stone arrangement is egg-shaped. Both forms can be generated by applying certain geometrical principles which have nothing to do with the sun or stars. However, I for one find Eddy's position more acceptable. For one thing, Plains Indian tradition supports an astronomical hypothesis for the wheels, and there is no similar tradition known to support the purely geometrical explanation. For another, there seems no compelling reason, in the absence of any supportive data, for extending a hypothesis about megalithic structures from a vastly different time and place to the Plains cultures.

The medicine wheels, understood as celestial observing structures, are quite different from the horizon calendar so commonly used in the Southwest (Stephan 1939:Map 4). Instead of using specific horizon features as foresights for marking important celestial directions, the builders of the medicine wheels included both fore- and back-sights within the structure. This seems to me to be particularly appropriate to a nomadic culture that might visit the site only during a single season. Everything one needs to know to witness a particular astronomical event is contained within the structure itself. Its users needn't have been particularly familiar with the distant landscape, but only had to understand the function of the various pairs of cairns.

At the Big Horn and Moose Mountain medicine wheels, both sunrise and sunset are observable at the summer solstice, an important redundancy at sites where cloudy weather is likely to be common in mid-June. In addition, the potential for observing the heliacal rise of Aldebaran just before the summer solstice heightens the probability of finding the day of summer solstice. Here, the stellar alignment substitutes for the "offset" which is apparently in use at Hopi Pueblo for obtaining greater accuracy.[1]

By contrast, in the Southwest the more sedentary Pueblo peoples make considerable use of a place-specific horizon calendar. In other words, by standing at a particular place and watching the cycle of the sun throughout the year, the Pueblo sunwatchers are able to tie their agricultural and ceremonial calendars to sunrise or sunset behind particular horizon features. Certain Hopi ceremonies, such as Wüwüchim, an initiatory festival which starts the new year; Soyal, the winter solstice ceremony; and Niman Kachina, the last kachina dance of the ceremonial season, are based on solar observations. Others, such as Powamu, the February bean-planting ceremony; Palulukonti, the ceremony of the horned water serpent; and the snake dance, are all regulated by observations of lunar phases. It is of some interest to understand how the Hopi reconcile the two calendars. The problem, of course, is that the solar year is about eleven days longer than 12 synodic lunar months. This means that every three years or so they must add a month to bring the two calendars into agreement.

In a very interesting study of this problem for the Hopi ceremonial-agricultural calendar, Stephen McCluskey (1977) has applied the methods of Fourier analysis to the dates of important ceremonies to determine how and when they add extra days to bring the two calendars into coincidence. This is an excellent example of the success of an interdisciplinary approach--the use of mathematical techniques originally developed to solve heat-transfer problems in physics applied by an historian of science to a Native American calendar. McCluskey has been able to show that with Wüwüchim each year the lunar and solar calendars are "first fully reconciled, which is appropriate for the month beginning the new year" (1977:190).

The horizon calendar is the simplest form of calendar; the observer merely needs a convenient place to stand and horizon features which can be used as mnemonic devices. They are, or have been, in use among all the Pueblo peoples. This is in sharp contrast to the Plains Indian tribes, who no longer employ medicine wheels for determining summer solstice, although at the beginning of the century they did apparently use the Big Horn wheel

---

[1]At Hopi, the sun is observed every day, and the announcement for the commencement of the summer or winter solstice ceremonies is made well before the solstice, when the sun is moving more quickly than at the solstices, thus making accurate observations possible.

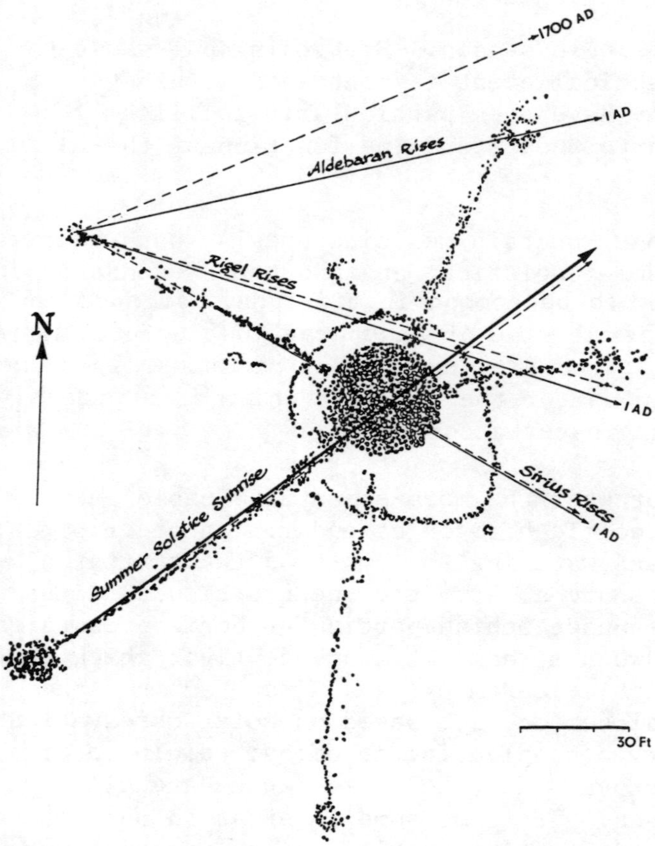

Fig. 8.3 Moose Mountain Medicine Wheel, Canada (Drawn by J. A. Eddy)

Fig. 8.4 Solar observation site near Holly House, Hovenweep National Monument

for ritual purposes (Simms 1903). Today, conservative Hopi and other like-minded Puebloans still use a horizon calendar for planting, for harvesting, and for dating ceremonies. Most of our data on these calendars derive from the Hopi and Zuni, although Florence Ellis, by careful search of the ethnohistoric literature and through her own studies, has shown (1975; personal communication 1979) that the Rio Grande Pueblos had (and still do have to a certain extent) a rather elaborate lunar-solar horizon calendar as well.

These calendars, as simple and obvious as they are to construct for a society whose detailed frame of reference changes but little over many years, must have prehistoric antecedents. The ethnohistoric literature contains many useful clues about what to look for, and in that literature perhaps the most significant and most often quoted bit of data is from Frank Cushing:

> Each morning, too, just at dawn, the Sun Priest, followed by the Master Priest of the Bow, went along the eastern trail to the ruined city of Ma-tsa-ki, by the river-side, where, awaited at a distance by his companion, he slowly approached a square open tower and seated himself just inside upon a rude, ancient stone chair, and before a pillar sculptured with the face of the sun, the sacred hand, the morning star, and the new moon. There he awaited with prayer and sacred song the rising of the sun. Not many such pilgrimages are made ere the "Suns look at each other," and the shadows of the solar monolith, the monument of Thunder Mountain, and the pillar of the gardens of Zuni, "lie along the same trail." Then the priest blesses, thanks, and exhorts his father, while the warrior guardian responds as he cuts the last notch in his pine-wood calendar, and both hasten back to call from the house-tops the glad tidings of the return of spring. Nor may the Sun Priest err in his watch of Time's flight; for many are the houses in Zuni with scores on their walls or ancient plates imbedded therein, while opposite, a convenient window or small port-hole lets in the light of the rising sun, which shines but two mornings in the three hundred and sixty-five on the same place. Wonderfully reliable and ingenious are these rude systems of orientation, by which the religion, the labors, and even the pastimes of the Zunis are regulated [Cushing 1967:40].

Williamson and Young (1979) and Williamson et al. (1977) have used the suggestion that sun-watching places will be marked by recognizable iconographic elements to identify several possible sun-watching sites. The most significant is at Holly House in Hovenweep National Monument, Colorado (Fig. 8.4). There, petroglyph elements with astronomical motifs combine with local natural features to produce calendrically-related displays of light and shadow. The petroglyph panel contains two spirals, a sun symbol, a twin-like figure, a possible plumed serpent, and two S-shaped figures. Except for the spirals and the *S*s, each of the panel elements has a meaning related to the sun in modern Hopi and Zuni tradition. The concentric circles represent the sun, the twin figures may represent the morning and evening stars (the Twins of Hopi and Zuni myth), and the plumed serpent is the guardian of springs and waterfalls. The plumed serpent, it should be noted, has a central place in the Hopi and Zuni winter solstice and spring equinox ceremonies. As we are finding out from recent related research, the spirals, considered to be representative of water, may also have been related to the sun in prehistoric times (Sofaer 1979a; 1979b).

The connection to the calendar can be described by the following sequence. The petroglyph boulder and another large boulder north of it form a narrow corridor which opens to the eastern horizon. From about February 20 to March 21, the rising sun shines into the corridor and illuminates the serpent and the twin-like figure. For a month, boulder C obscures the sun, but around April 10 and for another four weeks or so, the sun symbol is illuminated briefly at sunrise. The sun is then obscured again until near the summer solstice, when the sun is far enough north to shine through a narrow slot and once more illuminate the sun.[2] In the sun's southward journey, the process is reversed. This sequence of dates is remarkably similar to those which the Hopi use. Williamson and Young suggest that the inhabitants of Hovenweep used these naturally-occurring boulders as a calendar device, and that they placed the petroglyphs on the panel in order to aid in determining the calendar.

Anna Sofaer and her colleagues (1979a) have reported on a prehistoric spiral petroglyph on Fajada Butte in Chaco Canyon which also seems to have served as a calendrical marker.[3] There, three vertical slabs form two slots which create two vertical daggers of light as the sun passes horizontally overhead near local noon. On the summer solstice, the eastern dagger of light passes through the center of the eastern and larger spiral. The western dagger is barely visible at this time. Later in the year, the western dagger of light becomes larger and more visible. At the winter solstice, the two light daggers bracket the larger spiral. This calendar construct is particularly interesting because it is the first known ancient example of marking the solstices at noon rather than at sunrise or sunset. Among historic Pueblo tribes, the Isleta (Parsons 1930a:292) are known to have marked the arrival of the summer and winter solstice suns with noon ceremonies. It would perhaps be fruitful to explore the possible formal relationship between the Isleta practices and the Chaco Canyon solstice marker.

It should be noted that the Cushing quote also mentions the use of small ports or loopholes for observing the sun at calendrically significant times. Williamson et al. (1977) have investigated several towers at Hovenweep which were most probably used in part for observing the sun. Erected in the late twelfth and mid-thirteenth centuries by a group of Anasazi related to the Mesa Verde people, these towers were originally two or three stories high and contained several rooms. They are either round, square, or "D" shaped and are usually situated on the very edge of the canyons or on large boulders. Part of one such building, Hovenweep Castle, was constructed in about A.D. 1200. Several generations later it was enlarged by the addition of several rooms. One of these rooms is apparently a solar observing room. While constructing the room, the Anasazi who built it incorporated two special ports or sighting holes into the walls, one aligned to the summer solstice sunset, the other to the winter solstice sunset (Fig. 8.5).

---

[2]Following the Santa Fe Conference, several participants journeyed to the site to witness a remarkable display of light daggers which occurs around and on the solstice. Though a solstice function for the site had been suspected, this was the first confirmation of such a possibility (Williamson 1979).

[3]A detailed account of the Fajada Butte solar marker can be found in the paper presented at the Santa Fe Conference and published in *Science* (Sofaer et al. 1979b).

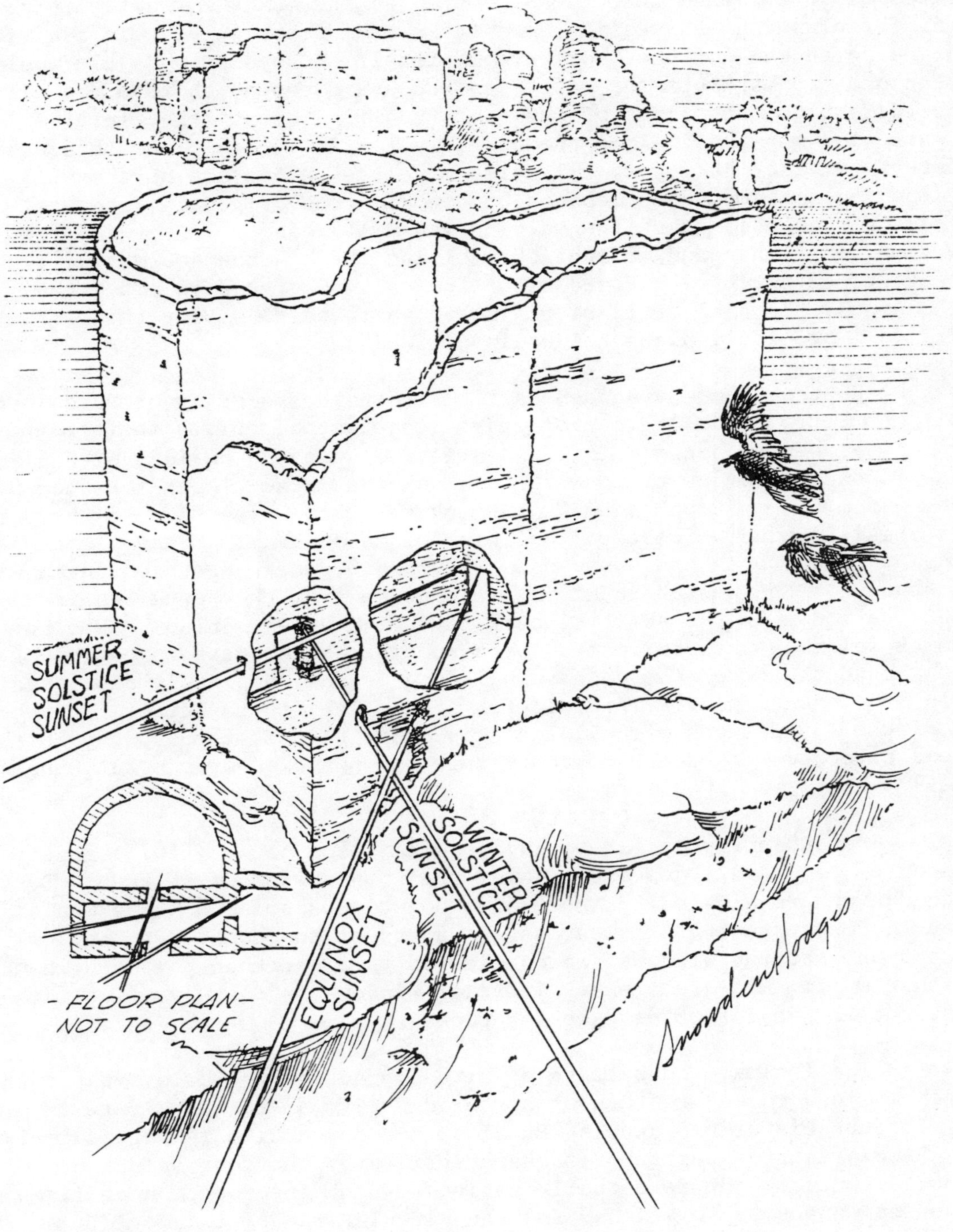

Fig. 8.5 Hovenweep Castle, Hovenweep National Monument (Drawing by Snowden Hodges)

In addition, the Anasazi designed the door to the outside so its jambs align to sunset on the equinox. The fact that the sunlight falls along a line between the port and a corner of a doorway or between door jambs leads us to believe that the Pueblo people of the area were using this building to make calendar observations 700 years ago. Hovenweep Castle incorporates clear evidence that these people had the means for determining a calendar for planting, harvesting, and for ritual observances. In addition, the inside north wall of the "solar room" itself at Hovenweep Castle would have been used to mark off the days. As the sun moves north from the winter solstice direction, the light of the sun streaming through the winter solstice port will move east along the north wall. As it does, the shadow of the western exterior door jamb will also move east along the north wall. After a period of time which depends on how wide the original winter solstice port was, the light of the setting sun will cease to enter the port. At the same time, the band of sunlight falling through the doorway will narrow and continue to move eastward along the wall. A few days after the true vernal equinox, the band of light will have narrowed to a very thin strip and will fall precisely along the equinox line.

Southwest of Hovenweep Castle, in another section of Hovenweep National Monument, a small group of ruins (Cajon) collaborate to form another rather unusual set of astronomical alignments. Here the buildings (Fig. 8.6) interact in such a way as to mark the equinoxes and the winter solstice by shadows. From the autumnal equinox through the winter solstice to the vernal equinox the left-hand, or western, building casts a shadow on the right-hand building at sunset. As can be seen in Figure 8.6, the eastern building will cast a shadow on the western building at sunrise from the vernal equinox through a period near the solstice back to the autumnal equinox. Whether the summer solstice was ever marked by a sunrise shadow-cast is not clear because the north end of the eastern building is now completely destroyed. It would be of considerable interest for archaeologists to investigate the extent of the foundations of building A in order to test the hypothesis that it indeed extended far enough north to mark the position of the sun's northerly extreme.

Jonathan Reyman (1976; 1977), Williamson et al. (1977), and Williamson (1977) have shown that two of the corner windows of Pueblo Bonito in Chaco Canyon are aligned to the winter solstice sunrise. These would function in very much the same way as the ports of the houses at Zuni, with the exception that the windows are considerably larger than ports (0.75 m rather than 0.1 m) and are clearly interior windows or doorways (i.e., they exist at the *inside* corner of exterior rooms).

The intimate connection of ritual practice to worldview and thence to astronomy is a marked feature of historic Native American culture as a whole. For the historic Southwest Pueblos, the connection is particularly clear. Pueblo sacred geography is characterized by the four solstice directions, a center which is ritually defined by the intersection of lines drawn between opposite solstices, and the circular rim of the world. At Zuni and among the Tewa Pueblos, the center of space and the center of time (the winter solstice) coincide--both are termed "the middle" (Ortiz 1972:143). The twin sons of the Sun, represented in the sky by the Morning and the Evening Stars, are venerated in Pueblo mythology as the helpers and protectors of the Pueblo people. Their dualism is one manifestation of several cosmic

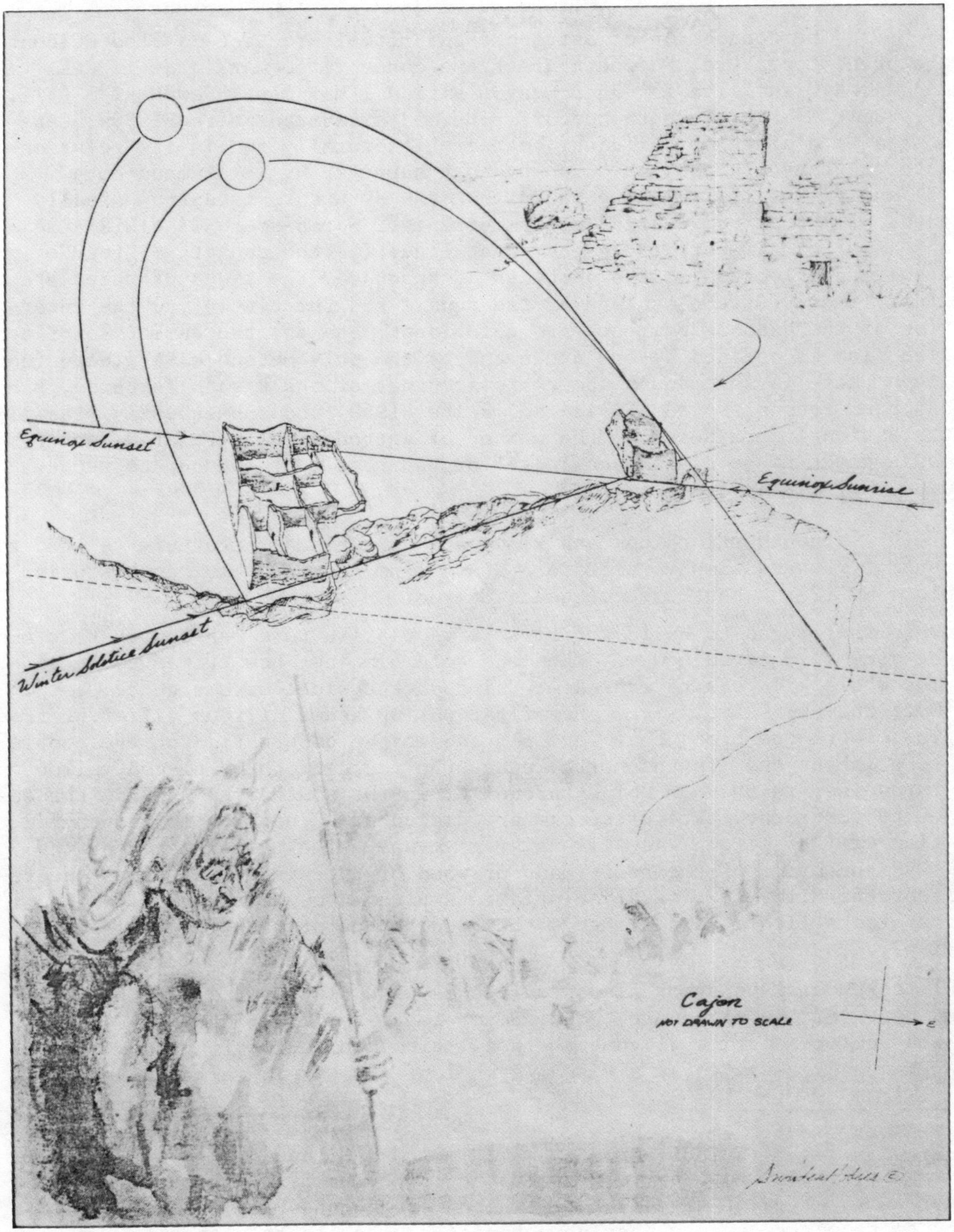

Fig. 8.6 Cajon Ruins, Hovenweep National Monument
(Drawing by Snowden Hodges)

dualities in Pueblo life (e.g., sun:earth, summer:winter, above:below).

The connection of astronomy and ritual practice is also evident in the prehistoric era. Nowhere are these concepts so marked as in Casa Rinconada, a great kiva in Chaco Canyon National Monument, New Mexico (Fig. 8.7). Built in the eleventh century A.D. and excavated in the 1930s, Casa Rinconada is a singular building, both architecturally and in its relationship to the canyon's pueblos. Two transit surveys, by Williamson et al. (1977) and Fisher (1978), have shown that the structure is astronomically aligned: (1) the "major axis" of symmetry is a north-south line; (2) the "minor axis" of symmetry is an east-west line; (3) the sockets of four original roof poles are located at the vertices of a square, the sides of which are cardinal lines; (4) the centers of the square and the circle and the intersection of the axes of symmetry are coincident; and (5) the summer solstice sunrise line is defined by one niche and by the only extant kiva window (door). Although there is some doubt about the accuracy of the kiva's features, since it has been reconstructed (Vivian and Reiter 1960) and subsequently stabilized by the National Park Service, the number of astronomically-related features seems too great to be either accidental or the result of inaccurate reconstruction by Vivian and Reiter.[4]

In understanding Casa Rinconada as a ritual structure related to astronomy, we need simply refer to a story from Acoma Pueblo, an historic pueblo some 125 km southeast of Chaco Canyon:[5]

> When they began to build the first kiva, Iatiku told Oak man that it must be done in a certain way. Then she told him just how it was to be done. The whole kiva was to represent Shipapu, the place of emergence. . . . When they built the kiva, they first put up beams of four different trees. These were the trees that were planted in the underworld for the people to climb up on. In the north, under the foundation they placed yellow turquoise; in the west, blue turquoise; in the south, red; and in the east, white turquoise. Prayer sticks are placed at each place so the foundation will be strong and will never give way. The walls represent the sky, the beams of the roof (made of wood of the first four trees) represent the Milky Way, wakaiyanistiawi'tsa (way-above-earth beam). The sky looks like a circle, hence the round shape of the kiva [Stirling 1942].

Pueblo ceremonial behavior also provides us with clues with which to understand parts of the structure. The number four, the coincidence of centers, and the summer solstice alignment are all elements of Pueblo ritual. Thus Casa Rinconada seems to be a visual symbol of the way in which the Anasazi

---

[4]Later in the summer of 1979, with the help of the staff of the Laboratory for Anthropology of the Museum of New Mexico, I located transit measurements of major kiva features which were made during excavation and prior to reconstruction. They do not differ significantly from my own recent transit maps. That it is important to verify the existence of features in a reconstructed building is exemplified by Reyman's experience (1978); he "discovered" a window alignment at Wupatki ruin which later turned out to be an artifact of reconstruction.

[5]Although according to Acoma legend the first kiva was circular, historic kivas at Acoma are rectangular.

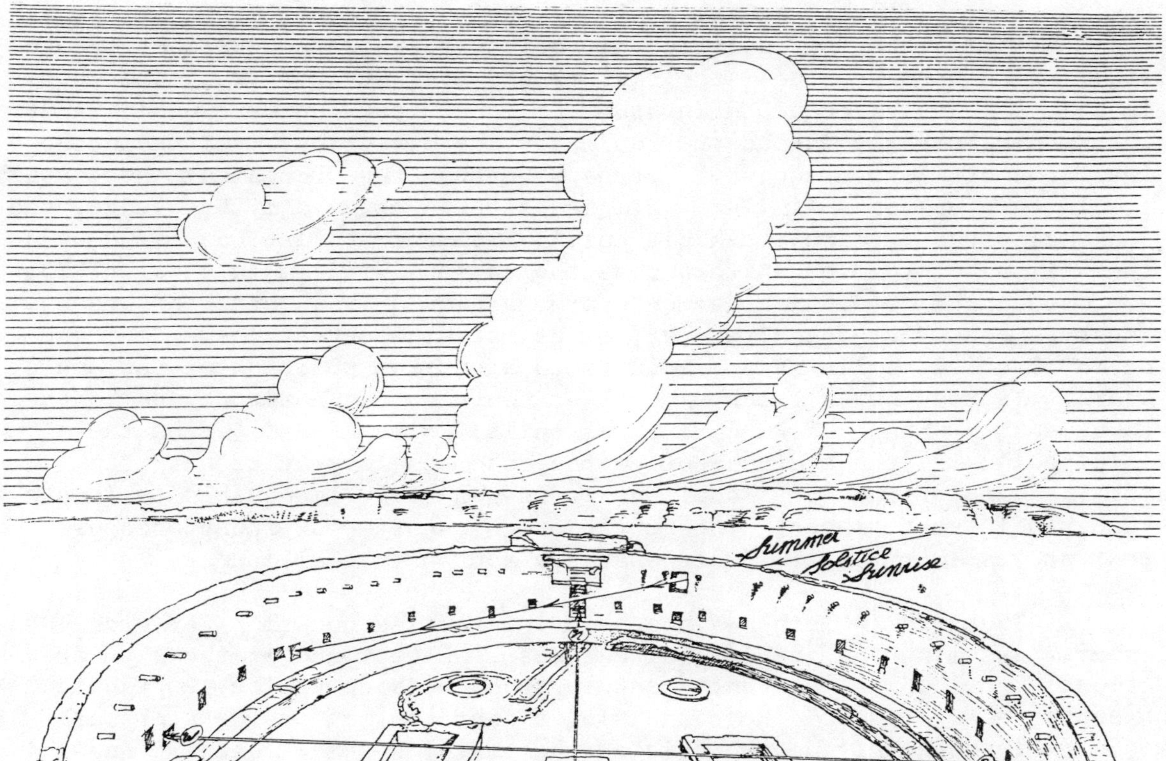

Fig. 8.7 Casa Rinconada, Chaco Canyon National Monument
(Drawing by Snowden Hodges)

of Chaco Canyon saw their world.

The astronomical features which cannot be explained in historic Pueblo ritual and myth are the alignments to the cardinal directions. When the Hopi priest casts sacred corn meal to the four directions in his ritual observances, it is the four solstice directions to which he gestures (Fewkes 1892:16). When the Zunis tell their story of Water Strider, who stretches his legs in the four directions in order to locate the Center (Cushing 1896: 415), they also clearly intend the four solstice directions, not the orthogonal cardinal ones of Western society. Perhaps further research among the Anasazi ruins will turn up other similarly aligned structures. To my knowledge, the only Anasazi structures that exhibit cardinal alignments are Casa Rinconada and the village of Pueblo Bonito across the canyon (Williamson et al. 1975; 1977). Orthogonal cardinal alignments are particularly interesting because they must be constructed indirectly from celestial observations. They cannot be constructed by lining up features to a horizon position of the sun or the stars. The sun never reaches north or south on the horizon, and in the eleventh century when Casa Rinconada was built, Polaris was 7° from the celestial pole. At the equinox, of course, the sun lies quite near the celestial equator for about a day, but it is generally impossible to locate true geographic east or west by observations of sunrise or sunset because the east or west horizon will not be at the appropriate altitude.

Thus, Casa Rinconada poses an interesting question for archaeoastronomy. Assuming the cardinal directions we have found in the kiva are intentional, how did its eleventh century architects determine geographical north-south and east-west? Methods using either the sun or the stars are possible. Given the strong emphasis on the sun, I suspect that the Anasazi must have used shadow-casting to locate the meridian and then drawn a perpendicular to it to find east-west. This, of course, must remain speculative until we find further clues. How singular are the cardinal orientations of Casa Rinconada and Pueblo Bonito? Is their apparent uniqueness merely a sampling selection effect, a function of their easy accessibility to archaeoastronomical studies, or is it truly a reflection of how the Chaco Canyon Anasazi differed intellectually from their Southwestern neighbors? The precision with which the purely geometrical features of Casa Rinconada were constructed suggests that the Chacoans had even developed *concepts* of the square, circle, and perpendicular lines which went beyond mere utility or the straight-forward desire to image their cosmology in stone. Can such a conclusion be supported by other Anasazi constructions?

These are all questions which will be answered, I hope, by the continued development of archaeoastronomical studies in the Southwest. Many new archaeological data are being gathered on the Anasazi, particularly from villages whose builders were related to the Chacoans. It would be of considerable interest to undertake a systematic investigation of all the Chaco-related villages which have already been mapped in order to determine how wide-spread their inclusion of astronomical alignments might be. Such an investigation might also provide the basis for determining how uniform their astronomical practices were. At this stage, archaeoastronomical investigations in the Southwest still lack the benefit of a systematic overall approach. Some areas, such as Hovenweep, have been thoroughly studied. Others, such as Chaco Canyon, have been studied in a haphazard way, but there still remains much to do. Thus, it will be important in the coming years to learn just how wide-spread and how uniform Anasazi astronomical practice was as a whole. The Anasazi clearly traded back and forth over hundreds of miles. In their trading efforts, did they also take along their ideas and share their knowledge of the sun and stars?

A similar question has intrigued North American archaeologists for years. To what extent did Mesoamerican trading contacts influence North American cultures? Although the archaeological record suggests numerous trading contacts (Kelly and Kelly 1974; DiPeso 1968; Wiegand et al. 1976) and some possible influence in the form of architectural details (Ferdon 1955), archaeoastronomy as yet provides nothing to support or deny a substantial cultural interaction.

Wimberly and Eidenbach (1977) and Eidenbach (1979) have found evidence of two potential observatory sites high in the Sacramento mountains of New Mexico. Wizard's Roost (10,360 ft.), and Wally's Dome (6500 ft.) southeast of it, are both constructed of dry-laid masonry. The latter site appears to possess summer solstice sunrise and sunset alignments, as well as some stellar alignments, while the former appears to have alignments to Sirius, Aldebaran, and Vega. Clearly both sites need further study, to determine the accuracy of the alignments and to discover their cultural affiliations (they are in the region inhabited by the prehistoric Mogollon, but as yet no datable artifacts have been found at the sites). What makes each of them particularly interesting is the fact that they lie along a line defined by the winter solstice sunrise and the summer solstice sunset. If they prove to be Mogollon upon further study, and if the indicated alignments appear intentional, they could provide another important link between Mexico and the Southwest. Schaafsma (1980) has already demonstrated that there was a movement of rock art iconography northward from the Mogollon country into the Rio Grande. Numerous dry-masonry circles and other structures were documented by Douglass (1912) and Renaud (1942) in early archaeological surveys of the northern Rio Grande area. Some of them seem to possess astronomical orientations. They should also be studied for their intrinsic interest as well as for their possible cultural affinities with the structures at Wizard's Roost and Wally's Dome.

With the possible exception of Casa Grande, for which there may be evidence of lunar alignments (Molloy et al. 1973), and the postulated stellar alignments at the Sacramento Mountain sites, Southwestern prehistoric astronomy seems to be entirely solar, with an emphasis on the solstices, and to date there is no known feature of Pueblo or Navaho astronomy which can be identified as a Mesoamerican invention. One might look for an interest in Venus, but although there is a strong emphasis among the historic Pueblos on the Morning Star and the Evening Star, they do not seem to identify them solely with the appearance of Venus. The Zuni, for example, time certain ceremonies by the Morning Star, but since the ceremony must occur at a specific day in the solar year any convenient Morning Star will do.[6] It is, of course,

---

[6]In her description of the Zunis, for example, Stevenson (1904: 130) notes that the New Fire ceremony began with the appearance of the Morning Star. On January 1, 1892, the date of this ceremony, Venus was an evening star. The only planets which could have been used as timing devices

quite possible that historically-recorded practices are remnants of earlier Mesoamerican astronomical notions which travelled north along with certain religious deities. Ellis's recent investigations (1976) into the striking similarities between Mesoamerican and Southwestern deities, including Morning and Evening Star, suggest that astronomical knowledge could have travelled as well. Clearly, a great deal of further ethnographic work is needed here in order to clarify the possibilities for archaeoastronomical research.

Another area of the North American continent where Mesoamerican influence has been suggested is along the Mississippi River, and in particular at Cahokia, the site of the largest Native American mound north of Mexico (Monk's Mound). Cahokia is a large, well-organized, and complex mound city of the Mississippi culture whose builders were an agricultural people who flourished between the ninth and sixteenth centuries A.D. The flat-top mounds which are so typical of Cahokia have led to suggestions that they are evidence of Mesoamerican influence. However, as Reed (1973) has argued, they can be explained quite well in terms of indigenous development. In this connection, however, archaeoastronomy may provide other clues. Near Monk's Mound, Warren Wittry has discovered a circle and several arcs defined by the remnants of post holes. One of the circles, dubbed an "American Woodhenge" by Wittry (1970), seems to have astronomical significance. Wittry first investigated the circles of Cahokia in the mid-60s, and at that time focused on the solstice and equinox alignments in the complete circle, Number 2. In more recent work (Norrish 1978), he reports unearthing more post holes which indicate the presence of several circles, each with similar radii but different post spacing and centers. Wittry believes that there are at least three and possibly four circles or arcs of circles in the immediate vicinity of the circle (labeled Circle 2) that he found in his original excavation of the site (Fig. 8.8).

In spite of this intriguing new development, Circle 2 remains the best understood astronomically. Several facts support Wittry's earlier hypothesis as to its astronomical nature. An unusual potsherd, found in a firepit superimposed on the rampway which was dug to aid in placing the winter solstice post, contains a design which could be interpreted as signifying the winter solstice. The central design, vaguely reminiscent of quartered-circle designs found in Southwest Pueblo rock art (Packard 1974:6, 7, 25), is surrounded by a circular broad band from which, on the lower right, leads a straight band (Fig. 8.9). Wittry understands the "open pathway" to signify the winter solstice sunrise, an interpretation which seems plausible since the sherd was found in front of the winter solstice post. Of more interest, however, is the suggestion that two of the posts in Circle 2 are aligned to Capella rise and set (A.D. 1150). Wittry postulates that the architects of the circle originally set it out using the Capella alignments, and placed the solstice and equinox markers afterwards; in this way, he can explain the placement of the observation post, which is five feet east of the circle's center. In A.D. 1150, Capella would have been seen to rise and set along lines drawn from the center of the circle, at which there is no remnant of a posthole. With the radius of the circle determined by the

---

were Mars and Saturn. Since it rose just before dawn, Saturn seems the more likely possibility. Starting the New Fire ceremony with the appearance of Morning Star *may* have a Mesoamerican origin, but more evidence is surely needed to assess that hypothesis.

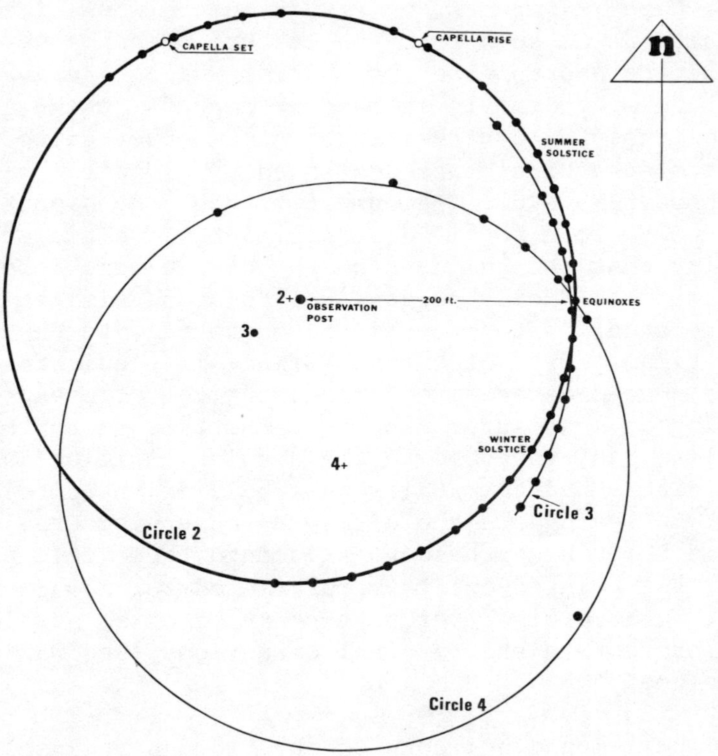

Fig. 8.8 Posthole arcs at Cahokia Mounds State Monument (Courtesy Warren Wittry)

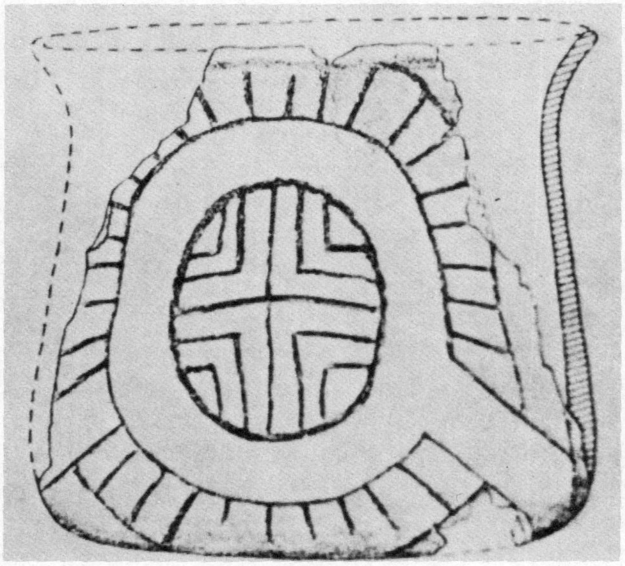

Fig. 8.9 Potsherd decoration, Cahokia Mounds State Monument (Courtesy Warren Wittry)

placement of the Capella markers, only pure chance would lead to posts at the solstice positions as seen from the center. Hence a new center must be chosen to make the appropriate angles from which to view the solstices. The problem with this hypothesis is that it depends on the circle having originally had 47 posts, a hypothesis which can never be tested because the westernmost part of the circle was destroyed by a "borrow pit" for construction of Highway I-55-70. The number 47 could be significant because 47 lunar months is very close to 51 "draconic months." This suggestion leads to the possibility that the builders of the circle were interested in predicting solar eclipses. While suggestive, this and similar hypotheses will require much more study. Ethnographic accounts fail to support any interest among Native Americans north of the Rio Grande in predicting eclipses, although there are some data on native reactions to eclipses (Schwartz 1925). I find it puzzling or even surprising that there is no north-south line in the circle, at least as presented in the drawing (Norrish 1978:8). If the builders of the circle had the skills to subdivide the circle nearly evenly, and the interest in laying out the circle by observing Capella, they most surely could have built a north-south axis into the circle as well. As Fowler (1973:19) has pointed out, several ridged-top mounds together with Monk's Mound define a cardinal set of axes at Cahokia, a relationship which confirms their interest in the cardinal directions (see Daniel-Hartung's paper in this volume).

The shell-mound builders of the Florida Peninsula also seem to have followed the annual cycle of the sun. Clark Hardman (1971), for example, has suggested that some of the shell mounds along the Crystal River in Florida may exhibit intentional astronomical alignments. Stelae (Bullen 1966) and significant mound features are placed in such a way as to suggest that the shell-mound builders observed the solstices and the equinoxes. This work has not been followed up or expanded upon to my knowledge, nor has anyone pursued a comprehensive search of the ethnohistoric literature about the mound builders (see, however, Daniel-Hartung in this volume). There are some intriguing possibilities, as the following observations of the Natchez Great Sun (chief) by Father Maturin LePetit (1699) attest: "Every morning the great chief honors by his presence the rising of his elder brother (the sun), and salutes him with many howlings as soon as he appears above the horizon . . . afterwards raising his hand above his head and turning from east to west, he shows him the direction he must take in his course." Other equally tantalizing direct references to solar observations exist, but they generally stop short of mentioning any interesting details. The literature, particularly the stories and myths which were recorded, may provide more clues and they ought to be studied.

Given the pattern of evident interest in astronomical practice among the prehistoric native population of the West and Midwest, it would be surprising indeed if that pattern failed among the Eastern Woodlands groups. The problem is that much of the evidence which might have illuminated the calendrical and observational customs of these tribes disappeared in the heavy settlement which followed the arrival of Western European and British colonists nearly 400 years ago. Furthermore, few observers of this period saw the native population as anything more than a people to be conquered and exploited. Those who did record their customs were generally uninterested in astronomical information and were certainly only vaguely aware of the

Indians' intellectual accomplishments. The rapid spread of the European colonists and their settlement of choice areas led to the destruction of many of the buildings and building foundations which might have provided us with clues about their astronomical interests. Further, the moist Eastern climate destroyed much of what the colonists and later developers did not. Here again, a comprehensive study of the ethnohistoric literature may be fruitful.

Although most of the wooden structures erected by Eastern native groups such as longhouses and dwellings have been irretrievably lost, some scholars have recently suggested that the numerous stone chambers, "standing stones," and rock carvings of New England, Pennsylvania, and New York may be remnants of an earlier North American culture (Cook 1978; Trento 1979). Some, along with Fell (1976), have rushed to the unfortunate and undocumented conclusion that these interesting stones and stone structures are the result of very early European and North African intrusions into the North American continent. Though few would deny that Europeans could have made transatlantic trips prior to the Vikings and Columbus, the data which Fell and his adherents present as evidence of such visits are poorly documented and of generally doubtful quality. In addition, the arguments they follow are composed largely of leaps of faith interspersed with unimpressive visual comparisons of New England and European similarities.

Nonetheless, the chambers are worth study. Some of the stone chambers and certain other stone features exhibit documented astronomical alignments (Dix 1977). Either they are fortuitous or they are intentional, in which case some explanation is in order. The problem is that such features have been largely ignored until recently, when several groups of amateurs "discovered" these artifacts of the New England hillsides. Because the professionals have, with few exceptions, been content to dismiss the potential of these structures right along with the clearly absurd claims of this new breed of diffusionists, little professional study or archaeology has been devoted to understanding them. Much more could clearly be done, and it would seem worthwhile to start by making a thorough archaeological investigation of structures and stone alignments which present the clearest astronomical orientations (e.g., see Dix and Mavor, this volume). In the light of what we know about the rest of North America, it certainly seems possible that these stone works, which have captured the imagination of hundreds of intelligent laymen, are the product of Native American populations who were here long before Europeans "discovered" this continent.

In spite of what we have done so far in investigating the wealth of astronomical practices of the aboriginal population of North America, we have surely just scratched the surface. In this brief summary I have sought to highlight some of the recent developments in North American archaeoastronomy, and indicate some interesting areas for further research. We seem to know more about the Southwestern and California groups than about the Eastern tribes, and we know very little about the astronomical knowledge of the Pacific Northwest. As Leonora Cope (1919) reminds us, they had calendars, and where there are calendars, there are certain to be regular observations of the sun and/or the moon.

What is surely clear is that in spite of the identical nature of astronomical appearances, individual tribes responded to them rather differently. Hence, they developed a multiplicity of astronomies, each emphasizing observations which best fit particular circumstances. Astronomy enjoys a special status among intellectual disciplines precisely because it relates to something very basic in our psyches--the need to make order out of chaos. The regular motions of the heavens served as a model of orderliness and dependability for mankind. The native tribes of North America observed this order and used it for ritual and for practical ends. Though their varied modes of observing and using the orderly progression of the celestial sphere often differed from other traditional peoples across the world, they shared with them an intense and lively interest in the sky.[7]

---

[7]The research on Hovenweep was supported by the National Geographic Foundation. Much of the research on Casa Rinconada and on general Native American archaeoastronomy was completed while I was a Smithsonian Fellow. Thanks are due to Snowden Hodges for his drawings, to Jean Monroe for typing and critiquing the original manuscript, and to Ed Krupp for a thoughtful reading.

9 Zodiac Ridge

Nev E. Autrey and Wanda R. Autrey

INTRODUCTION

Intensive investigation and mapping of a circular rock alignment was initiated by the authors during the fall of 1977 under the direction of the Pima Community College Department of Human Resources. This circular alignment, known as Zodiac Ridge (BB:5:27, Arizona State Museum), is located in the lower bajada of the Tortolita Mountains, northwest of Tucson, Arizona. The rock circle is approximately 40 meters in diameter and involves more than 800 stones of a coarse granitic structure. Most of the stones mapped are embedded horizontally in the soil; however, numerous upright stones are found in the perimeter circle. Near the center of the circle is an alignment of approximately 30 large stones oriented true east-west. It was this alignment that prompted investigation of the site.

The present paper deals with the archaeological survey and hypothesized astronomical significance of Zodiac Ridge, including the possible environmental conditions and agricultural practices utilized during its period of use by Native Americans. We also present ethnographic data from several Southwestern tribes to lend cultural support to our hypotheses. We will argue that a knowledge of horizon-based astronomy existed in the pre-Hispanic culture of the people who built Zodiac Ridge, and that these people used the rock alignments for sighting astronomical bodies as they appeared on the horizon.

MAPPING PROCEDURE

Realizing that hypotheses of an astronomical nature come under critical scrutiny from both astronomers and archaeologists, we sought a rapid yet accurate method of surveying. We elected to use a method previously employed in the mapping of circular sites in Great Britain (Atkinson 1946: 106-107).

We determined the center point of the circle from diametric measurements of the perimeter of the circular stone alignment. This center-point was not suitable, as vegetation and an animal habitat would not allow a convenient setting up of the transit and tape over the point. Therefore, we chose a datum near the center of the circle which allowed visual and linear access to all stones included in the survey. We set the transit up over this fixed, central datum point and magnetically established an approximate true north reference with a 13.5° declination correction for Tucson (Kissam 1971:75). The true north reference point was later corrected through two independent solar shots. The datum and north reference became control points for our surveying transit and tape. We held a plumb line over the marked center of a stone (determined from horizontal visual area measurements), sighted it with the transit, and recorded the azimuth from the vernier reading. We then drew the tape from the datum to the marked center of the stone and recorded the distance. This method proved extremely rapid in the field as all angles and distances could be taken in a continuous clockwise motion from one central point. After converting the transit and tape readings from polar to rectangular coordinates, we plotted the location of each stone on a map. Rectangular coordinate points allow easy mathematical calculations of bearings from point to point, and provide a data base for computerized applications.

We determined elevations in ten-degree intervals around the horizon (Table 9.1). Although the site is flanked east and west by two mountain ranges, the obstruction to a 0° horizon ranges from a maximum of +4°56' to a minimum of -0°32', which offers optimum conditions for horizon-based observations.

The resultant plot is a surface map (Fig. 9.1) of all visible stones within the survey area. The shaded area indicates the stones investigated for astronomical significance. The chance of establishing a significant astronomical alignment utilizing the perimeter stones is almost 1:1. Therefore, we decided early in the project that only alignments involving interior and exterior stones would be presented at this time.

TABLE 9.1

HORIZON ELEVATIONS, 0° TO 350°, ZODIAC RIDGE (BB:5:27)

Azimuth	Horizon Angle	Azimuth	Horizon Angle
00°00'	+0°51'	180°00'	-0°32'
10°00'	+0°49'	190°00'	-0°26'
20°00'	+0°34'	200°00'	+0°15'
30°00'	+0°30'	210°00'	+1°24'
40°00'	+0°27'	220°00'	+1°40'
50°00'	+0°36'	230°00'	+2°49'
60°00'	+1°00'	240°00'	+2°49'
70°00'	+1°29'	250°00'	+4°56'
80°00'	+2°24'	260°00'	+3°30'
90°00'	+3°02'	270°00'	+3°00'
100°00'	+3°49'	280°00'	+3°12'
110°00'	+4°30'	290°00'	+3°12'
120°00'	+4°29'	300°00'	+2°03'
130°00'	+3°42'	310°00'	+2°13'
140°00'	+4°08'	320°00'	+1°32'
150°00'	+3°42'	330°00'	+1°17'
160°00'	+2°29'	340°00'	+1°04'
170°00'	+1°42'	350°00'	+0°56'

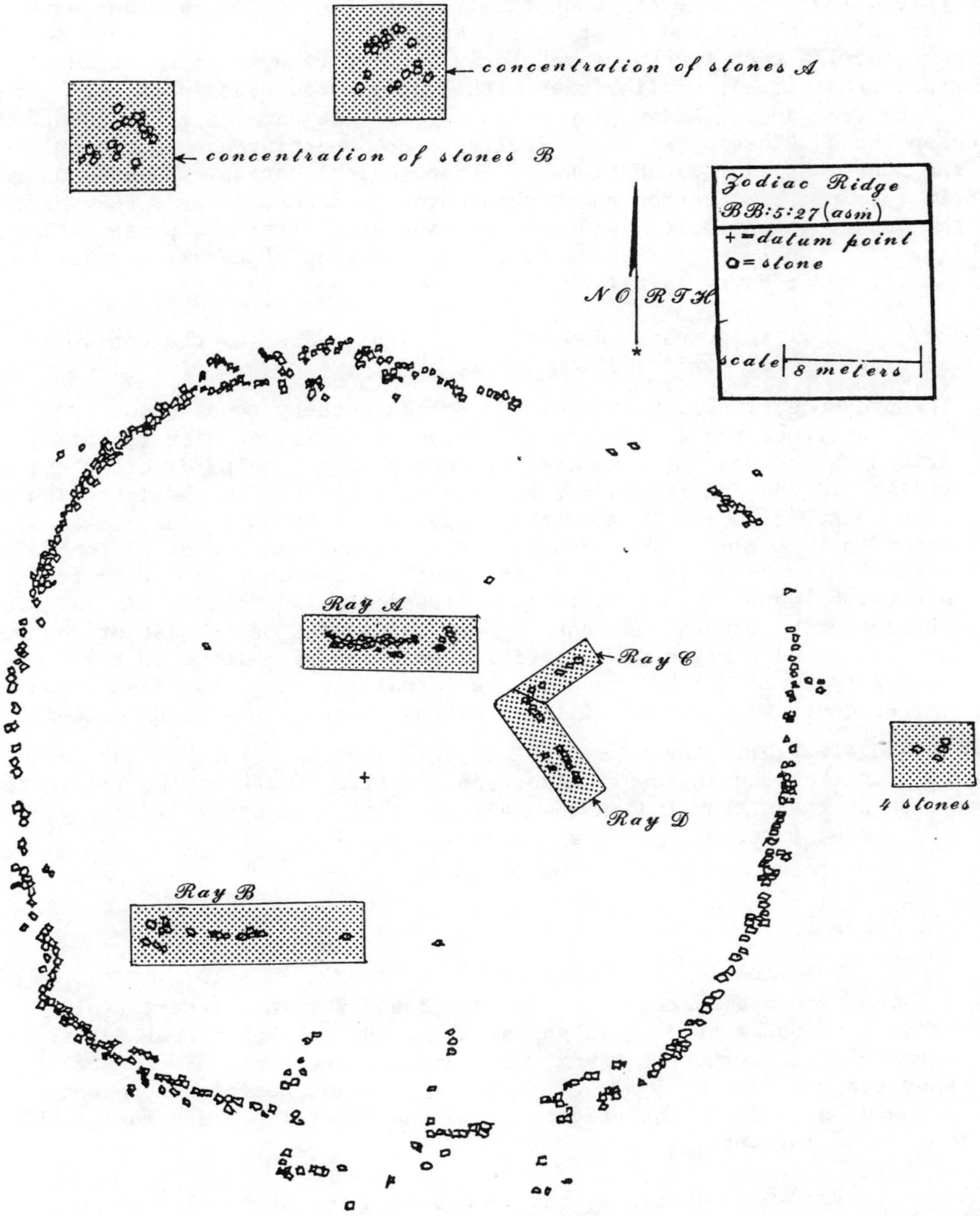

Fig. 9.1 Zodiac Ridge general site map. Shaded areas indicate regions investigated for astronomical alignments.

DATING

In order to investigate the possible astronomical nature of this site, an approximate time period for its construction was necessary for data interpretation. No date has been firmly connected to Zodiac Ridge site; however, using the method of stylistic cross-dating of ceramics and architecture (styles tend to change regularly through time, forming regionally distinct traditions), similarities with nearby dated sites when correlated with data from Zodiac Ridge give us an approximate date of A.D. 1300-1500 (Tucson Phase, Classic Period). Stylistic cross-dating has long been used in the Southwest as a foundation for chronological dating. Based upon recently obtained radiocarbon and archaeomagnetic dates, Haury's re-evaluation of the Snaketown chronology did not substantially alter the phase dating established in the 1930s on the basis of cross-dating of ceramic styles (Gladwin et al. 1938; Haury 1976).

Hewitt (private communication 1979) addresses the dating of Zodiac Ridge as follows:

> The presence of Gila Polychrome, a Salado pottery type, associated with surface structures at the nearby Indian Town village site (BB:5:25, Arizona State Museum; 2 km west of Zodiac Ridge) helps date the use of upright foundation stones for surface architecture in the Tortolita region to A.D. 1300+. This pottery type is not known in the Tucson Basin prior to this time. The presence of a few painted sherds of Tanque Verde Red-on-brown found at Zodiac Ridge, a common type dating from A.D. 1200-1450, further supports a Classic Period date for Zodiac Ridge. Surface architecture undergoes a parallel evolution in most of the Tucson Basin and periphery, appearing during the transition to the Classic Period (A.D. 1100-1200) and becomes formalized as compound units and large, contiguous-roomed villages by the Tucson Phase (A.D. 1300-1500).

Zodiac Ridge exhibits the same type of surface architecture as the large compound unit at Indian Town. Also, the practice of inhumation rather than cremation at the nearby Indian Town village site is another indicator of the Classic Period (Stephen, personal communication 1979).

BEARING PLOTTING

From Aveni's astronomical tables (1972a:531-540; 1972b), we calculated the rise/set bearings for Zodiac Ridge of solar, lunar, stellar, and supernovae events as they would appear along the horizon throughout the year. We investigated solstice, equinox, maximum moon declinations, 20 stellar, and 4 supernovae events. Only 20 of the 25 stellar positions presented by Aveni could be used for investigation, as the remaining 5 are not visible at N 32°30'8" latitude.

If Native Americans had indeed used the stone alignments as a calendar or observatory for celestial events, then an observation point or points were needed. Looking at Figure 9.1, we see that in the interior of the circle there are four distinct features: two east-west rock alignments (Ray A and Ray B), and two rays of rocks (Ray C and Ray D) which form a 90° angle. At the east end of Ray A is a concentration of rocks which we have designated Observation Point A; the same applies to the concentration of rocks at the west end of Ray B, which we have designated Observation Point B. These two observation points would be ideal for observing the sunset and sunrise, respectively, at the equinoxes. On the autumnal equinox in 1977 and 1979, we observed the sunset and photographed the phenomenon in the field along Ray A from Observation Point A (Fig. 9.2--September 1979). If the sunrise and sunset of the equinoxes could be observed from these two observation points, then other celestial events could possibly be observed from these points as well.

ALIGNMENTS

Table 9.2 presents the data used for the investigation of astronomical alignments at Zodiac Ridge. This table was prepared from Aveni's, using A.D. 1000-1500 for computation and plotting purposes. We plotted rise/set bearings from each observation point on the surface map. We found no supernovae reference; however, solar, lunar, and five stellar correlations were found. The following text presents the hypothesized astronomical alignments located at Zodiac Ridge calculated for A.D. 1400 (the median date of the Tucson Phase, Classic Period).

Equinoxes (Fig. 9.3)

For horizon-based astronomy, the equinoxes occur twice a year with the sun rising and setting approximately due east and west. The vernal or autumnal sunrise could be observed from Observation Point B along Ray B. The azimuth of this alignment is 89°43'. From Observation Point A, equinox sunsets could be observed along Ray A which has an azimuth of 269°38'.

Winter Solstice Sunrise (Fig. 9.3)

The sunrise as observed from Observation Point A rises through the gap existing at the intersection of Ray C with Ray D. The azimuth of the two rocks between which this gap occurs is 188°58' for the northern one and 126°08' for the southern one. The winter solstice would occur at 121°42'. At Zodiac Ridge, as the sun proceeded southward from the fall equinox, it would rise over the rocks present in Ray C.

Summer Solstice Sunrise (Fig. 9.3)

From Observation Point B the sun would reach its summer home and appear to rise through the gap in Ray D. The azimuth of the southern edge of this gap is 64°04' and the northern edge is 59°22'. Sunrise at the summer solstice would occur at 62°16'. Solar calculations for Zodiac Ridge show that the variance of the sun's rise azimuth per 100 years is 0°01'.

South Extreme Moonrise (Fig. 9.4)

The moon completes its entire north-south cycle along the horizon in just one lunar month. Unlike the sun, whose extreme positions along the horizon (the solstices) are perceptibly constant for many years, the moon appears at a range of north-south extremes. The complete lunar cycle from maximum extreme north and south excursion to minimum extreme north and south

TABLE 9.2

RISE (OR SET) AZIMUTHS FOR PROPOSED SOLAR, LUNAR, AND STELLAR ALIGNMENTS AT ZODIAC RIDGE

	Horizon Elevation	A.D. 1000	A.D. 1100	A.D. 1200	A.D. 1300	A.D. 1400	A.D. 1500	Variance per 100 yrs.
SOLAR								
Summer Solstice Sunrise								
First Gleam	1°00'	61°52'	61°52'	61°53'	61°54'	61°55'	61°56'	0°01'
Center	1°00'	62°02'	62°03'	62°04'	62°05'	62°06'	62°07'	0°01'
Lower Tangent	1°00'	62°12'	62°13'	62°14'	62°15'	62°16'	62°17'	0°01'
Winter Solstice Sunrise								
First Gleam	4°30'	121°24'	121°23'	121°22'	121°21'	121°20'	121°19'	0°01'
Center	4°30'	121°36'	121°35'	121°34'	121°33'	121°32'	121°31'	0°01'
Lower Tangent	4°30'	121°47'	121°46'	121°45'	121°43'	121°42'	121°41'	0°01'
LUNAR								
North Extreme Moonrise								
Minimum North								
First Gleam	1°29'	69°09'	69°10'	69°11'	69°12'	69°13'	69°14'	0°01'
Center	1°29'	69°19'	69°20'	69°21'	69°22'	69°23'	69°24'	0°01'
Lower Tangent	1°29'	69°29'	69°30'	69°31'	69°32'	69°33'	69°34'	0°01'
Maximum North								
First Gleam	0°48'	55°59'	56°00'	56°01'	56°02'	56°03'	56°04'	0°01'
Center	0°48'	56°09'	56°10'	56°11'	56°12'	56°13'	56°14'	0°01'
Lower Tangent	0°48'	56°20'	56°21'	56°22'	56°23'	56°24'	56°25'	0°01'
South Extreme Moonrise								
Minimum South								
First Gleam	4°30'	115°38'	115°37'	115°36'	115°35'	115°34'	115°33'	0°01'
Center	4°30'	115°48'	115°47'	115°46'	115°45'	115°44'	115°43'	0°01'
Lower Tangent	4°30'	115°59'	115°58'	115°57'	115°56'	115°55'	115°54'	0°01'
Maximum South								
First Gleam	4°00'	128°30'	128°29'	128°28'	128°27'	128°26'	128°25'	0°01'
Center	4°00'	128°41'	128°40'	128°39'	128°38'	128°37'	128°36'	0°01'
Lower Tangent	4°00'	128°52'	128°51'	128°50'	128°49'	128°48'	128°47'	0°01'
STELLAR								
Set								
Capella	1°23'	324°41'	325°01'	325°21'	325°41'	326°01'	326°20'	0°19'
Rise								
Spica	3°57'	99°16'	99°56'	100°36'	101°15'	101°56'	102°34'	0°40'
Rigel	4°01'	104°15'	104°01'	103°47'	103°33'	103°19'	103°04'	0°14'
Vega	0°30'	42°42'	42°39'	42°36'	42°33'	42°30'	42°27'	0°03'

Fig. 9.2 Sunset, autumnal equinox 1979, as observed along ray A from observation point A

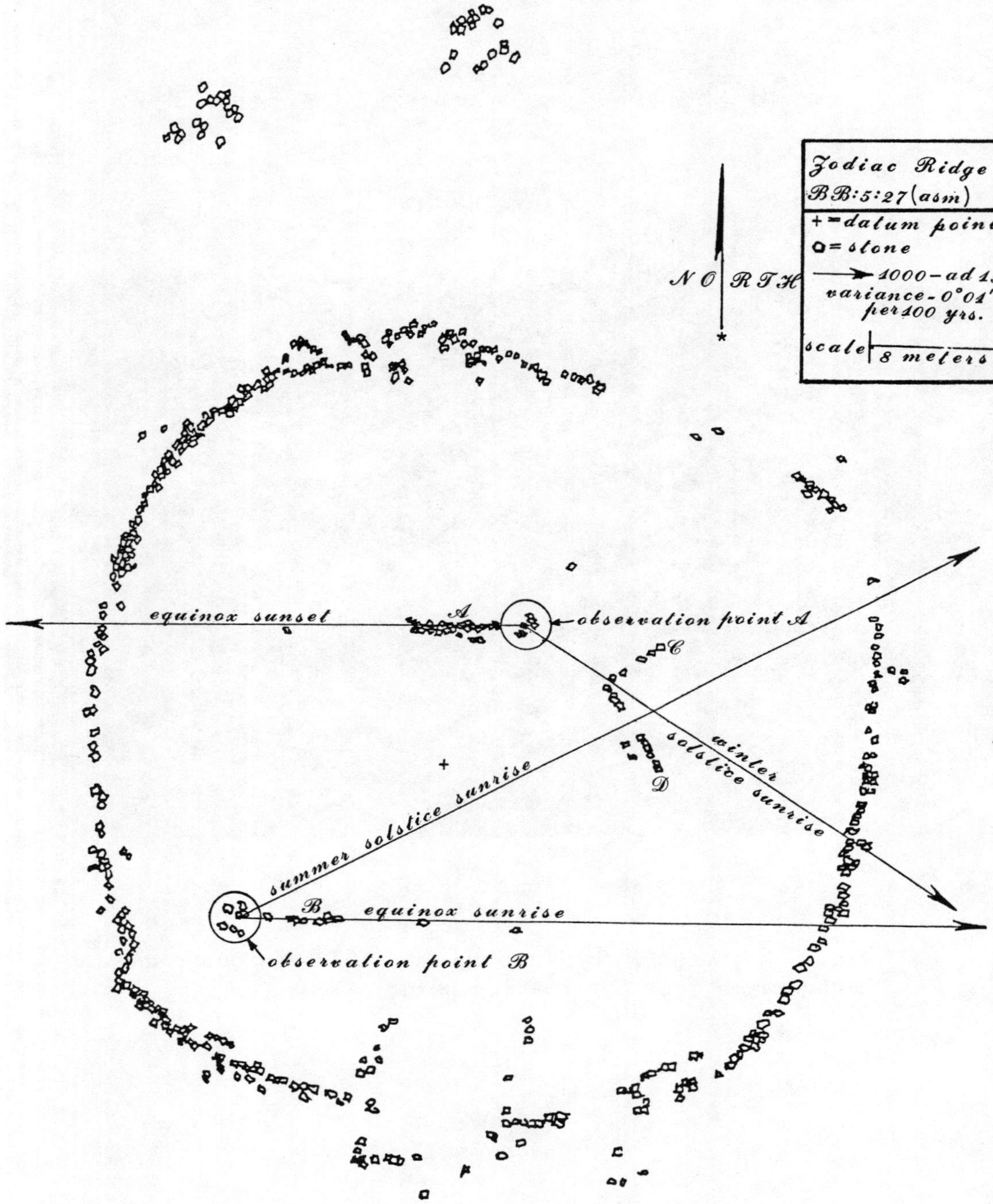

Fig. 9.3 Solar alignments at Zodiac Ridge, A.D. 1000-1500

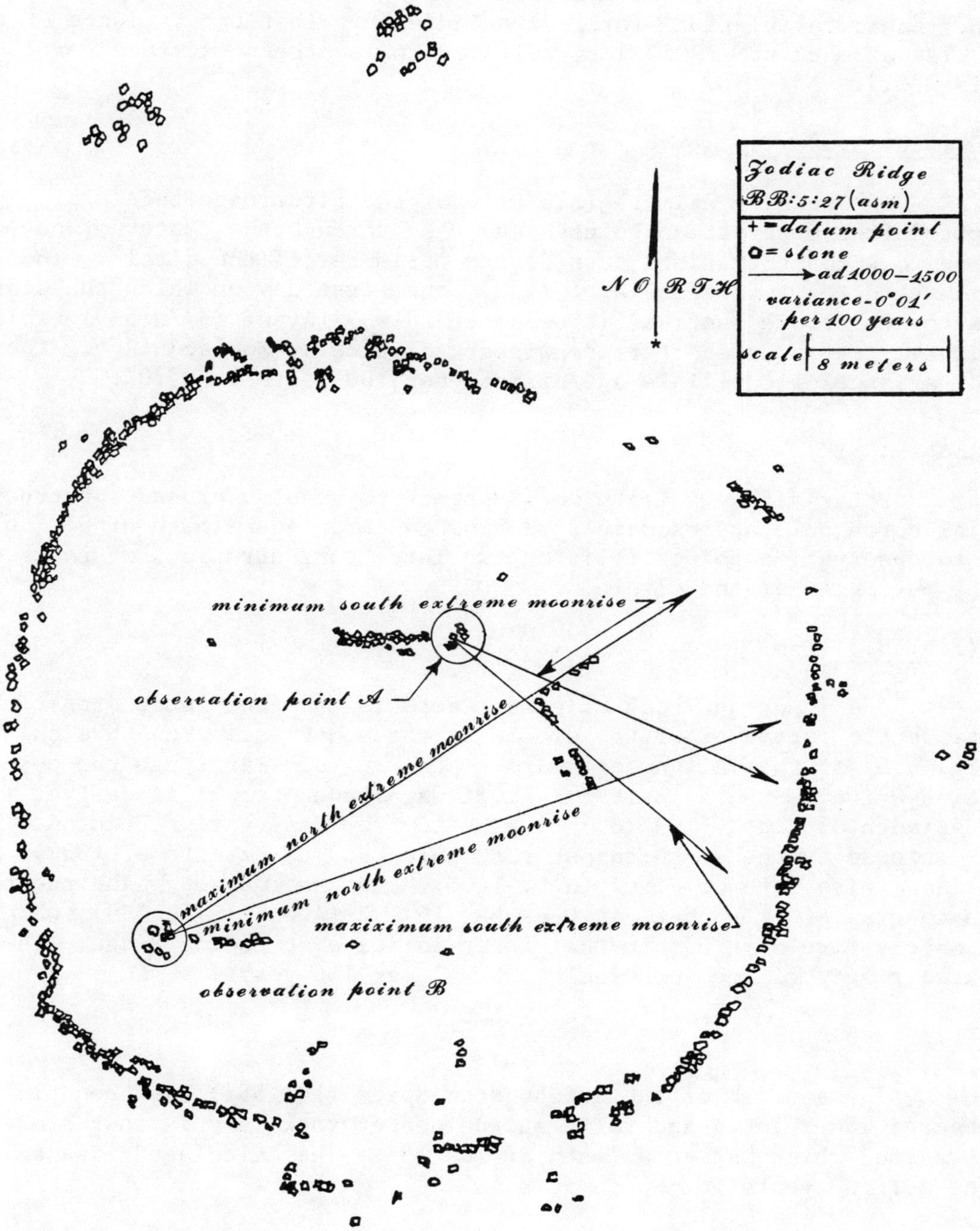

Fig. 9.4 Lunar alignments at Zodiac Ridge, A.D. 1000-1500

excursion and back takes approximately 18.6 years. The south extreme declinations fall in a straight line with Ray C and at the southern extreme edge of Ray D as observed from Observation Point B.

North Extreme Moonrise (Fig. 9.4)

The maximum and minimum northern extremes for the moon have been plotted from Observation Point A and fall along the interior portions of rays C and D. Lunar calculations for Zodiac Ridge show that the variance of the moon's rise azimuth at its maximum northern and southern extremes per 100 years is 0°01'.

Capella (Fig. 9.5)

The setting azimuth (326°01') of the first magnitude star Capella was plotted from Observation Point A and goes through the center of one of the large rock concentrations outside the perimeter of the circle. The heliacal set date of Capella (A.D. 1400), the first day on which the star is seen to set before sunrise, is December 28 $\pm$ 3 (Aveni 1972a:539), which is approximately seven days after the winter solstice at Zodiac Ridge. The calculated variance of Capella's set azimuth per 100 years is 0°20'.

Polaris (Fig. 9.5)

Polaris at A.D. 1500 would appear to revolve around the true celestial north pole approximately 3° from center. An azimuth of 357° plotted from both observation points falls on two large concentrations of rocks outside the perimeter of the circle.

Rigel (Fig. 9.5)

The important rock alignment here is the four large stones that lie outside the perimeter of the circle to the southeast. The physical layout (Fig. 9.6) of this alignment bears a striking similarity to the constellation Orion, with the belt and first magnitude star Rigel west of it. If the azimuth of Rigel (103°18') is plotted from Observation Point A, it closely intersects the southernmost rock which has an azimuth of 102°41'. The heliacal rise date of Rigel (A.D. 1400), the last day on which the star is seen to rise after sunset, is December 16 $\pm$ 3 (Aveni 1972a:539), which is approximately five days before the winter solstice at Zodiac Ridge. The calculated rise variance for Rigel is 0°14' per 100 years.

Spica (Fig. 9.5)

The azimuth of the bright star Spica (101°55') has been plotted from Observation Point A and falls on the center rock in this four-stone configuration, which has an azimuth of 102°33'. The calculated rise azimuth variance per 100 years for Spica is 0°40'.

Vega (Fig. 9.5)

From Observation Point B, the first magnitude star Vega would appear to rise at an azimuth of 42°30', directly over the Observation Point A. The calculated variance of Vega's rise azimuth per 100 years is 0°03'.

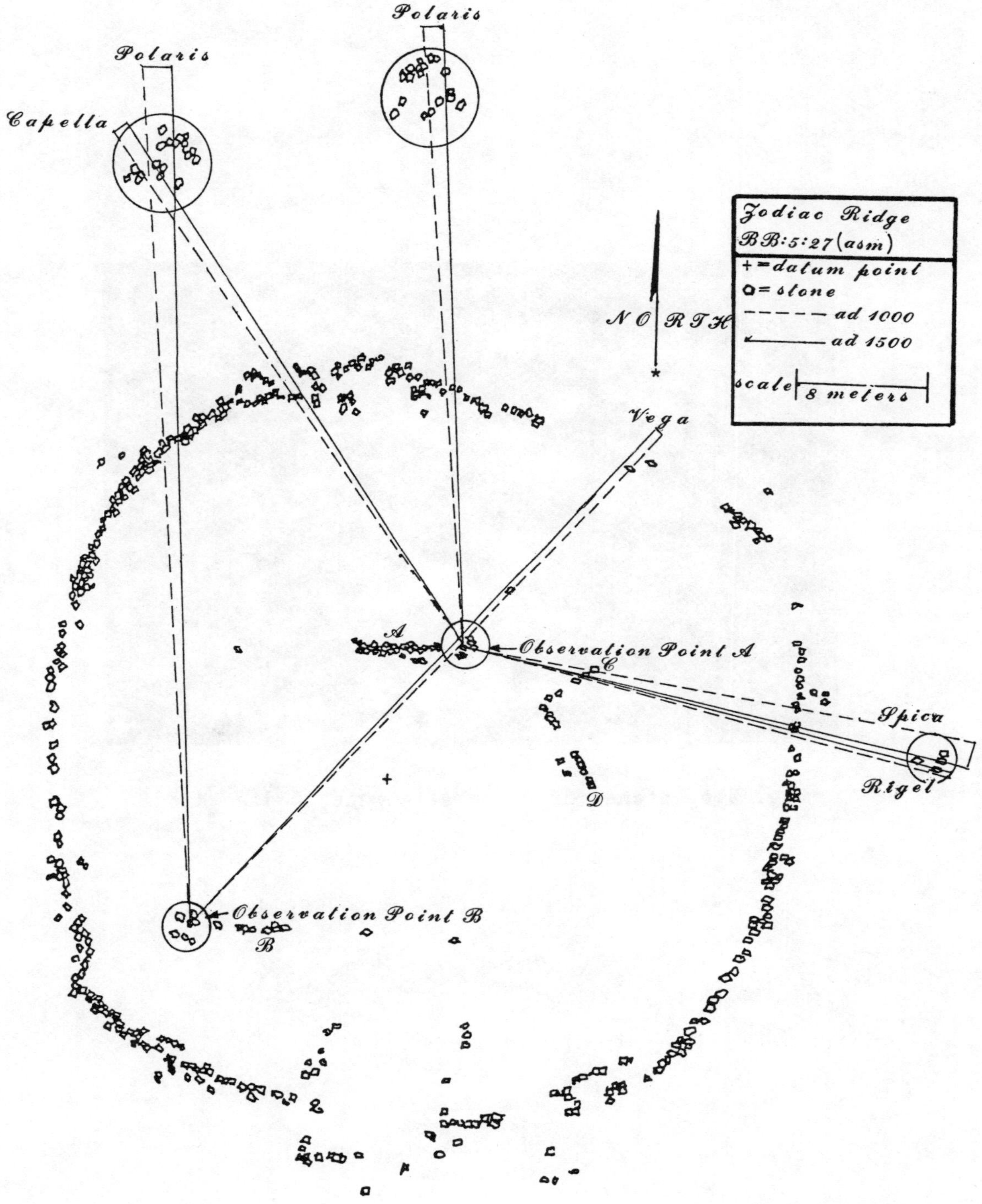

Fig. 9.5 Stellar alignments at Zodiac Ridge, A.D. 1000-1500

Fig. 9.6 Stones of Orion alignment, Zodiac Ridge

WHO BUILT ZODIAC RIDGE?

Most archaeologists agree that the predominant group prehistorically inhabiting the Tucson Basin was the Hohokam, with abandonment of the area by A.D. 1450, leaving behind the Papago Indians we know today. We consider the Hohokam in the Tucson Basin to have been a group that practiced various subsistence strategies that included floodwater and dry farming (Doyel 1977: 103) coupled with hunting and gathering, a subsistence pattern similar to that of the modern Papago.

The occurrence of Tanque Verde Red-on-brown sherds at Zodiac Ridge places it within the time period of Hohokam domination of the Tucson Basin and its periphery (A.D. 1200-1450). No one hypothesis for the origin of the Classic Period Hohokam settlements is generally accepted; however, the two strongest hypotheses for the Tucson Basin propose that during the Classic Period northern Pueblo (specifically Salado) influence was present and felt by the Hohokam in varying degrees, and that this cultural presence influenced the shift in ceramic and architectural styles (Hewitt 1977:8).

ETHNOGRAPHIC SUPPORT

We have correlated Southwestern archaeological and ethnographic evidence of celestial observations with the discussion of specific alignments at Zodiac Ridge, and have also looked at documented agricultural practices that would perhaps be benefited by knowledge of certain celestial events. We have to look to the Pueblo, Papago, and Pima peoples, known to have inhabited the area or at the very least to have culturally influenced groups residing in the Tucson Basin, for comparison. Documented astronomical practices of these groups hopefully have aided us in better understanding and explaining the construction of Zodiac Ridge.

The archaeological evidence, the appearance of compound walls, and polychrome pottery tend to support the hypothesis of a Pueblo intrusion into southern Arizona (Schmidt 1928; Gladwin 1929; Haury 1945). It is well documented that historic Pueblo Indian religious ceremonies involve celestial observations (Ellis 1975; Parsons 1939). The study by Williamson et al. (1977) of Anasazi solar observatories in the Chaco Canyon area substantially supports celestial observations by other prehistoric Pueblo peoples as well.

The Papago delegated certain individuals to observe celestial movements in relation to their bearing upon planting time. It is believed that in ancient times among the Pima, the headmen called the men together about six days before planting was to begin, and the person in charge of calendrics told the people when the time had arrived to begin planting (Castetter and Bell 1942:143). The following quote from Castetter and Bell would seem to indicate that there was at least some reliance on horizon-based solar positions (1942:142-144): "The Pima depended on the relation of the sun to the gap of the Sierra Estrella, the chief purpose being to determine the latest period at which it was possible to plant crops and have them mature before frost."

Knowledge of the equinox occurrence is important because this event can be used to mark the changing of the seasons and thus be used to divide the year into an agricultural and non-agricultural part. There has been no reference to equinox alignments in the medicine wheels investigated by Eddy (1977). These wheels often occur in areas where sedentary agriculture was not or could not be practiced. Equinox alignments have been documented in Mexico at the Caracol (Aveni 1977:16) and at the Big House in Casa Grande (Wilcox and Shenk 1977:44-45), both located in areas where agriculture was practiced. In the Tucson Basin area, the equinoxes are significant events to a sedentary agricultural community. Determination of the last freeze, coincident with the vernal equinox, and the first freeze possible shortly after the autumnal equinox, is essential to crop survival and yield. The two equinox alignments at Zodiac Ridge are the major support for the remaining alignments; inception from the observation points at these two east-west alignments is universal among all alignments documented at this site.

Williamson et al. (1977:203-217) state that there were Ansazi solar observatories in southeastern Utah with likely summer solstice alignments, and Anasazi winter solstice sunrise alignments occurring at a sun-watching station on the north wall of Chaco Canyon, east of Wijiji. Dozier (1966) has demonstrated the importance of the winter solstice ceremony and Ortiz (1969) has described some of the summer solstice ceremonies being practiced today among the Tewa. Among the Hopi, the Winter Solstice Ceremony starts several days before the actual celestial event with priests and participants entering the kivas for prayer and preparation (Talayesva 1972: 451-455). At Zodiac Ridge, as the sun proceeded south from the fall equinox, it would rise over the rocks present in Ray C and could thus be used to time ceremonial events leading up to the winter solstice. It is perhaps a meaningful correlation that Rigel and Capella could have been used to mark the beginning and ending of winter solstice ceremonies, had they been held by the peoples that built Zodiac Ridge. This is supported by the fact that in A.D. 1400 Rigel rose approximately five days before the winter solstice and Capella set approximately seven days after the winter solstice. This correlation may be further supported ethnographically by accounts of Pueblo winter solstice ceremonial practices of beginning approximately one week before and ending approximately one week after the winter solstice.

In the Southwest, the only previously documented moonrise alignment is at the Casa Grande site (Molloy 1977; see also Evans and Hillman, this volume); if the moonrise alignments at Zodiac Ridge can be substantiated through excavation, there will be two sites in the Southwest with moonrise alignments. Both sites were built by agriculturally-based groups, as were the sites with moonrise alignments in Great Britain (Hawkins 1965; Hadingham 1975). The absence of moonrise alignments among Eddy's medicine wheels tends to support the idea that the occurrence of moonrise alignments may be directly related to a sedentary agricultural subsistence. However, it must be mentioned that an understanding of the link between the maximum northern and southern declination of the moon and agriculture is not known.

Our review of the ethnographic literature reveals no direct reference to the stars Capella or Vega. We did not find Polaris alignment documentation in previous investigations of Native American astronomical sites. There are, however, many documented Native American legends involving Polaris (Scott 1942) and the constellation of Orion, including Rigel. Parsons (1939:613) states that among the Pueblo Indians, Orion is used to time ceremonial activity during the night. It is also interesting to note that in curing ceremonies among the Navajo, the proper ceremony for someone ill is prescribed by specialists who view the refracted colors of first magnitude stars through crystals or glass (Coe 1975:6). A worn, six-sided crystal was recovered by Pima Community College (Hewitt and Stephen personal communication 1980) during the test excavation of the equinox alignment at Zodiac Ridge. Zodiac Ridge is the only known observatory in the Southwest possessing both stellar and solar alignments.

ENVIRONMENTAL/AGRICULTURAL SCENARIO

We present the following discussion on environmental conditions because we believe that Native American groups substantially increased the economic vitality of their culture when they established calendrics and possessed the ability to determine rainy seasons and late and early frosts for agricultural purposes. This, coupled with exploitation of the environment through hunting and gathering techniques and their timing, could provide a very effective economic base.

The Tortolita area lies in the region known as the Sonoran Desert, and its topography is characteristic of the Basin and Range Region. The country is arid with eroded fault-block mountains separated by waste-filled basin regions. The chief climatic feature of the area is low and erratic precipitation. Rainfall distribution in the Sonoran Desert is usually concentrated in two distinct periods: summer and winter. The summer season extends from May to October and is characterized by violent thunderstorms forming over the mountains, which result in torrential downpours of short duration in the desert basins. The winter season is from November to April and is characterized by active storms of two to three days' duration, moving inland eastward out of the Pacific. In past time periods, however, periods of drought (A.D. 1000-1300) alternated with periods of increased effective moisture (A.D. 1300-1500).

Agricultural practices in the desert can be seen as functions of two environmental variables: (1) growing seasons determined by late and early frosts which vary geographically according to latitude and elevation, and (2) availability of water determined by access to perennial or intermittent streams for channeling or runoff, or dependence on summer rains for floodwater farming.

Doyel (1977) states that in the Santa Cruz Valley area the evidence suggests that prehistoric agriculture was diversified: various dry farming and floodplain/ground-water systems were utilized rather than canal irrigation. Each agricultural method was employed where appropriate. We do not know the principal method of agriculture connected with the Indian Town-Zodiac Ridge complex. No evidence of canals has yet been found here. This is not surprising, as canal irrigation in the Tucson Basin/Santa Cruz Valley would have generally been limited to the river floodplain areas. It is possible, however, that in periods of increased moisture (A.D. 1300-1500) runoff could have been channeled from intermittent streams. Another possibility is that crops were planted in the streams and on slopes at various localities (after the Hopi system) to ensure crop yields of some fashion in either light or heavy rain periods. It is equally feasible that in times of peaceful coexistence with other groups, floodwater farming could have been accomplished approximately ten miles due east of this complex in the Canyon del Oro drainage.

In times of great aridity and little mountain snowfall, one planting could be done at the start of the rainy season, near the summer solstice, along the ephemeral streams of the bajada (in the immediate vicinity of Indian Town and Zodiac Ridge) and the intermittent Canyon del Oro drainage. During times of slightly increased moisture, planting could be started after the vernal equinox in the Canyon del Oro drainage, which runs effectively during late March, April, and May from snowmelt in the mountains. During periods of moderately increased moisture, an early planting could be accomplished along the ephemeral streams of the bajada, as well as in the intermittent Canyon del Oro area.

The important aspect of this environmental/agricultural scenario is that the periods of crop planting are determined by increased or decreased effective moisture. When two plantings are possible, determination of the last freeze coincident with the vernal equinox is essential to crop survival. If available moisture only allows one planting, determination of the beginning of the rainy season, near the summer solstice, becomes essential. Where ancient agricultural communities had sufficient water sources to make two plantings, as the Pima Indians did, utilizing both irrigation and summer rains, predictions of early and/or late freezes become paramount. Accurate determination of these dates via celestial observations would certainly offer a selective advantage in agricultural processes.

ARCHITECTURAL CONSIDERATIONS

Using horizon landmarks and the related position of the sun and stars for timing ceremony and agriculture was and continues to be one of the preferred methods of the Pimas, Papagos, Hopi, and other Pueblo groups. However, what of problem areas where there are no convenient landmarks to designate horizon points, or where the horizon is visually confusing? The area around the Big House at Casa Grande is flat desert, and there is little in the way of horizon landmarks. Nevertheless, in this area where two agricultural plantings were possible, calendrics must have certainly played its part. Molloy has stated that in the Big House, portholes were used to establish solar and lunar events (Wilcox and Shenk 1977:44-45). These Native Americans obviously solved their horizon problems by the use of an architectural sighting device.

We believe Zodiac Ridge falls within this category as well. The horizon around Zodiac Ridge is confusing. The eastern horizon is a long mountain range comprised of many peaks and crevices with little to distinguish one from another. The majority of hypothesized alignments at this site offer an eastern orientation and, because of the confusing nature of the horizon in this direction, we believe the construction of Zodiac Ridge was necessary as an aid to help distinguish the peaks and crevices associated with the important (critical) events for this area. More specifically, a series of alignments on the ground, when used in conjunction with the mountainous horizon, provided an efficient device for following and permanently marking celestial events and establishing calendrics.

PRELIMINARY EXCAVATION

Test excavation at Zodiac Ridge was conducted by Pima Community College in January 1980. Three one-meter-square test pits and east-west Ray A were investigated in this initial excavation (Figs. 9.7 and 9.8). Sterile floor surface was 20 cm below the surface level.

The preliminary findings suggest east-west Ray A was a free-standing wall structure. The bottom surfaces of the stones associated with this alignment rest 15 cm below the surface. There appears to have been some wall fall toward the south of this structure. Two postholes (60 cm deep, 25 cm in diameter) were located just south of the wall. These postholes are approximately 3 meters apart and are oriented parallel to the alignment of stones. An earthen hearth was excavated north of the wall (east end). Radiocarbon-datable material was retrieved from this feature for analysis.

Tanque Verde Red-on-brown sherds, a rubbing stone with heavy hematite deposits, and a worn six-sided crystal were recovered from the excavation.

CONCLUSION

Much of what occurs in archaeoastronomy is ultimately beyond the realm of proof. We can use our magic computers to determine azimuths and dates for important events in prehistoric times, but we cannot observe the site or device when it was being used by the peoples who built it. A professor once told us that the only way he could accept astronomical alignments would be if we were to excavate the skull of a priest looking down an alignment--and the skull had sun symbols painted on it. This is an exaggeration, but it nevertheless bears a seed of truth. The east-west alignment at Zodiac Ridge has been excavated and there are no skulls or sun symbols, only the remains of a free-standing wall and possible ceremonial objects. Further excavation may provide additional support to our hypothesis, or show cause for reconsideration.

The acceptance or non-acceptance of Zodiac Ridge as a prehistoric observatory lies with the reader. We believe that ethnology among Southwestern Indians supports the use of a horizon-based astronomy in calendrics, and that calendrics produces a more efficient basis for exploiting the environment through agriculture and the timing of hunting and gathering expeditions. We also believe that Zodiac Ridge is a physical expression of astronomical knowledge as it existed in the Tortolita Mountain region during its occupation by Native Americans.

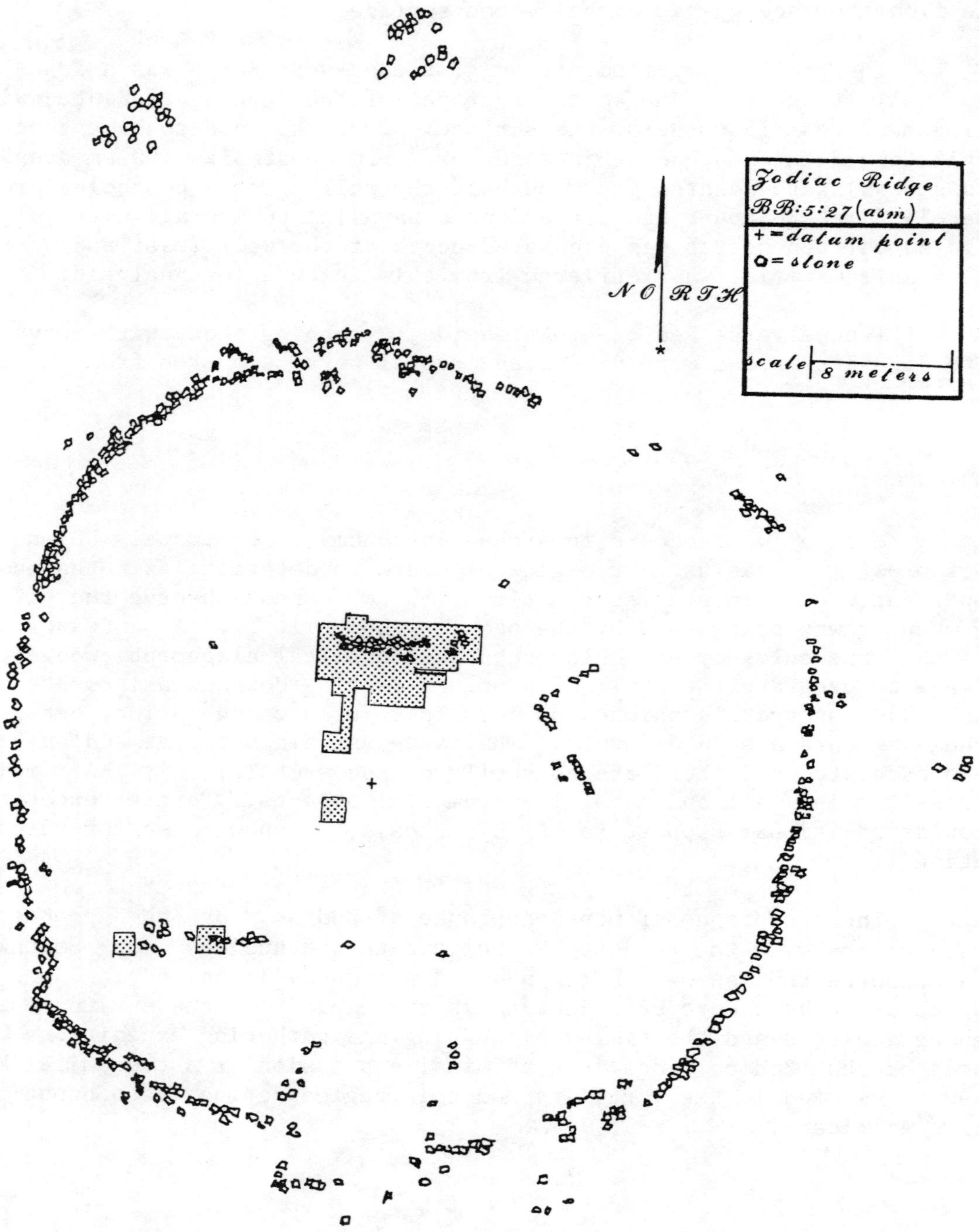

Fig. 9.7 Zodiac Ridge Site Map--Excavations

Fig. 9.8 Equinox alignment--
Ray A after excavation

10

Archaeoastronomy at a Selection of Mississippian Sites in the Southeastern United States

Ann L. Daniel-Hartung

INTRODUCTION

There have been many pseudoscientific articles concerning the "wonderful secret knowledge our ancestors had that has been lost" (MacKie 1977:7). Such topics as prehistoric computers, ley lines, lost continents, and contact with visitors from space have gained much popular support. This support has resulted in prejudice against reputable work in archaeoastronomy since it has been seen by some as part of pseudoscience. Though archaeoastronomy is hardly pseudoscience, some of its practitioners are doing field work without sufficient knowledge of the culture or cultures that occupied the area. As a result, they have postulated alignments for cultures which may not have had a sophisticated knowledge of astronomy or have had an interest in those particular celestial bodies. In order to remedy this situation, a systematic, logical approach to archaeoastronomy must be developed, with concentration on the evidence and theories that best fit the realities of the specific area and time being considered.

Four major problem areas in archaeoastronomical research have been identified (Reyman 1975a): (1) an inadequate conceptual scheme or theoretical approach; (2) an insufficient control of the relevant ethnohistoric, ethnographic, and/or archaeological data; (3) the failure to formulate specific field problems, hypotheses, and test implications; and (4) the lack of a consistent, systematic procedure for conducting fieldwork, coupled with the possibility of unsuitable field equipment. To avoid these problems I followed a research procedure based on the use of ethnographic, ethnohistoric, and archaeological records; on-site measurements; and accurate mapping, all applied in support of a stated hypothesis. For this particular research, the hypothesis to be tested was as follows: that evidence of Mississippian knowledge of astronomy should appear in architectural alignments found in the remaining structures of this culture, the mounds. This hypothesis is based on possible Mesoamerican-southeastern United States contact; if the contact was made, a transfer of celestial knowledge may have occurred.

In the southeastern United States during the period from approximately A.D. 700 to A.D. 1600, a cultural system, labeled Mississippian because of its apparent geographical origin, spread throughout the region. Here occurred the most complex development of aborginal sociocultural organization north of the civilizations of Mesoamerica. In fact, this development is often attributed to diffusion from Mesoamerica involving such traits as human sacrifice, mound and plaza construction, iconographs, ceramic complexes, cultigens, games, and the form of personal ornaments:

> There is some evidence that the Mississippian development was stimulated by the introduction of concepts, ceremonial attitudes, and practices from Mexico. It was based on such improved agricultural procedures as the marked use of the flint hoe and probably of improved strains of corn, which resulted in large populations and a more sedentary societal organization [Griffin 1971:248-249].

Those improved strains of corn, specifically the eight-row Eastern Complex corn introduced into the Southeast about A.D. 800, were well adapted to the growing season of the southern United States (Stoltman 1978:724). Corn was popped, made into hominy, or ground to make a bread; "perhaps the most cheering and heartwarming use the Indians made of maize was the production of alcoholic beverages" (Paul Weatherwax, as quoted by Heizer 1973:98). Heizer (1973:30ff) suggested religious ceremonies as the origin of agriculture--seed offerings scattered over a field from which it had been gathered to appease the gods. It is possible that these ceremonies were continued and expanded as intentional cultivation developed. According to this hypothesis, the ceremonies would be transmitted to other culture groups along with the seeds, because they became an integral part of the planting and perhaps the harvesting of the maize. It seems reasonable to suppose that such a process brought Mesoamerican ceremonialism *and* associated astronomy into the Southeast, along with truncated pyramids, vessels with tripod feet, plumed serpents, monolithic hatchets, seated human figures, sculptured idol heads, spool-shaped ear ornaments, and long ceremonial swords shaped from flint.

To discover if Mesoamerican contact included transmission of celestial knowledge, I analyzed five major ceremonial centers for evidence of astronomical interest. These centers--Cahokia, Illinois; Moundville, Alabama; Etowah, Georgia; Kincaid, Illinois; and Angel, Indiana--were the most influential centers of the Mississippian period (Fig. 10.1). A sixth center, Spiro, Oklahoma, has been destroyed, and insufficient information remains to study this site. What remains of the five extant ceremonial centers is the pattern of earthen man-made mounds. Other man-made structures have been destroyed, either by the elements or by man. Dwellings and ceremonial structures built of wood, cane, and grass are not durable in the climate and soils of the southeastern United States. Evidence of these structures, such as decayed posts remaining in the ground, was not always recorded during early excavations. Often these post molds were not recognized; excavators were more interested in the recovery of artifacts than in reconstructing the culture. For these reasons, it is now impossible to study the remains of structures for evidence of astronomy.

BACKGROUND DATA

I have made a thorough search of ethnohistoric, ethnographic, and archaeological documents to provide a cultural background for an interest in astronomy. There are no recognizable written records from this time period; what is available are the narratives of DeSoto's men as they traveled through the Southeast during the mid-sixteenth century, and studies made in the late nineteenth and early twentieth centuries of native peoples remaining in this region. Several chroniclers (the Gentlemen of Elvas 1907; Garcilaso 1951; Bourne 1904) recorded what they saw as they traveled across the Southeast. The French and English explorers unfortunately left few narratives of their

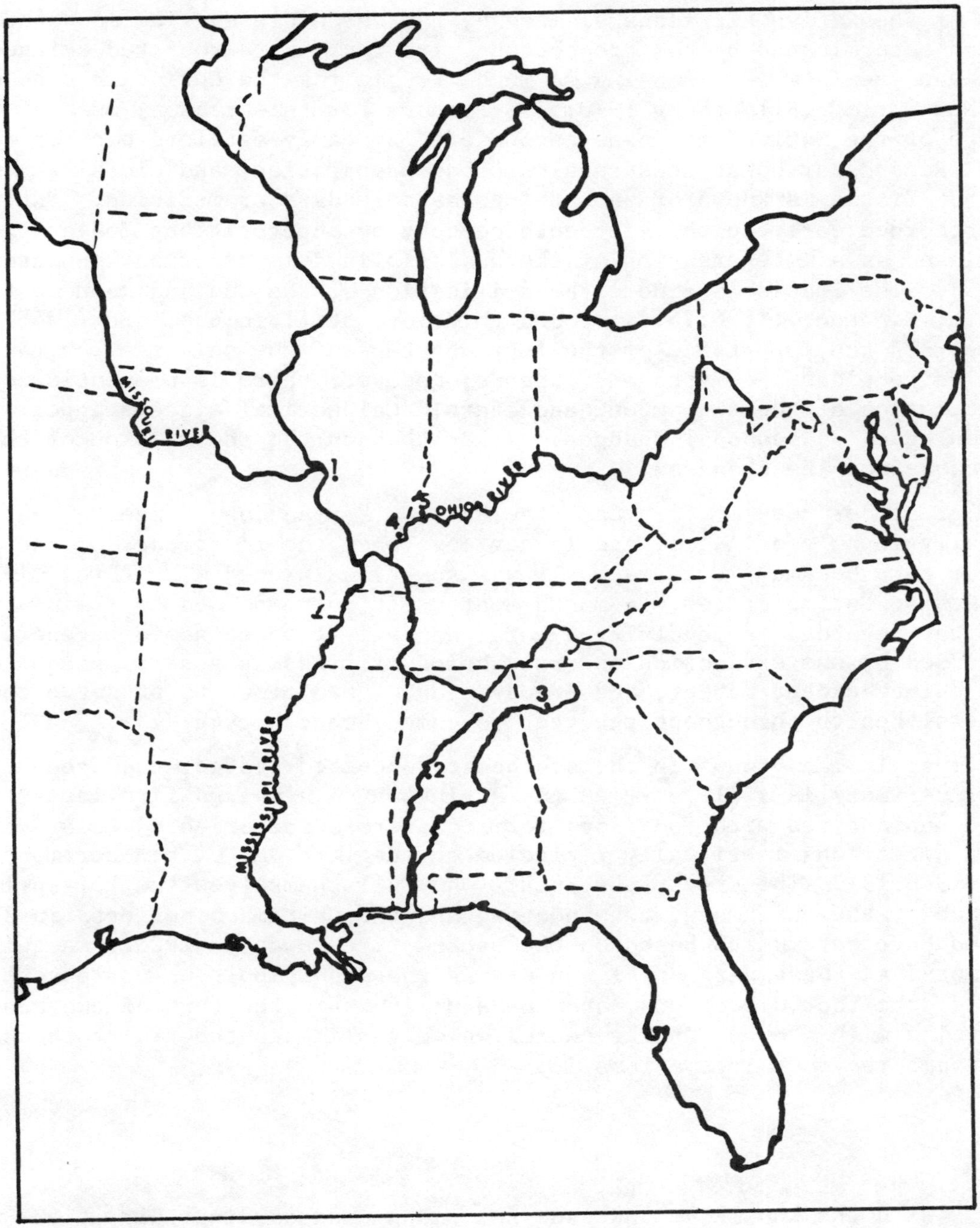

Fig. 10.1 Location of Southeastern Mound Sites:
1) Cahokia, Illinois; 2) Moundville, Alabama;
3) Etowah, Georgia; 4) Kincaid, Illinois;
5) Angel, Indiana

travels in this area, though the French recorded their encounters with the Natchez, who lived along the southern end of the Mississippi River. Early studies of this region indicated that the mound-builders were not the ancestors of the Native Americans visited by the chroniclers. The mound-builders "had been exterminated by the treacherous, ignorant, murderous red-skinned savages who even now were causing so much trouble for the Christian settlers of the New World" (Silverberg 1970:5). Perhaps for this reason, much of the knowledge of the Native Americans encountered by early settlers was not recorded. Records of house construction, food preparation, and clothing can be found, but little is known of such things as calendars or medicine. "Naturally, a tale recorded early in the sixteenth century by superstitious Indians would be dismissed as a mere creation of the undisciplined imagination" (Swanton 1946:755). The Spanish did note the deification of the sun and moon in the Southeast; Swanton (1946:761ff) found this idea still in use. He (1946:767) also recorded the importance of the four quarters of the universe, or the four cardinal directions, in myths and legends; however, there is no mention of the apparent motion of the sun, moon, and stars. Calendrical systems appear to have been based on seasonal changes, rather than on the recognition of the sun as the cause for the changes.

> The great ceremony of the year, the busk or "green corn dance," occurred usually in July or August and in any case when the first ears of the flour corn became fit to eat. It was sometimes preceded by three minor feasts or "stomp dances," a month apart. It corresponded to the new year, and was regarded as involving a moral as well as an economic regeneration, typified by the extinction and relighting of fires, a general pardon of all crimes except murder, and preparation of medicines to preserve the general health throughout the year to come [Swanton 1946:775].

Today, among Indian groups in the southeastern United States, the Green Corn, or Busk, ceremony is still carried on, during which a sacred fire identified with the sun and fed with four logs forming a cross and oriented toward the cardinal directions is ritually rekindled on the last day of the ceremony (Howard 1968:19). The Creek, the Yuchi (who call themselves the "offspring of the Sun"), the Chickasaw, the Choctaw, and the Chitimach believed in a sun-deity and have ceremonies based on the sacred fire (Swanton 1928). A quite common motif at these ceremonial centers is the sun symbol, which represented the world, the four directions, and the sun; it is in the form of two concentric circles with a cross in the center which replicates the rim of the sacred fire pit and the four logs across it.

FIELDWORK

With the knowledge that the sun, moon, and cardinal directions were important in the Southeast, I visited each site to take sightings on the markers, in this case the mounds. Because of the poor condition of the mounds, a Brunton compass provided sufficient accuracy for the necessary sightings. Erosion and man and animal activities have created indistinct lines to be used for sighting--the sides of the mounds are no longer (if they ever were) straight and level, horizontally or vertically. I made magnetic corrections according to the Federal Aviation Agency sectional aeronautical charts.

Sightings were taken in both forward and reverse directions. For example, if a ramp faced east, I took sightings in a northerly and a southerly direction along that side of the mound, as well as from the top center of the ramp. The horizon was scanned for prominent peaks which could have been used as foresights for alignments. Possible alignments using the sides of the mounds were not considered because natural and man-made modifications have changed their shapes. The center of each mound and the top center of each existing ramp were used as possible markers. The height of and distance to horizon features were measured during the field work and then included in the analysis.

I then made maps of mound locations from these data, using topographic maps and aerial photographs obtained from the Agricultural Stabilization and Conservation Service of the Department of Agriculture. From these maps I made measurements involving combinations of mounds, taking into consideration topographic obstacles such as intervening mounds and heights of the mounds involved. Possible alignments using the sides of the mounds or the diagonals were not considered because the mounds have been significantly modified so that the sides can no longer be considered original. The center of each mound and the top center of each ramp, if one exists, were used as possible markers.

I calculated azimuths for each of these alignments, and then compared them with the azimuths for the visible planets, stars, and the sun and moon, taken from astronomical tables provided by Aveni (1972a). What became apparent in the analysis is a concern for the cardinal directions, reflected in site orientation and individual mound orientations; other alignments are questionable due to modification of the mounds.

CAHOKIA

Cahokia is the largest prehistoric site in North America north of Central Mexico. Monks Mound, the largest mound in the United States, and the third largest prehistoric man-made structure in North America, is here. Near it Wittry (1961) discovered a pattern of post pits during an archaeological salvage project in the 1960s; some of the post pits found in one excavated tract formed three complete circles. Wittry has interpreted the pits found in another tract as forming arcs or parts of four circles (Wittry 1964). One of these proposed circles, number two, has been the focus of astronomical research at Cahokia. The western part of this circle has been destroyed, so the total number of post pits will never be known; however, three post pits on the eastern side mark cardinal directions and two pits may be associated with solsticial alignments. During recent field work Wittry uncovered more pits along the circumference of circle number two, some of which he interpreted as Capella markers and eclipse predictors (Norrish 1978). There are a number of unanswered questions concerning these post pits. Do they indeed form circles? For each post pit there is an uncertainty; for each three pits, determining an arc, there is an uncertainty as to the center of the circle described by the arc. Thus, for each three pits, there is the possibility of many different center positions. Why would four, or more, be built? All that would be necessary for an observatory are the appropriate foresights and a backsight. If these were incorporated in a circle, only one circle would be necessary. Any casualty to the markers could be repaired without constructing another circle. To determine if the post pits do form circles and if there are intentional alignments, more data must be made available--through excavations and publication.

Monks Mound has a long axis running 5° east of north; similar orientations have been found in surrounding mounds, houses, and the eastern stockade (Reed 1977). This may have been magnetic north at the time of its original construction, or the orientation may have been in relation to nearby Cahokia Creek. No evidence of a compass has been recognized here, so this question remains unanswered.

Buildings in one excavated tract were arranged in rows running north and south, with the long axis running east and west; buildings in the other excavated tract were also arranged in north and south directions, but the long axis was neither north-south nor east-west (Benchley 1974:36).

The mounds were not randomly scattered, but were constructed in a pattern which may represent community organization. The majority of mounds cluster along a natural ridge forming an east-west axis for the site; this is also the highest and driest land in the area (Fowler 1978:462). Each of these mound clusters has its own plaza and platform and burial mounds, forming suburban areas within the metropolitan city limits. There also are special mounds, ridgetop in shape, that mark the city limits in three of the cardinal directions (Krupp 1977). The area where the fourth mound may have been has been disturbed. I found no apparent alignments in the buildings or mounds in the settlement planning other than orientation toward the cardinal directions. Intermound and structural alignments are inconclusive at present. Location of structures on the mounds is not well known because of the limited field work and the amount of time required to excavate 100 mounds. The relationship between mounds may be lost as a result of the destruction caused by modern urbanization.

MOUNDVILLE

Moundville is considered the second largest ceremonial center of the Mississippian period. There are eighteen mounds surrounding a plaza and the two largest mounds within the plaza (Fig. 10.2). If a north-south line is drawn through Mound A and Mound B, and a series of parallel lines are drawn through this along the axis of the winter solstice sunrise, the mounds along the edge of the plaza alternate between those with large platforms and no burials and those with relatively smaller platforms and rich-accompanied burials (Peebles 1978): Mound C and Mound D have burials; Mound R and Mound E have no burials; Mound Q and Mound F have burials; and so forth (Fig. 10.2). However, a "'fudge factor' of at least 5 degrees is necessary to make the mounds on the edge of the plaza fall into solstice and equinox alignments when viewed from the centrally placed Mound A" (Peebles 1975:69).

The shapes of the mounds have evolved through time; in the last 70 years, one mound has been removed, others altered, and many ramps destroyed. In the intervening years the mounds have been reconstructed or repaired and their original shape may have been altered. For this reason, I found it nearly impossible to determine the original intention of possible alignments; because of the structural modifications, no man-made markers were identified. The site itself appears to have been constructed with an orientation along the cardinal directions. The exception is Mound A but, because of its size and location, it must have had a special meaning. The plaza is oriented along a north-south axis, not in parallel with any geographical phenomenon, and the

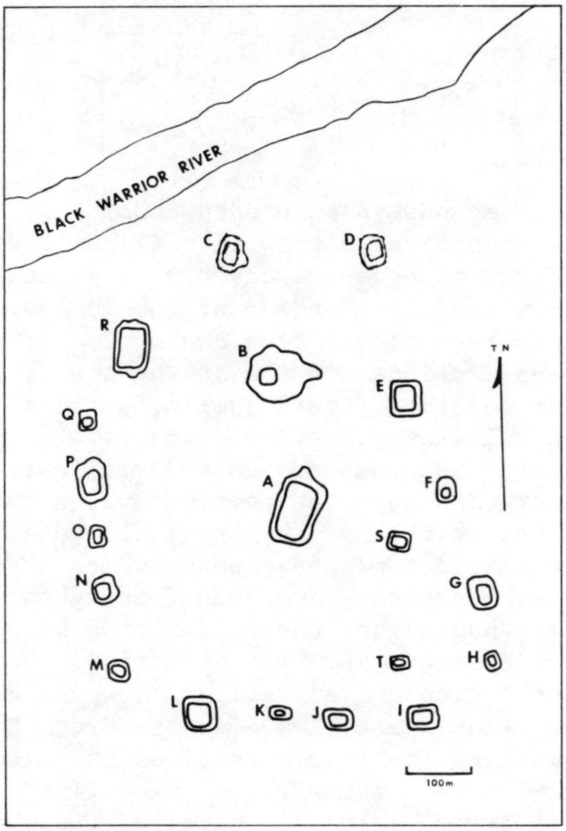

Fig. 10.2 Map of Moundville mounds

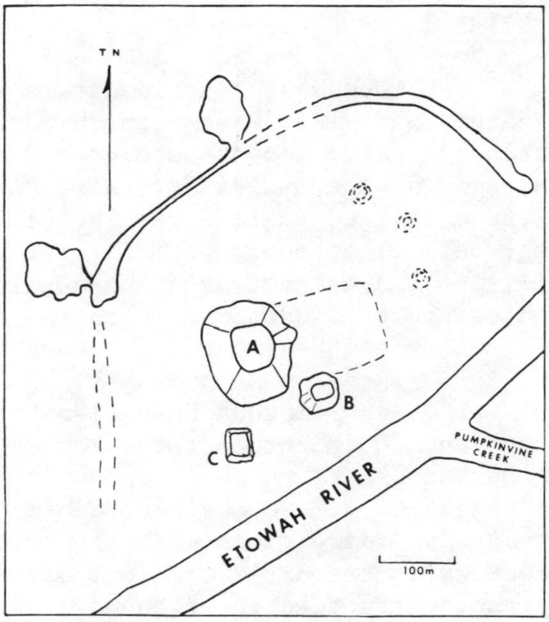

Fig. 10.3 Map of Etowah mounds

mounds around the plaza reflect this pattern.

ETOWAH

Etowah now consists of two large mounds in an eroded condition, one reconstructed mound, and evidence of a moat and borrow pits (Fig. 10.3). Three small mounds at the eastern end of the site were excavated to ground level some time back and not rebuilt. Mound A, the largest of the mounds, was farmed until recently; the surface has been plowed to a depth of approximately twenty centimeters. A local legend mentions a notch at the top of the ramp on Mound A aligned toward the summer solstice (Lewis Tumlin, personal communication); because the top has been cultivated, any evidence of a notch has been destroyed. Mound B has been tested and found to be a domiciliary mound, perhaps the residence of the leader. Mound C has been excavated and rebuilt, so its present orientation may not be that of the original mound. From the summit of Mound A there are two distinct mountain peaks which stand out on the horizon to the east. These two peaks may have been foresights; if the top center of the ramp of Mound A is a backsight, there is a possible alignment with Castor. With cultivation, erosion, and reconstruction, the center of the ramp is now difficult to determine and any alignment would be suspect. Present day Cherokee and Creek tribes participate in the Green Corn ceremony, including the fire with logs marking the four directions and dancing around a sacred mound (Howard 1968:19). Both these groups have lived in the Etowah area; it is possible that this ceremony is a survival of Etowahn practices. A ramp on the east side of Mound A marks one cardinal direction, the only evidence at Etowah of an interest in astronomy.

OTHER SITES

The mounds at Kincaid are in such a state of deterioration that their shapes and sizes have been obscured. This region is subject to periodic inundation, and the mounds have provided refuge for man and animals; for this reason, some of the mounds have structures and roads on them. When the site was first excavated there were nineteen mounds; after excavation and extensive farming, only eight mounds remain visible (Fig. 10.4). What may have been the village area has not been thoroughly excavated, so the orientation of structures there is unknown. With the data available today, it is difficult to find evidence of astronomy, or even a knowledge of the cardinal directions.

The Angel mounds have also been altered through cultivation and construction. Only three of the mounds remain for analysis; the other eight are either too small to be useful or are not identifiable now without excavation (Fig. 10.5). Houses were consistently oriented with corners directed to the cardinal directions east of Mound A, while to the west of the mound the house walls faced the cardinal directions (Black 1967:501ff). The semi-subterranean structure under Mound I was oriented toward the cardinal directions, as were the sides of Mound E and Mound D. This concern for the cardinal directions is consistent with other Mississippian ceremonial centers and with the concept the Mississippian people may have had of a quadripartite universe.

The significance of the orientation of Mound A and Mound F of more

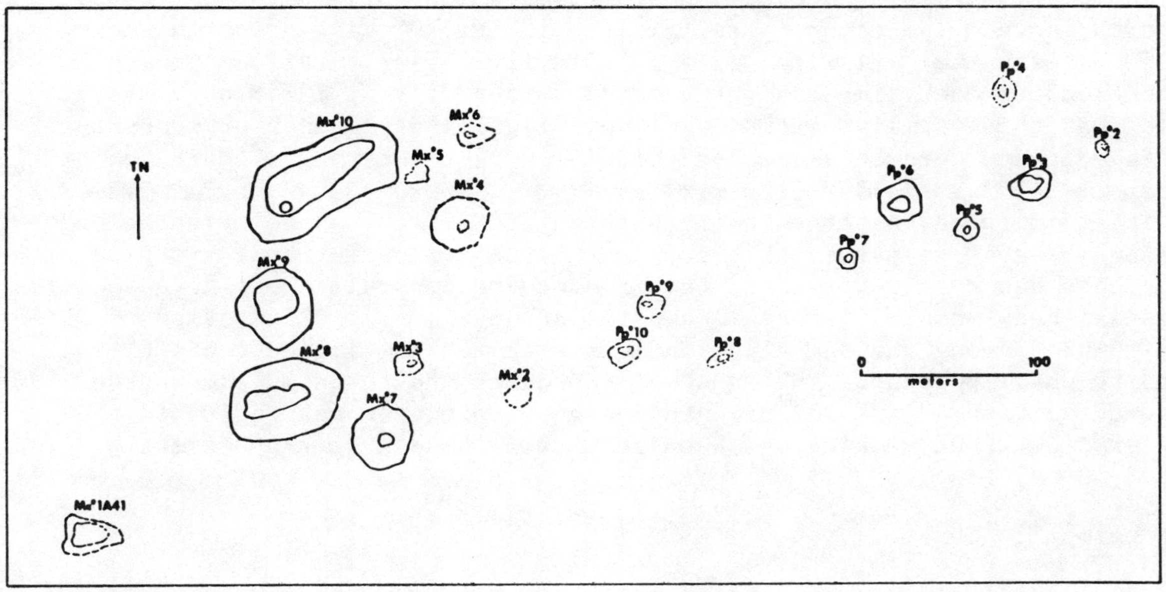

Fig. 10.4 Map of Kincaid mounds. Extinct mounds are indicated by hatched lines.

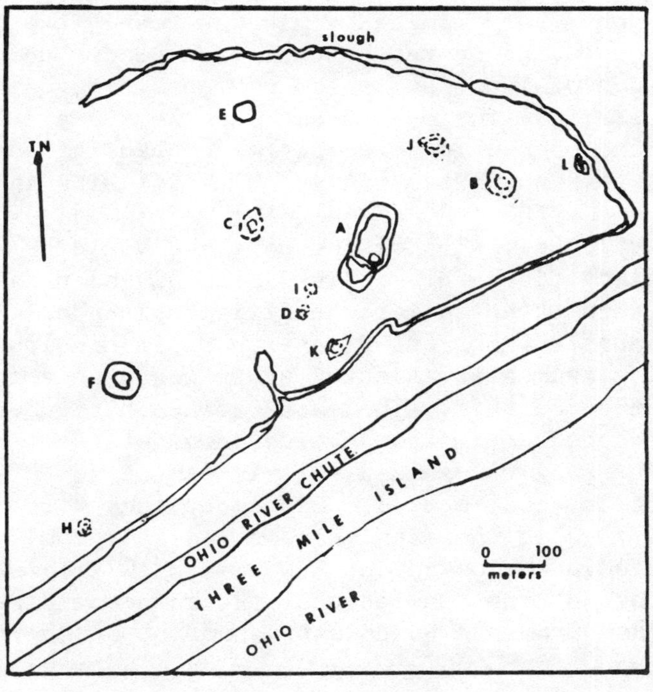

Fig. 10.5 Map of Angel mounds

than 20° east of north is not known. There is no obvious celestial body that might be in alignment, and the river flows in another direction. There are, however, site alignments in Mesoamerica which fall in the 20° to 30° east-of-north range. The majority fall in the 15° to 20° east-of-north range, which suggests the "existence of a 17° 'family' of orientations though no axial trend through time has yet been discerned" (Aveni 1975:166). Is it possible that this 'family' marked some specific celestial body or occurrence? Teotihuacan and ceremonial centers within 100 kilometers of Teotihuacan fall within this 17° east-of-north category (Aveni 1977a:5-7). If there was a continuity of purpose, is Angel a northern outpost of this site orientation? We will not be able to pursue this idea until the reason for this grouping of alignments has been determined; that reason--perhaps alignment to a specific celestial body--could then be looked for at Angel. The orientation of primary Mound F along the same line indicates that this alignment of 20° east-of-north was important. The mound of the chief and the temple mound deliberately constructed with a parallel orientation must have been meaningful to the builders, but that meaning has been lost or is hidden at the present.

CONCLUSIONS

The ethnographic and ethnohistoric data indicate an emphasis on the sun and moon in the religion of this region, and this interest has been recognized at Mississippian centers. Because agriculture was important in the development of large ceremonial centers, apparent solar motions and the corresponding seasons were significant. An interest in the sun and a quadripartite division of the universe were unifying concepts at these major centers. Astronomical knowledge on a very limited scale was found at each of these sites in varying degrees, depending on the size of each center and the intervening destruction. Knowledge of lunar and solar motions is fundamental and is not necessarily the result of diffusion. Observing the sun and the moon are natural activities, and ones which need not have been introduced from another culture. Therefore, astronomy does not support diffusion into the Southeast.

A quadripartite division of the universe would be a shared trait if based on the same orientation; the Mississippian division is based on the cardinal directions, and the Mesoamerican division is not. These two facts argue for an independent development of astronomy in Mississippian culture. If so, the recognition and observation of stars would be a logical development, but there is no evidence of this. The catastrophe that ended the Mississippian florescence may have occurred in the developmental stage. In the Southwest, historically, only the Zuni recognized Polaris as the north star (Reyman 1971: 123). The Pawnee Indians had four directions but they were not correlated to north; Wedel (1977) calls them semi-cardinal directions. There is evidence of cardinal alignments in Chaco Canyon, New Mexico (Williamson et al. 1977: 203ff), but whether or not the knowledge of the cardinal directions was transmitted to the Southeast from the Southwest cannot be determined.

Cultigens came from Mesoamerica, but burial practices and other traits may have been local developments, based perhaps on Adena-Hopewell antecedents. Although contacts may have occurred, they do not seem to have involved the transmission of celestial knowledge. More research is required to demonstrate the presence of Mesoamerican elements in the astronomical knowledge of the Mississippian people.

11 Two Possible Calendar Sites in Vermont[1]

Bryon E. Dix and James W. Mavor, Jr.

INTRODUCTION

Many unusual stone structures (including chambers, walls, standing stones, and perched boulders) located in the northeastern United States have been recorded through historic times to the present. The purpose of this paper is to present a brief overview of the New England material, followed by a brief description of one possible calendar site (Calendar II) and a more comprehensive analysis of another (Calendar I) in the central Vermont region. Our method of investigation has been one of passive surveying, locating all features within a site with respect to one another. We also identify the relationships between the intrinsic site features and the surrounding horizons and other topography. This method of analysis was developed in order both to help preserve and to document the subtle nature of the sites.

All of the sites studied are at elevated locations, 300-500 meters high, near the White River in central Vermont. The topography of the area is one of glacially-eroded ancient hills, with mountains of up to 1200 meters in height to the south and west. The geology of the area is best described as that of a contact zone between metamorphic and igneous rock.

The Vermont stone chambers appear to be organized in groups, with one group in southern Vermont and two in central Vermont. The two chamber groups in central Vermont are centered around the Calendar I and Calendar II complexes.

Standing stones, of one to three meters in height, are associated with these chambers, sometimes on top. Many cairns (stone piles) have also been recorded in and around the sites. One cairn field near the Calendar I site features a cluster of over 70 cairns on an eastern slope. In addition, many unusual marked stones have been found in and around the sites. Most of the marks are deeply incised, linear, and are organized in such a way as to suggest some kind of notation. Some of the markings are assembled in the form of specific symbols. Literally hundreds of these marked stones have been noted, some of them on astronomical sight-lines.

---

[1]We are grateful for the assistance of Betty and Bob Sincerbeaux, Lucinda Gerhart, and Mary H. Mavor; we also wish to thank William Lovering for graphics and Harold McCarthy, owner of the Calendar I bowl, for permission to survey his property. We appreciate the helpful comments of Richard Brinckerhoff and Ray Williamson.

CALENDAR II SITE

This site is located near Mount Ascutney in central Vermont, a 1000-meter-high mountain which dominates the horizon. The topography of the site involves ancient rolling hills with infrequent, distant mountains of substantial height to the south and west. The site contains a group of five stone chambers. The center of the site is located on the eastern side of a long rolling hill at an elevation of 475 meters. The southwestern horizon is formed by a mountain ridge with an elevation of 518 meters. The site may best be described as an assortment of intact and ruined stone structures some of which are historical or are connected to historical constructs. The structures include a stone chamber, standing stones, walls featuring large upright stone slabs, and curiously marked stones. The site was shown to Dix in 1974 by Elizabeth Sincerbeaux of Woodstock, Vermont, who has been instrumental in locating over 40 chambers in the central Vermont area.

An aerial view of the site center (Fig. 11.1) reveals a complex arrangement of stone walls surrounding a large stone chamber and platform. The inside dimensions of the chamber are 3 m x 6 m. Observations at the 1975 and 1976 winter solstices confirm that the sun rises in a notch on the southeast horizon along a line coincident with the long axis of the chamber (Fig. 11.2). The chamber is roofed by nine stone slabs of up to 4.5 m in length, some of which weigh over three tons. The chamber has an open port adjacent to the back wall which penetrates the roof. Near the entrance to the chamber is a large well-like stone feature, over 1 m in diameter, with a large slab roof supported by stone beams.

CALENDAR I SITE

The possibility of ancient astronomical observatories in New England raises unique questions about techniques of exploration, survey, and interpretation, as well as about the nature and origins of the people who might have built them.

We provide evidence here that supports the theory that certain man-made and natural features of this Vermont site conspire to form a number of unambiguous alignments which can be associated with astronomical events. We do not assume that astronomical activity would have taken place all at once; it may have occurred during a long period, with the design added to or changed over time.

Brief History of Fieldwork

One of us (Dix) first visited the site in 1974 and observed that the natural bowl could have been used as a solar calendar. An observer could, from a single location within the bowl, have sighted features of the natural peaks and valleys of the eastern and western horizons that were in approximately the correct locations to serve as foresights for all six of the principal solar events: the rising and setting at the winter and summer solstices, and the vernal and autumnal equinoxes (Dix 1978a; Fig. 11.3). This central location, which is unmarked today and which we call the *astronomical center*, was located initially through the architectural insight of Elizabeth Sincerbeaux and approximately confirmed by the use of a sighting device on an

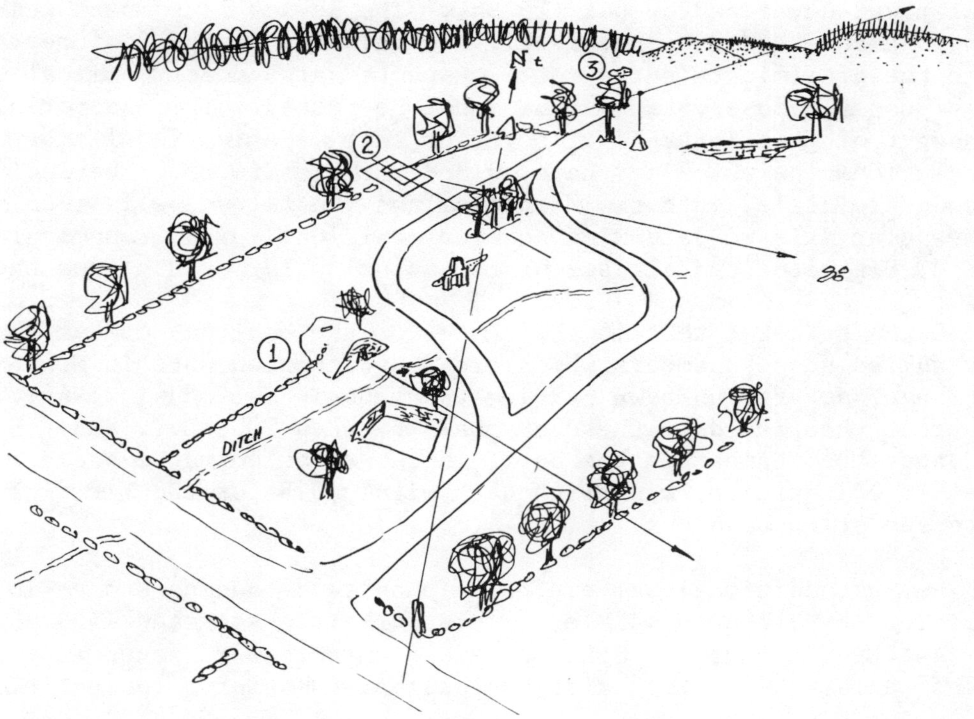

Fig. 11.1 Site sketch of Calendar II showing observation areas: (1) Chamber (winter solstice sunrise)
(2) Platform (winter solstice sunrise, equinox sunrise)
(3) Rock outcropping with markings (summer solstice sunrise, sunset).

Fig. 11.2 Winter solstice sunrise at Calendar II, from Chamber.

equatorial mount developed by Dix (1978b). The device is compact and can be set up to track the sun conveniently and quickly. These features are important in the preliminary survey of a suspected archaeoastronomical site where many possible observation points must be tested. The equatorial mount can be thought of as a portable planetarium that aids in visualizing the spatial relationships which may have guided the architects. The declination and latitude are dialed into the mechanism and the telescope is then swung around the polar axis to determine horizon positions for astronomical events. In addition, Dix witnessed all six of the actual solar events from the center.

Dix also noted that in the center of the bowl was one of the enigmatic dry-walled stone chambers which are found in abundance in New England and which have, as yet, unknown origins and functions (Goodwin 1946; Trento 1978); current theories date these chambers between 2000 B.C. and A.D. 1840. Dix noted that the chamber in the bowl, which has an entrance passage with a quite precise orientation, was aligned to within 0.5° of the first gleam of the rising sun at the equinox.

During additional surveys, Dix found two standing stones in situ which appear to have been possible horizon markers associated with observations from the center of the bowl. Other possible markers were found at the intersections of walls, which might also have served as observation positions.

In 1978 and 1979, we made a precision transit survey in order to classify the important features of the site, identify and map accurately homogeneous sets of markers and all possible horizon alignments associated with them, and verify the natural foresights; we also witnessed the major solar events. Our purpose was to collect evidence for or against the use of the site as an astronomical observatory or calendrical setting device.

The precision surveying of the site has been time-consuming because of dense overgrowth, even though much of the area has been pastureland within the past 50 years. The use of high intensity lights at night, to verify the positions of markers in wooded terrain, has been a very helpful technique. Actual horizon profiles were identified through observations of the site during all seasons. The most productive work was done in the early spring when the leaves were off the trees and the brush down. The apparent use of stone foresights leads us to the hypothesis that ground level was the horizon reference, rather than tree-top level. This convention is used in calculating the horizon declinations.

Description of Calendar I

Calendar Site I presently includes about 20 square kilometers and seven stone chambers. Seven of the twelve chambers which are included in the Calendar I and II complexes are known to have orientations that correlate with major solar calendrical events.

The center of Calendar I is a natural bowl (Fig. 11.3). An observer standing within the bowl has a near horizon with distinct features on three sides and a varied distant horizon to the south. The west ridge, as viewed from the bowl, is mostly barren and the horizon is outlined by a stone wall. The east ridge consists of two peaks separated by a valley. The ridge is densely wooded and the ground horizon is obscured in many places, even in

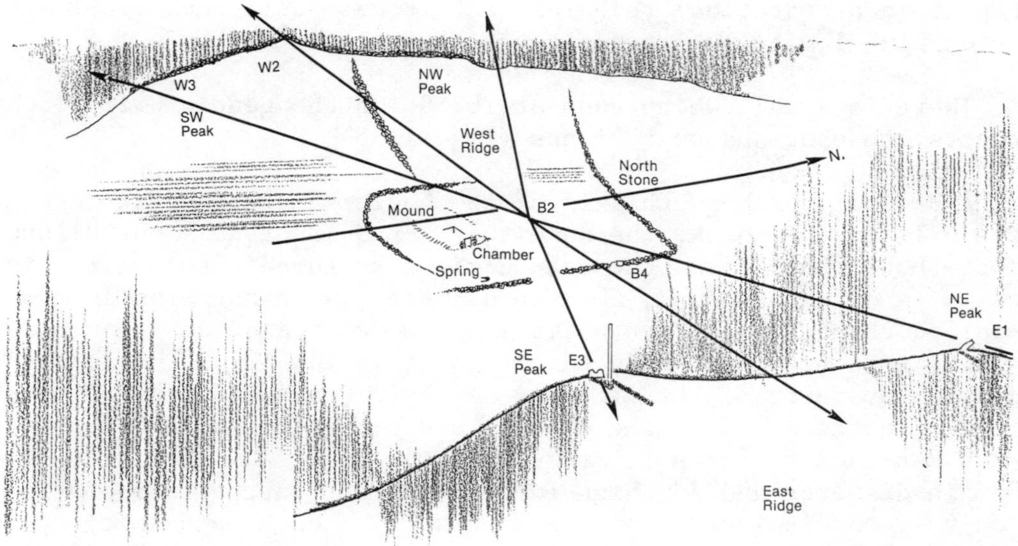

Fig. 11.3 Sketch of Calendar I bowl showing principal features, including sight lines from unmarked astronomical center, B2, to natural peaks and notches.

Fig. 11.4 Standing stone at position E3. Notch ½ meter in width is precisely on the horizon as seen from W2.

winter. A standing stone (E3) is located very near the summit of the south peak (Fig. 11.4). The position of this stone has been precisely determined relative to other significant features of the site. Excavations around the stone (report in preparation) revealed that it was positioned vertically on a foundation slab 1 m below present grade.

There is a wall which runs at 107.9° on the gentle hill to the north that connects the east and west ridges (Fig. 11.5).

A natural north-south axis of the bowl is marked by the crest of the gentle hill on the north, the summit of a prominent hill one kilometer south of the bowl which forms the main horizon feature in that direction, and the center of a circular mound, 60 m in diameter, within the bowl. One of the dry-walled chambers has been built into the east side of the mound (Fig. 11.6). There is a prominent pointed orthostat in the 107.9° wall; this lies on the axis of symmetry of the site and has been designated the north stone (Fig. 11.5). To the south of the chamber and along this axis are several natural and man-made features in dense forest. They include a steep-sided mound 20 m in diameter and 7 m high (which deserves further exploration), grouped cairns, and isolated standing stones. The close proximity of one cairn to another in the groups makes a land-clearing rationale difficult to accept.

A stone wall on the eastern side of the bowl runs at 356.5° and ends at the wall previously mentioned, forming a corner (Fig. 11.5). Some 51.2 m south of the corner of the east wall there were once three large orthostats, two of which remain; the center one has been removed within the past few years. The southernmost of the remaining orthostats is marked with a grid symbol (Fig. 11.7). The center stone, now removed, was also marked. A second grid symbol is located on a stone in the east wall 25 m south of the corner (Fig. 11.8).

There is an unusual stone row which lies along a line normal to the west ridge wall connecting W1 with the chamber (Fig. 11.5). Distributed along the length of the row are several low heaps of stones. The stone row ties the chamber to the site architecturally, and relates it particularly to the west ridge wall, which is the western horizon. The latter includes observation points W1 and W2, which are two of the most important astronomical features of the site (Fig. 11.9).

The central mound is partially surrounded by a wall consisting of contiguous straight-line sections and a curved section.

An upright stone, E1, stands on the east ridge at the position of a foresight for the summer solstice sunrise alignment as seen from the astronomical center of the bowl. We have mapped this stone precisely relative to other significant features of the site.

There is a second chamber 600 m southwest of the one in the bowl. This chamber is positioned along the bank of the stream, whose source lies at the bowl chamber. There is a third chamber 2.5 km due south of the bowl, and a fourth chamber 2 km to the southeast. There are three additional chambers within 5 km of the bowl which are currently being related spatially and temporally to the center of the site.

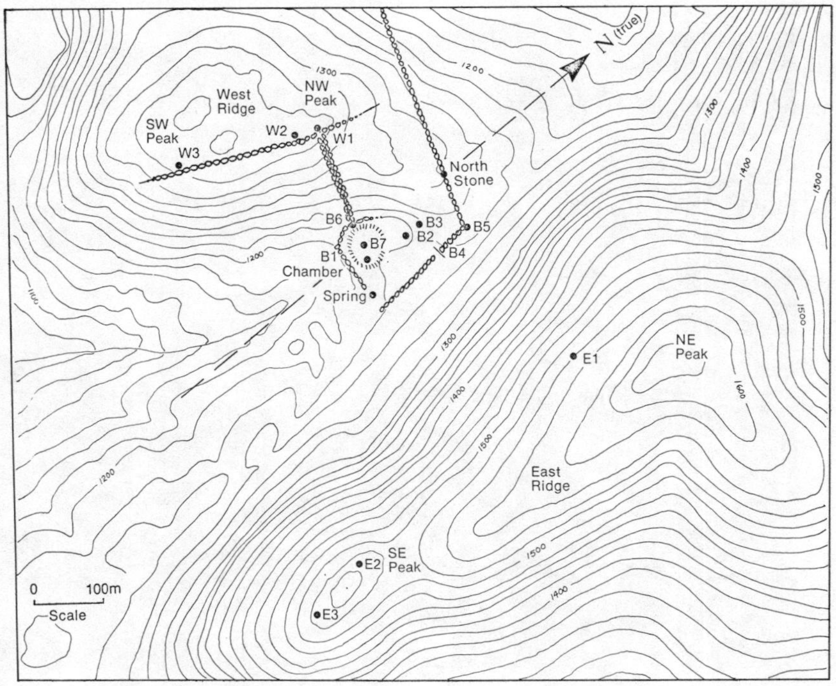

Fig. 11.5 Calendar I contour map: contours 20 feet (6.10 meters). Numbers on contours are feet above mean sea level. Stone features used as alignment markers are shown.

Fig. 11.6 The chamber entrance at Calendar I faces east. (There is no historical record of a slab roof on the chamber proper.)

Fig. 11.7 4 by 5 grid on east face of orthostat in east wall 1.4 meters south of point B4

Fig. 11.8 Grid on east face of stone in east wall 25 meters south of corner B5

One feature of the site that was important in terms of its occupation is an all-weather spring located 46 m southeast of the chamber in the bowl (Fig. 11.5). This is the source of a stream which eventually finds its way to the White River and thence to the Connecticut and Long Island Sound.

There are three rectangular stone enclosures on top of the mound in the center of the bowl. They are flush with the ground and may be the remains of the foundations of historical buildings. The center of the group lies on the axis of the bowl, and the longitudinal axis of each enclosure is aligned in the same direction as the chamber, toward the equinox sunrise. A building foundation is located on the northwest periphery of the mound (Fig. 11.5).

There are two small standing stones at location B3. One is oriented east-west and the other north-south (Fig. 11.5). It seems quite natural to stand within the enclosure formed by these stones to observe the horizon. They may have been placed there in order for a person to position his feet precisely for observation.

The man-made features studied were constructed from local materials and present a complex arrangement of stone walls, chambers, standing stones, and cairns. These particular architectural forms embody geometric and astronomical concepts which appear to fall outside orthodox farmyard practices.

Results of Field Work

The quantitative results of the field work reported in this paper are confined to an archaeoastronomical survey of the Calendar I site. The evidence that astronomy was practiced at the site is based primarily on statistical probability and is essentially circumstantial. A comprehensive archaeological investigation that would establish the date and culture connected with the site would certainly be a valuable contribution. However, the locations which we have identified as architecturally-likely ritual sites have not exhibited sufficient conventional archaeological clues to identify them as either important colonial or Amerindian settlements.

Our method has been first to identify a site on the basis of celestial and terrestrial spatial clues, and then employ conventional archaeological techniques. The investigation of the spectacular solstitial sun patterns through rock slab settings on the top of Fajada butte in New Mexico's Chaco Canyon followed a similar procedure (Sofaer et al. 1979b).

The results of the field survey to date are presented numerically in Tables 11.1-11.5 and graphically in Figures 11.3, 11.9, and 11.10. Azimuths and altitudes were measured using transits capable of one minute resolution or better, and declinations were calculated from the astronomical triangle (Thom 1971). True north was determined by sun sights to an accuracy of $\pm$ 0.02°. The error range for each measurement reported includes the dimensions of observation points and foresights and uncertainties in refraction and parallax. In the tables, S indicates that the results are entirely from the field survey, T indicates calculation using the contours on topographic maps as well as original surveys, and W means that an astronomical event was witnessed. Though some witnessed events indicate an upper limb tangent to the horizon today, all the tabulated solar and lunar alignment declinations are

TABLE 11.1

POLAR AND SOLAR ALIGNMENTS FOR STATISTICAL ANALYSIS
(see Fig. 11.9 and Table 11.2a)

Alignment No.	Alignment	Azimuth (deg)	Apparent Altitude (deg)	Calculated* Declination (deg)	Astronomical Feature
1. S	B7-N1	0.00±0.05			North-South
2. S	B1-E3	138.13±0.10	10.70±0.10	-23.51±0.20	*Winter Solstice Sunrise*
3. T	B3-W3	230.20±0.20	5.52±0.20	-23.25±0.40	*Winter Solstice Sunset*
4. STW	B1 Axis	104.66±0.30	15.05±0.20	+ 0.20±0.50	*Equinox Sunrise*
5. STW	B4-W1	260.95±0.20	7.72±0.30	- 1.19±0.50	*Equinox Sunset*
6. T	W2-B5	66.33±0.10	7.94±0.30	+22.46±0.40	*Summer Solstice Sunrise*
7. T	B6-W1	288.37±0.10	15.97±0.10	+24.12±0.20	*Summer Solstice Sunset*

*Includes effects of parallax and refraction where required. Compare with uncorrected event declinations.

S = Survey, T = Topographic Map, W = Witnessed.

TABLE 11.2a

HOMOGENEOUS SET OF MARKERS (WERE OR MAY HAVE BEEN MAN-MADE) FOR USE WITH ALIGNMENTS 2-7, TABLE 11.1

B1	Chamber, observation point on top of chamber.
B3	Two small flat standing stones at right angles to one another.
B4	Centerline of gap in stone wall between two large orthostats, one having a grid on its face.
B5	Corner of stone wall which runs west and south from this point.
B6	Intersection of stone row W1-B6 with short wall on western boundary of central mound.
W1	Intersection of stone row W1-B6 with west ridge wall.
W2	Peak of west ridge, at summit and corner of stone wall.
W3	South peak of west ridge at corner of stone wall, obscured.
E3	Notched standing stone at horizon as seen from W2.

TABLE 11.2b

ADDITIONAL OBSERVATION POINTS OR FORESIGHTS

B2	Unmarked astronomical center of bowl.
B7	Center of mound into which chamber is built. Lies on north-south axis of site at intersection with line W1-B1.
N1	North Stone. A large, flat, pointed stone in north wall.
N2	Natural notch in northeasterly skyline as seen from W2.
E1	Northeast peak as seen from B2.
E2	Southeast peak as seen from B2.
E4	Horizon for sight from B3 toward E3.
E5	Actual summit of southeast peak, seen from W2.

TABLE 11.3

OTHER SOLAR AND LUNAR ALIGNMENTS
(see Table 11.2)

Alignment No.	Alignment	Azimuth (deg)	Apparent Altitude (deg)	Calculated* Declination (deg)	Astronomical Feature
8. ST	W1-E3	130.35±0.05	4.91±0.05	-24.10±0.10	*Winter Solstice Sunrise* Acronical Rise Rigel, 2660 B.C.
9. S	B7-E3	137.24±0.30	10.31±0.20	-23.39±0.50	*Winter Solstice Sunrise*
10. ST	B5-Wall	107.85±0.20	18.37±0.30	+ 0.48±0.50	*Equinox Sunrise*
11. T	B5-B3	224.21±0.05	2.62±0.30	-28.44±0.40	*Lunar Major Set***
12. ST	B5-N1	287.85±0.20	7.23±0.20	+18.44±0.40	*Lunar Minor Set***
13. ST	B3-E4	144.57±0.10	9.29±0.20	-27.10±0.30 -27.93±0.30	*Lunar Major Rise***
14. ST	W2-E5	123.00±0.30	5.58±0.20	-18.28±0.50 -19.00±0.50	*Lunar Minor Rise***

*Includes effects of parallax and refraction where required. Compare with uncorrected event declinations.

**Major and Minor standstills of 18.6 years cycle.

S = Survey, T = Topographic Map, W = Witnessed.

TABLE 11.4

SOLAR ALIGNMENTS TO NATURAL PEAKS FROM ASTRONOMICAL CENTER, B2
(see Fig. 11.3)

Alignment No.	Alignment	Azimuth (deg)	Apparent Altitude (deg)	Calculated* Declination (deg)	Astronomical Feature
15. STW	B2-E1	75.45±0.20	19.15±0.20	+23.46±0.40	*Summer Solstice Sunrise*
16. STW	B2-E2	138.50±0.20	11.00±0.20	-23.46±0.40	*Winter Solstice Sunrise*
17. STW	B2-W2	259.30±0.10	10.97±0.10	- 0.03±0.20	*Equinox Sunset*
18. ST	B2-W3	233.60±0.30	6.50±0.30	-20.40±0.60	

*Includes effect of refraction. Compare with uncorrected event declinations.

S = Survey, T = Topographic Map, W = Witnessed.

TABLE 11.5

ADDITIONAL ALIGNMENTS

Alignment No.	Alignment	Azimuth (deg)	Apparent Altitude (deg)	Calculated* Declination (deg)	Astronomical Feature
19. SW	W2-E3 (See Fig. 11.5)	128.04±0.04	4.78±0.02	-22.80±0.06	Acronical *Rise Rigel* at Winter Solstice 2424 B.C.
20. S	B1-Axis	284.66±0.40	9.11±0.05	+16.85±0.50	*May 6, Aug. 7, sunset*
21. S	B7-W2	279.05±0.05	12.70±0.10	+15.25±0.20	*May 6, Aug. 7, sunset*
22. TS	B1-B5	23.35±0.20	6.22±0.20	+47.05±0.40	
23. S	W2-N2	14.84±0.10	1.03±0.03	+44.81±0.30	
24. T	B1-B4	35.34±0.10	11.91±0.40	+45.94±0.50	
25. T	B3-B5	44.21±0.10	13.12±0.50	+41.36±0.60	
26. ST	N1-W2	232.50±0.20	5.56±0.20	-21.84±0.40	
27. ST	B3-B4	95.21±0.10	15.89±0.40	+ 7.28±0.50	
28. S	B4-B3	275.21±0.10	5.79±0.20	+ 7.65±0.30	
29. S	B3-W2	253.45±0.10	8.68±0.10	- 5.65±0.20	
30. S	B4-W2	257.74±0.10	9.07±0.10	- 2.41±0.20	
31. S	B5-W2	246.33±0.10	6.27±0.10	-12.38±0.20	
32. T	B1-W3	242.60±0.50	7.24±0.50	-14.08±1.00	

*Effect of refraction included where required. Compare with uncorrected event declinations.

S = Survey, T = Topographic Map, W = Witnessed.

for the center of the body at the foresight. It is not generally known whether the center, upper limb, or lower limb was used as a reference. Therefore, there is an unavoidable uncertainty of 0.50° in interpreting most solar or lunar alignments. The resolution of contours on the U.S. Geological Survey topographic map (U.S. Geological Survey 1944) is 20 ft. The contours for horizon features were found to be accurate within this limit, but topography within the bowl of Calendar I was sufficiently unresolved to require independent measurements of elevations.

Astronomical refraction is included in the values of declination for alignments and follows Thom (1971). It is fortunate that most of the elevations of the alignments of Calendar I are sufficiently large (5°-20°) that astronomical refraction is less than ten minutes of arc, so that the usual uncertainty of several percent does not cause significant error, even for precise alignments. The sighting ranges from the center of the bowl to the ridges or from ridge to ridge (250-650 m) and are a good balance between long distances required for accuracy and short distances to keep terrestrial refraction to a negligible value. Parallax is appreciable for lunar alignments and is considered. The elevations are almost all much greater than the extinction angles of the stars considered.

Table 11.1 first presents the north-south axis of the site, an alignment from the north stone (N1) through the center of the mound (B7) and continuing to a natural summit to the south. Following this in Table 11.1 are the six major solar alignments associated with a homogeneous set of nine markers (defined in Table 11.2a), for alignments to the east and west ridges, selected by including all surveyed, prominent features of the site which were or may have been man-made. This selection is, in part, subjective.

For a set of n markers, there is a maximum of

$$2\ C_n^2 = 2\ n!\ /\ 2(n-2)! = n(n-1)$$

directions that can be obtained from straight lines connecting the markers of the set. For each pair of markers, there can be two associated horizon alignments having opposite directions and different elevations to the horizon from the marker furthest from the horizon. In some cases, a marker is the horizon foresight. In cases where it is clear which is the observation point and which is the foresight, the possible number of alignments can be reduced, but this has not been done here in order to be conservative and to counteract any bias in selection of the markers. The number of possible alignments associated with the set of nine markers is n(n-1) = 72.

All of the six major solar directions have been discovered to be associated with this set, as a result of an analysis of some 32 alignments. The six possible astronomical events considered in Table 11.1 are the following: (1) winter solstice sunrise and (2) sunset; (3) equinox sunrise and (4) sunset; and (5) summer solstice sunrise and (6) sunset.

What is the probability of discovering six solar alignments using a random array of markers? As shown above, the number of directions associated with a set of nine markers is 72, and there are six solar directions to be considered. There are six alignments which have been identified as astronomically significant within less than 1° of declination, and there are 360 possible

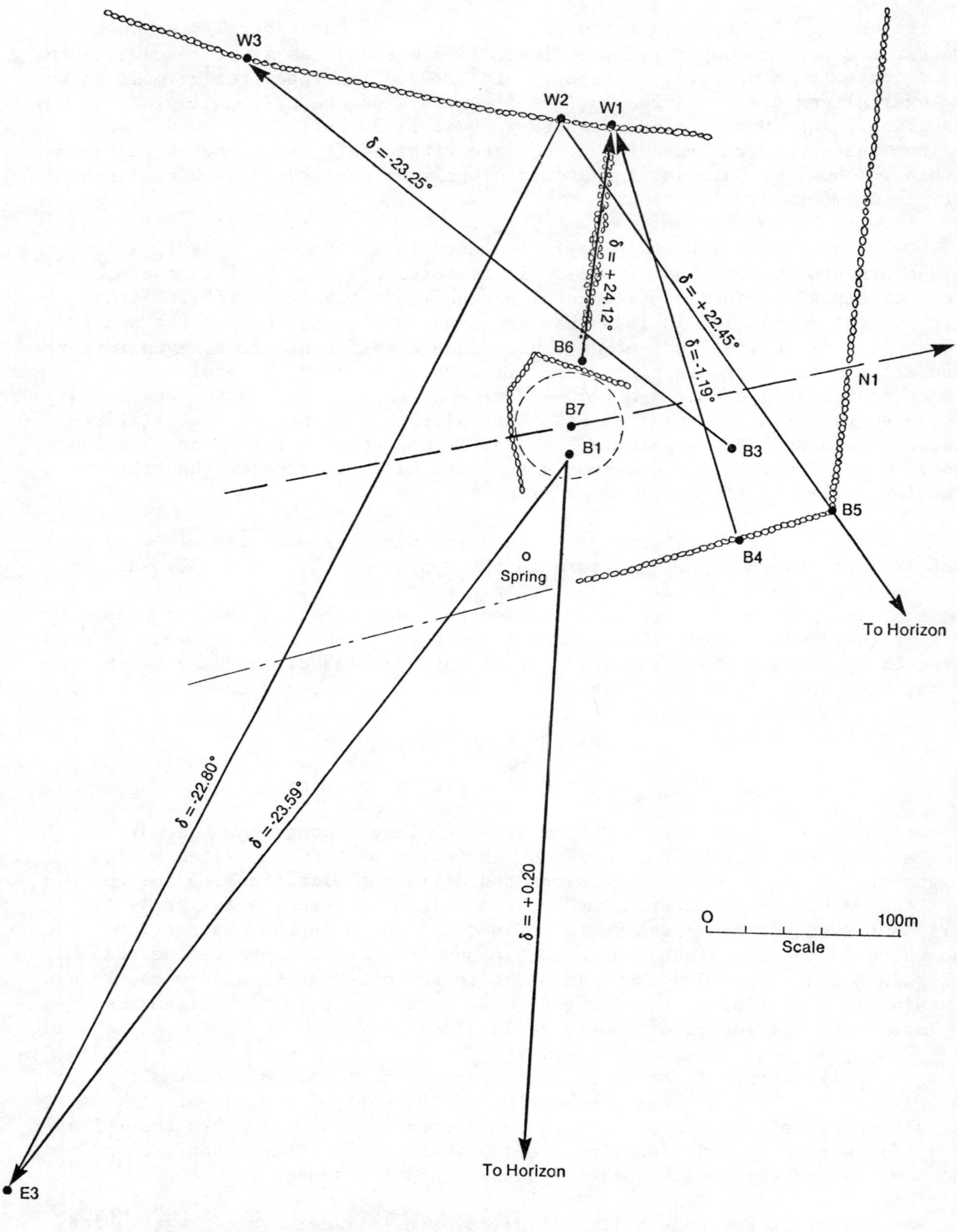

Fig. 11.9 Six solar alignments associated with homogeneous set of nine markers (see Table 11.2a, p. 121); acronical rising alignment W2-E3 also shown.

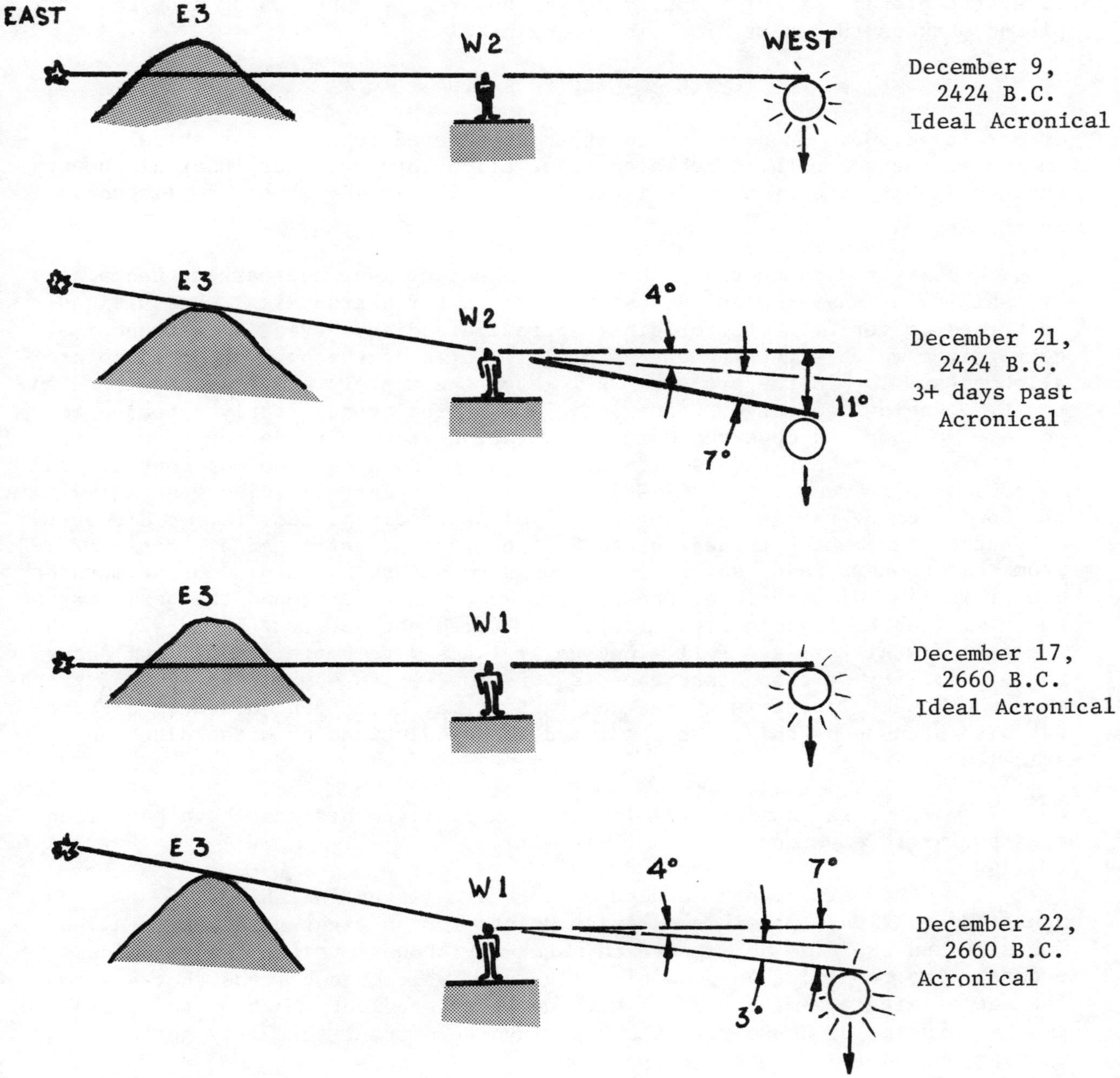

Fig. 11.10 Acronical rising of Rigel at winter solstice

directions if the minimum accuracy of the results in azimuth of $\pm 1°$ is used (Table 11.1). The total number of ways to pick 72 directions out of a possible 360 is given by C^{72}_{360}. The number of ways to pick six directions out of an acceptable six is $C^{6}_{6} = 1$ and the number of ways of picking 66 meaningless directions out of a possible 354 meaningless directions is C^{66}_{354}. Therefore, the probability of identifying six solar alignments using nine randomly placed markers (Uspensky 1937) is approximately

$$P = (C^{6}_{6})(C^{66}_{354}) \;/\; C^{72}_{360} = 0.54 \times 10^{-4}$$

or one in 18,500. In addition to those considered for the statistical analysis, there are three repeated solar alignments and four lunar alignments associated with the markers in Table 11.2a. These alignments are listed in Table 11.3.

After finding that a homogeneous set of man-made markers had a high probability of having been purposely oriented for astronomical observations, we turned to verifying the original approximate discovery of an astronomical center from which sights to the natural features of the bowl produced solar alignments. To aid the process, we located the markers which we had found on the U.S. Geological Survey topographic map of the area, greatly expanded in scale. We compared observed horizon elevations for every degree of azimuth for the east and west ridges with elevations calculated from map contours and found good agreement after adjusting for local topographic discrepancies within the bowl. Using the map and several trial astronomical centers, we discovered a location (B2) which turned out to be the *original astronomical center of 1974*. From this center, major solar events could be sighted within 1° of azimuth or better in line with three of the four peaks which rise around the periphery of the bowl (Table 11.4 and Fig. 11.3). The foresight markers E1 and W2, which are part of this series, fell along solar lines with better than 1° of accuracy. The probability of this occurrence is very low, which shows that the topography of the Calendar I bowl is unique and that it may have been chosen as a calendar site because of this. We confirmed the predictions by witnessing the actual events.

The total number of solar and lunar alignments that have been identified so far is eighteen (Tables 11.1-11.5).

Other alignments between markers are listed in Table 11.5. Alignments 20 and 21 are possible calendar points for an eight-part year. Alignments 18 and 29-32 do not appear to match any known important astronomical events. The remaining 40 possible alignments which can be constructed from the set of nine man-made markers have not been studied. Most of these require the penetration of dense forest cover, involve distant southerly horizons outside of the bowl area, or are spurious.

It is difficult to distinguish between intentional and accidental stellar alignments unless there is a consistent date for several stars or there is an acronical or heliacal event. We have found evidence for the recording of an acronical rising of Rigel; this is discussed below. However, no consistent pattern of star alignments has otherwise been found, so they are not tabulated. At this stage of the work, any stellar alignments must be treated with extreme caution in inferring either an intent or a date.

The high probability of a non-random placement of man-made markers in conjunction with the knowledge that eighteen solar and lunar alignments have been identified provides support for the theory that the site could have been a complex observatory or calendar site.

We observed the winter solstice sunrise from W2, looking toward marker E3 (Fig. 11.4) on December 22, 1978, and measured the declination of the alignment with high precision (Table 11.5 and Fig. 11.9). There are four other winter solstice alignments reported in the tables (Nos. 2, 8, 9, and 16). We surveyed No. 8, from W1 to E3, with a precision comparable to that employed from W2 to E3. Three of these alignments use an interesting marker stone at the horizon as a foresight. The stone (E3) has a notch just large enough to be observed when backlighted by the sun, even at a range of 640 m. The apparent frequency of winter solstice alignments to this stone caused us to wonder whether or not there might have been a stellar event at the winter solstice recorded at Calendar I.

Winkler (1972) describes an ideal acronical rising of a star as occurring when the star and the sun are on the level horizon simultaneously and the star is in the general direction opposite to that of the sun. The first day on which a designated star can be seen after sunset can be used as a calendar point. If we consider only stars of first magnitude or brighter plus the Pleiades, a heliacal or acronical rising at a solstice or equinox is a rare astronomical event. It has occurred annually during periods of about 200 years on less than fifteen occasions in the past 6000 years. In Egypt during the early third millenium B.C., the Nile in June was used to regulate the lunar calendar (Parker 1874) and was the basis of Egyptian chronology. In the British Isles, the heliacal rising of the Pleiades about 2000 B.C. was apparently noted and has come down to us as Beltane or May day.

At the latitude of Calendar I there was a special type of acronical rising in which the sun and the star Rigel rose near the same point on the horizon at the winter solstice. Since 10,000 B.C., only two first magnitude or brighter stars have risen precisely at the position of the marker E3 as seen from W2: Sirius in 3050 B.C. $\pm$ 20 and Rigel in 2424 B.C. $\pm$ 20 (Hawkins and Rosenthal 1967). On December 5, 2424 B.C., both Rigel and the sun rose precisely over the position of the marker E3 (as seen from W2, the prominent western peak). On December 9, Rigel rose on a level horizon precisely at the time of sunset, an ideal acronical rising. The marker E3, which plays such an important role here, was excavated and found to have been set up on a foundation. However, Rigel would not have contrasted sufficiently with the twilight at sunset on December 9, and could have been seen only after some days had passed, when it rose well after sunset. Aveni (1972), following the work of Lockyer (1894), states that a first magnitude star in the east becomes visible when the sun has sunk seven degrees below a level horizon, and he has confirmed Lockyer's observations under the best conditions at Hamilton, New York. There was a single day each year during which the rising star was first visible on the eastern horizon just after sunset. This annual event would have continued over a period of years until the precession of the equinoxes caused the star to rise too far away from the marker E3 for it to be useful. The width of the notch in the marker, 0.52 m, subtends an angle of 0.05° at the observation point W2 (Fig. 11.4). Rigel would have risen within the notch for ten years. Thom and Thom (1978) have concluded that a notch is the most effective horizon feature for observing the first appearance of the upper limb of the sun (first gleam).

The sun set on December 21, winter solstice eve, three minutes later than on December 9, and Rigel rose earlier by 3.93 x 12 = 47 minutes. Thus, the phase difference from an ideal acronical rising on the solstice eve was 44', which corresponds to a difference in altitude of 7°. Rigel rose at an altitude of 4° as seen from W2, and the sun set on a level horizon. Therefore, the altitude of the sun on December 21 at the time of the rising of Rigel was -7-4 = -11°. This means that the sun was 3° or five days past acronical (first day Rigel was visible). Variable atmospheric conditions permit a tolerance of three days and alignment accuracy another day, so the first rising of Rigel would have occurred between December 12 and 20.

This alignment (W2-E3) might have been useful at the time of the acronical rising to signal the onset of the winter solstice over a period of years. Perhaps the marker E3 was set up for this purpose. Rigel is of magnitude 0.11 and is usually the brightest star in the constellation Orion.

But the sun did not rise over marker E3, as seen from W2, on December 22. Its rising azimuth had shifted by 0.52° to the south during the period from December 5. Therefore, the alignment W2-E3 was not useful for observing the solsticial rise with precision. This troublesome circumstance led us to wonder why the architects did not go one step further and set up the markers so that both Rigel and the sun rose over the same foresight from the same observation point on December 22. With this in mind, we investigated alignment W1-E3 and found that the circumstances were right for such a combination of events. W1 is an intersection of two walls located 25.9 m down the west ridge wall. The alignment, however, corresponded to an earlier date, 2660 B.C. On December 22, 2660 B.C., Rigel and the sun rose precisely over E3 as seen from W1. The conditions for a useful acronical rising were also met here precisely. On December 22, the sun had sunk exactly 7° below the level horizon when Rigel rose over E3; the ideal acronical rising had occurred on December 17. Figure 11.10 illustrates the events.

As a result of the above analysis, we have two dates associated with the acronical rising of Rigel: a precise date of 2660 B.C. and another less precise date. It is possible that these dates define the period during which this acronical rising was useful in signaling the winter solstice.

A complex combination of circumstances would have been required to record this event. The three required markers are all in place with the necessary precision. First, the azimuth and elevation required for an acronical rising of Rigel along alignments W2-E3 and W1-E3 had to be precise enough to identify the star. Second, the azimuth and elevation required for a solstitial sunrise along alignment W1-E3 had to have been precise and of the same date as the stellar alignment. Rigel would have been visible at its rise only briefly and a marker in a precise location would have been very helpful, if not necessary, to alert an observer to its rising.

Eddy (1974) used the heliacal separation of 11° to establish a probable date for the Bighorn Medicine Wheel coincident with the heliacal rising of Aldebaran at the summer solstice. His astronomical date was confirmed by dendrochronology. At Calendar I, we have a similar basis for dating except that the rising is acronical rather than heliacal. We have man-made markers at locations for observing the rising of Rigel and the sun at the same point on the horizon on the day of the winter solstice in association with an acronical rising, an event much rarer than a general heliacal or acronical rising at the winter solstice. Thus, we observe that markers exist that may have been related to a highly unique and sophisticated astronomical event between 2660 and 2424 B.C. This was the only acronical rising of a star of first magnitude or brighter near the sun's azimuth at the winter solstice in the past 10,000 years. It is likely that it was observed in northern latitudes.

The date 2660 B.C. is too early for comparison with known archaeo-astronomy data in the western hemisphere. It falls within the Archaic period of eastern North American peoples. Astronomy has not been identified as a practice of the cultures associated with this period, but Fitzhugh (1978) reports a number of architectural and environmental features in the same locations in the Maritime Archaic settlements of coastal Labrador, a correspondence which invites comparison with the Vermont sites. These features include boulder alignments, rock-capped mounds, stone chambers, standing stones, and cairns on high coastal prominences. He cites evidence for a "sacred space" concept throughout the Maritime Archaic period. Further research to locate possible astronomical alignments is warranted in this northern region.

The limited excavations of astronomical markers at Calendar I have produced artifact assemblages, including red ochre, which (while not diagnostic) are not inconsistent with an Archaic provenance (report in preparation).

Since our work at Calendar I was carried out, we have found a site in Massachusetts where there are four independent, unambiguous astronomical alignments from the interior of a stone chamber to horizon cairns a mile away. They provide a consistent astronomical date of A.D. 710 (Dix and Mavor 1980). One of these alignments marks the heliacal setting of the Pleiades at the equinox, and a second, later example of the possible recording of the acronical-heliacal behavior of stars in New England using stone markers.

In addition, in expanding our work at Calendar I we have found that a May day sunset is marked at a standing stone on the horizon near a hilltop, as viewed from a chamber outside of the bowl (Dix, Mavor, and Potts 1980). The stone has been excavated (report in preparation) and, like marker E3, was found to be balanced on a foundation slab 1 m below present grade.

CONCLUSIONS

As of the present writing, we have identified eighteen possible solar and lunar alignments, plus a north-south line, using man-made and natural features of the landscape at the Calendar I site in central Vermont. All six major solar directions have been found to be associated with a homogeneous set of nine man-made markers. If these markers were situated in a random manner, the probability of the discovery is one in 18,500. In addition, the six major solar events are associated with alignments from a single point within a natural bowl to the natural peaks and notches on the horizon surrounding the bowl. We suggest an acronical rising of Rigel at the same azimuth as the winter solstice sun between 2660 and 2424 B.C. as an event which could have been associated with the alignments of man-made markers and which may date a phase of the use of Calendar I by man. The discovery of a heliacal setting recording at another site in New England seems to corroborate this conclusion.

We conclude that there are special sites in Vermont which can be described by a set of architectural and environmental features, some of them quite subtle. Astronomy is but one of them. The sites, their markers, and their structures may be historical or prehistorical or both, and their designs may have evolved over several millenia. We recommend careful, interdisciplinary study of the sites.

12 Documentation of Some Lunar and Solar Events at Casa Grande, Arizona

John H. Evans and Harry Hillman

The Casa Grande National Monument is located on the northern edge of the small city of Coolidge, Arizona, about midway between Phoenix and Tucson. Although small in area, it encompasses eighteen separate ruins ascribed to the classic period Hohokam culture: i.e., ca. A.D. 1175 to A.D. 1475. Perhaps the three major features of the monument are two compounds, about 200' by 400' in size, and a ball court. Only one compound has been excavated and studied, but a major feature of this excavated compound is the "Big House" that is the origin of the name of the monument.

This structure is about 60 ft. in length from north to south and about 40 ft. from east to west. The walls are made of puddled adobe up to four ft. thick, erected on a platform mound that is several feet above the plaza. To the north and south are two rooms whose long axis lies in an east-west direction. Between these two rooms are three others whose long axis lies in a north-south direction. Originally there were several floors, so the outer rooms all had two stories while the central room had three stories rising above the platform mound. There are openings piercing the walls which archaeologists classify as doorways, large windows, observation holes, and vent holes; we will omit the vent holes from our discussion. Several doorways have been blocked up, occasionally leaving a small rectangular opening at the top; these are not understood.

Frank Pinkley (1918), National Park Service archaeologist and administrator, first suggested that some of the openings in the Casa Grande may have served the Hohokam as astronomical viewing ports. This suggestion is reinforced by Pima oral accounts that Casa Grande "was inhabited by a cruel and bitter man that looked out through holes in his house to salute the sun." In 1969, assistant ranger D. H. Kayser and John Molloy, a graduate student in anthropology, decided to test this hypothesis. They measured azimuth sight lines using a hand-held sextant and working from a ladder. On the basis of these and other measurements they and additional collaborators proposed (1973) that some of the openings in the walls could indeed have been used to view the setting sun on the summer solstice, the sun at the equinoxes, and the moon at its extremes inside and outside the solstice points. They observed the summer solstice, the equinox events, and a lunar event.

Because the moon was so oriented in 1978 that its rising and setting points were at their extremes inside the solstice points, the authors decided to observe and document astronomical events occurring at Casa Grande. What was observed and photographed is explained here with the aid of Figure 12.1.

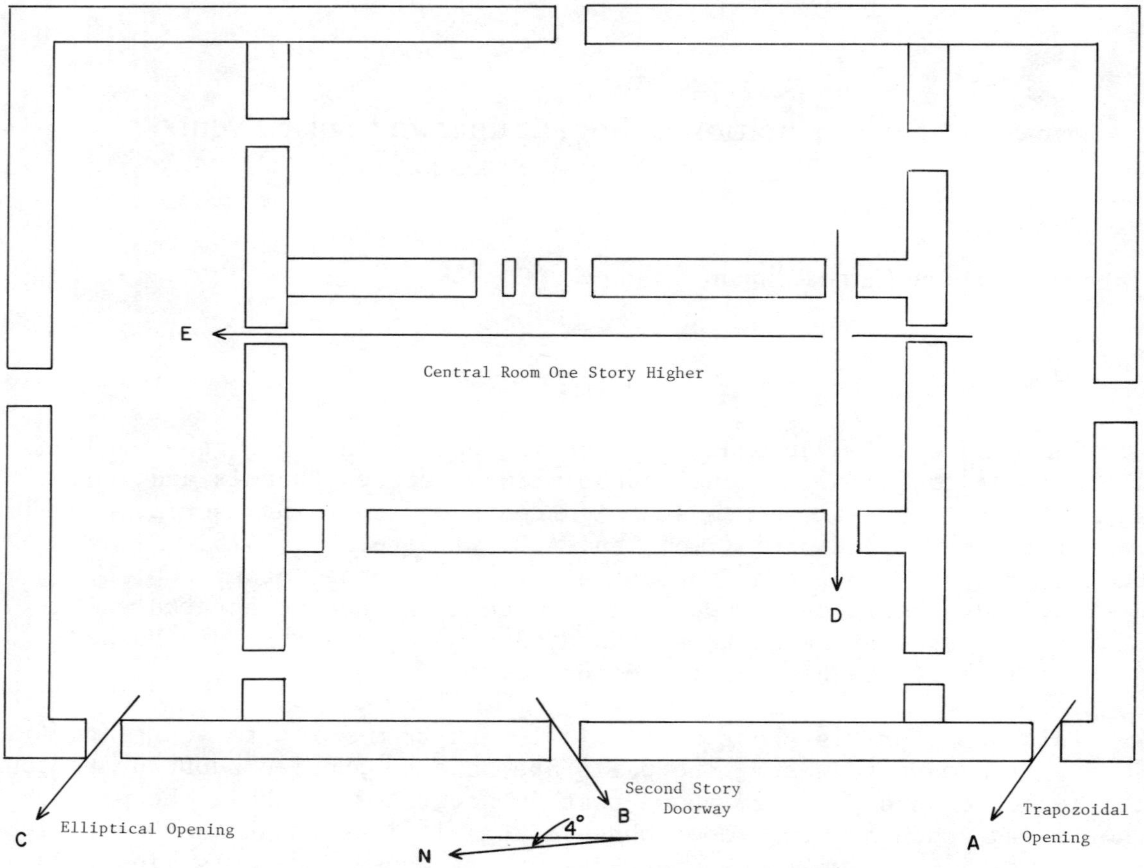

Fig. 12.1 Site plan, Casa Grande, Arizona

In August, September, October, and November of 1978, the moon set about 6.5° inside the solstice points. Since opening A, when viewed from the inner left to outer right window jambs, isolated nearly a 1°-wide strip of the horizon about 6° inside the solstice point, it clearly selected the lunar events and excluded the solar ones. The same type of event was repeated two weeks later along the direction marked B; this coincidence is much less marked, however, due principally to the eroded edges of the west doorway.

Openings A and C are at a very convenient viewing level above the second-story floor location (indicated by wall markings). Opening C is an excellent summer solstice marker. It is an elliptically-shaped hole through a wall about 91 cm thick. With one's head back a slight distance and placed so as to view from inner left to outer right, a narrow slot is formed that just frames the setting sun on the summer solstice. When one stands back out of the way, a spot of light moves along the south wall of the north room. Unfortunately, the plaster is eroded rather badly, so it is not known if there was a marker on this wall or not. There is a carving of a human figure on this south wall, but it is too low for the light to fall on it and it may be of modern origin.

The pair of openings at D serve as equinox markers, though with some difficulty. As the sun rises, a spot of light formed by the opening in the east wall appears on the west wall; as the sun moves up from the horizon and to the south, the spot of light moves down and to the north on the west wall. At a particular point in time the light disappears into the hole through the west wall.

The hole in the east wall is about 10.2 cm in diameter and does not go straight through the 76 cm-thick wall. As it goes from inside to outside, it slopes downward toward the horizon and toward the north. The hole in the west wall is about 11.5 cm in diameter and is not concentric with the hole in the east wall. It is also angled through the wall; it is level in the horizontal plane but slants off to the south when one looks through the hole from east to west. The net effect of the offset and slope is extremely interesting; if one is observing from the extreme west room, one's head must be placed in just the right position to see the eastern sky through both holes. There are only two days a year when conditions are such as to permit the light from the rising sun to pass through both holes; this is near each equinox. The alignment functions very well two days after the fall equinox and two days before the spring equinox, in each case about 45 minutes after sunrise. This event has been photographed on both faces of the west wall so that one sees the spot move down the wall and into the opening. A second photo was made through the hole at this moment; a brief, brilliant flash of light emerging on the western side of the upper story room can be seen.

In addition to our poor understanding of the equinox situation, there are many other openings whose functions can not even be guessed. For example, the holes marked by arrow E in the north-south direction do not seem to relate to any known celestial events. And are the four holes in the east wall of the upper room just redundant equinox markers? There are other openings in the east wall (not shown) that are expected to function in 1987 when the moon rises at its maximum outside of the solstice points. And there are still other openings (again not shown) that are not understood at all. The authors have since returned to Casa Grande and have made many more measurements on the azimuthal angles of these openings and horizon elevations; these data are now being analyzed for correlations with celestial objects and events around A.D. 1300.

13 Living the Sky: Aspects of Mescalero Apache Ethnoastronomy

Claire R. Farrer and Bernard Second

INTRODUCTION

Cosmology is important to each of us, even when we do not consciously attend to it. The very way in which our perceptions are decoded seems to be determined, in large measure, by our culture (Deregowski 1972). We shall demonstrate that perceptions, at least for the Mescalero Apache, rest upon a cosmological base but are simultaneously filtered through the Apachean culture.

We do not have to go to the Apache to illustrate the cosmological basis of perceptions, however. If we are mainstream Americans, we are a linear, hierarchical people, in keeping with our biblical idea that there is a heaven above, a hell below, and our world in the middle. Our notions of hierarchy also surface in our belief that there are those both greater and lesser than each of us. Often our ideas and values are so much a part of us that we resist being shown that their basis lies in a religious belief system.

No one of us, whether Mescalero Apache or mainstream American, is always consciously aware of the beginnings or the assemblage of our belief system, nor does each member of a culture have an equally sophisticated knowledge of the underpinnings of the belief system (see Wallace 1970:3-38). We usually have religious specialists who keep us informed of our responsibilities and privileges.

Whatever belief system holds our allegiance, we organize our world, our perceptions, our feelings, our sense of proper and improper behavior, at least in part around our cosmology. And no matter what the belief system, some of us are more erudite than others both in possessing knowledge of the cosmology and articulating it. We walk through life on a cosmological path that has been mapped for us by our deity or deities or (for some of us) by our own inventive minds.

Even when we do not believe the cultural cosmology, we are oftentimes forced to live it; for example, consider Sunday "blue laws." Whether we are church-going (or even believing) Christians or not, we are unable to purchase certain items or engage in specific events on Sundays if we live in an area where blue laws are in effect.

While our present data are from the Mescalero Apache Indians of the mountains in southcentral New Mexico, we wish to suggest that cosmology is significant for all of us. It is underfoot as well as being in and of the sky.

Furthermore, we suggest that there is differential access, not only to that cosmology, but also to interpretations of it. We do not expect the proverbial man on the street to have the understanding of Christianity that we expect from religious scholars; nor should we expect each Mescalero Apache to have a total understanding of the symbolic significance of Apachean religion and ritual. Our data are primarily from specialists about esoterica--but it is esoterica that influence the lives of all who dwell at Mescalero.[1]

The contemporary Mescalero Apaches live on a reservation of approximately 720 square miles in southcentral New Mexico. This area is considered to be the heartland of the area they once roamed: from Austin, Texas, to the Rio Grande River in New Mexico on a east-west axis, and from southern Colorado to Chihuahua City, Mexico, on a north-south axis. Despite their propensity for moving about in a large area, they had a Plains adaptation at the time of contact. Indeed, Coronado's expedition found them (or rather the Querechos, who are generally believed to be ancestral to present-day Apaches) on the Plains, the Llano Estacado (Winship 1964). Apache stories told today aver that even while on the Plains, advantage was taken of the mountains for the flora and fauna they offered.

Authorities differ on when the Apache arrived in the Southwest, but most agree their arrival predated that of the Spanish by a relatively short time (e.g., see Forbes 1971:xi-xx). It is generally agreed by both scholars and Apaches that they migrated southward from what is now Alaska and northwestern Canada. They share their Athapaskan linguistic stock with Indians still in the north as well as with the Navajos of the Southwest.

The Mescalero Apache Indian Reservation was created by treaty in 1873. In 1904-05, the remnants of the Lipan Apache joined their "cousins" on the Mescalero Reservation, and in 1913, almost 200 Chiricahua Apaches also moved to Mescalero from Fort Sill, Oklahoma. Today, the tribal count hovers around 2000 (the 1974 tribal count was 2174) and includes not only the descendants of the original 468 Mescalero who moved to the reservation but also the descendants of the Lipan and Chiricahua. There has been little

---

[1]These data have been gathered over a sixteen-year period by the senior author, and represent a lifetime of learning, beginning with an apprenticeship at seven years of age, for the junior author. Farrer began trying to understand in 1964 when, as an Anglo housewife from Alamogordo, New Mexico, she talked with an Apache housewife from Mescalero during a girls' puberty ceremonial. Within 36 hours of that first conversation, she had witnessed one of the more arcane events occurring among the contemporary Mescalero. These many years later she is still learning, most often from Second, a Singer of the girls' puberty ceremony and a religious specialist. Data collection has proceeded aperiodically and continues to the present.

While this is a joint effort, Farrer was primarily responsible for writing; Second was responsible for critical reading and provided many of the data. Farrer received dissertation support in 1974 and 1975 from the Whitney M. Young, Jr., Memorial Foundation, of which she was a Fellow, and from the Mescalero Apache Indian Tribe. A Weatherhead Resident Fellowship at the School of American Research in 1977-78 allowed her time to think through much of the material gathered earlier and to ask more sensible questions of Second. Grateful acknowledgment is tendered each.

Puebloan intermarriage through the years but there has been a considerable amount of intermarriage with Navajos, Jicarilla Apaches, and the Apaches of Arizona. However, we discuss only the Mescalero in this paper.

Our interest in collecting ethnoastronomical data on the Mescalero is not a pioneering effort; however, since academic interests seem to follow cycles and archaeoastronomy and ethnoastronomy are currently in fashion, there has been little published to date in the area. Morris Opler, the dean of Apachean scholarship, has not had astronomical matters as a primary concern, as a glance at his substantial bibliography indicates;[2] however, he does discuss beliefs relating to sky lore in passing (e.g., 1965:355 and 1969:91). Gifford has published names for seasons, the phases of the moon, certain constellations, as well as information on day and month counts (1940:153-156). Our contribution is to provide a context for the contemporary record and show how the sky is *lived* on an everyday basis, for the Mescalero can truly be said to be "living the sky."

METAPHORS IN MESCALERO RITUAL

CRF: Tell me about the beginning time. What happened then?

> BS: When my grandparents started off these stories, they started it off with the word *niaguchilaada*, when the world was being made. At that time there was nothing in the universe except--for the Great Spirit, God. And He--*He* made the world in four days. First came Father Sun and Mother Earth; then the sky elements and Old Man Thunder and Little Boy Lightning. Next came animals and on the fourth day came man, the Apaches.[3]

The universe thus described verbally is given visual representation as a circle bisected both horizontally and vertically. It forms a visual metaphor (see Fig. 13.1) that, together with the verbal metaphor, organizes much of Mescalero life and ritual; it is what has elsewhere (Farrer 1977; 1980) been termed the "base metaphor" of the Mescalero. The base metaphor holds that the world is constituted, and rests upon a foundation, of circularity, balance, the number 4, and directionality. The metaphor is readily apparent in the brief narrative of the beginning time: The number 4 is of primary importance for the world was created in four days; but it was also created in four steps that are seen as being in balance with each other. The first two days, when the beings that allow and sustain life came into existence, are balanced by the last two days, when animals and people appeared. The balance is further predicated on the animate (last two days) and the inanimate (first two days); each requires the other for harmony and completion.

---

[2]Melody (1977) lists over thirty of his publications; none of these has astronomy as the focus.

[3]In all quoted material a dash (-) indicates a pause in speech; each dash approximates one second of time. Elipses (. . .) indicate material omitted from the original narration. Quotations are from Second, unless otherwise credited, and are from either audio tape or hand-written notes.

The verbal/visual metaphor can be (and is) explained on several levels: it is first and foremost a depiction of the universe; it is a "blueprint" for a four-day/four-night ritual, the girls' puberty ceremony; it is how people should behave; it is the four stages of life (infancy, childhood, adulthood, old age); and it is how the world was made as well as how it functions. It is more as well.

The visual metaphor is said to be "the same thing" as a crescent moon and a four-pointed star painted on a dancer's body. And both the visual metaphor of Figure 13.1 and the designs it generates are said to be "the same thing" as a narrative concerning natural elements and the goodness of life in the mountains. The verbal/visual metaphor also generates a base metaphor for life--how one goes about "living the sky."

It is not necessary to be a religious specialist, or to speak to one, to see the metaphor in operation. It is on public display during the staging of the girls' puberty ceremony that is held during the first week of July each year at Mescalero. While there are several areas of activity during a ceremony, the Holy Lodge, a twelve-pole tipi, is the focus of the most sacred aspects of the event. And that Holy Lodge is in the form of the visual metaphor (Fig. 13.2).

The Holy Lodge has four poles for its essential structure. These four poles are called ". . . the Four Grandfathers who hold up the universe for us. . . ." These Grandfathers are said to be "the same thing" as the visual metaphor and "the same thing" as the beginning-time narrative. When viewed as a schematic, it is immediately apparent that the Four Grandfathers describe the figure that is the drawing of the universe (Fig. 13.1). Each primary pole (Grandfather) of the Holy Lodge is placed at a cardinal point: the First Grandfather in the east, the Second in the south, the Third in the west, and the Fourth in the north. They are always spoken of in that order, in a "sunwise way." These four, together with the other eight poles that are eventually added and the covering of the resultant tipi, form a circle with a Grandfather at each cardinal direction, as illustrated in Figure 13.3. The lighter dots represent the additional eight poles that form the finished Holy Lodge.

> The main ceremonial lodge is made of twelve evergreen fir trees. These poles . . . represent eternal life for us. And the twelve represent the twelve moons of the year. . . . The four main structure poles [the Four Grandfathers] . . . correspond to the four directions of the universe, the four seasons, the four stages of life--for in the natural world everything is based on four. . . .

The First Grandfather, who stands in the east, represents the moon and stars. The Second Grandfather, who stands in the south, represents the sky elements (wind, rain, lightning, clouds, thunder, rainbows, mountains). The Third Grandfather, who stands in the west, represents the animals. And the Fourth Grandfather, who stands in the north, represents humanity, for "since man is a frail being, it takes all the other three to hold him up." By the erection of the Holy Lodge's foundation poles, the creation of the universe has been recapitulated in structural terms paralleling the verbal ones.

When the Holy Lodge is being erected, it is done so in the sunwise way. The easternmost pole is the strongest; it is the pole of the heavens.

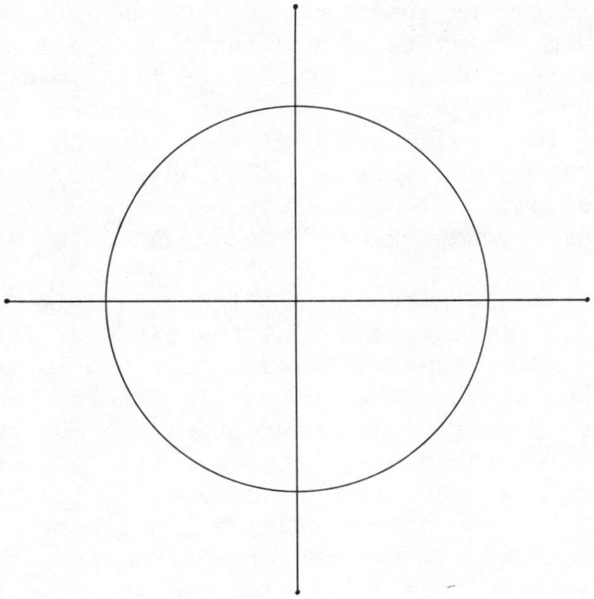

Fig. 13.1 Mescalero Apache depiction of the universe

Fig. 13.2 Holy Lodge four-pole tipi

Resting upon it (in order) are the elements, the animals, and human kind. East and south (the symbolic areas of the inanimate) balance west and north (the symbolic areas of the animate). But there is also a balance in the east and west, where the sun rises and sets, and in the north and south, through which the sun passes. (We shall also see below that the cross-cutting of symbolic fields carries over to body painting as well.) Animate balances inanimate; four is ubiquitous; directionality permeates all; and the circle unites them all--the base metaphor has been demonstrated again.

The original visual metaphor of Figure 13.1 is formed by the points for each of the cardinal directions and by the circular form said to be inherent within the four directions; this circular form is said to be described by the sun as it circles through the heavens in a solar year as well as in its daily motion from east to west. The inherent balance of the world dictates that if the sun moves from the east to the west (passing through the south) in the daytime, then it must move from west to east (passing through the north) during the night. Thus, the circle is seen as an integral part of the cardinal directions. The Holy Lodge is the structural correlate of the base metaphor and the visual metaphor: by connecting east and west and then connecting north and south a cross is formed, while the area inside the poles becomes the circular floor of the tipi. Furthermore, the Grandfathers are said together to balance the other eight poles constituting the Holy Lodge; this demonstrates that the base metaphor provides an organizing principle for the Mescalero through the generation of balance, circularity, directionality, and the number four with its various factors.[4]

By observing the natural world, the Mescalero construct a rational explanation of the way in which life should be managed. These observations have the sun as their focus; however, it is not to be supposed that the Mescalero are sun worshippers: "The sun to us . . . is the *physical representative* of the Great Spirit, God: the Power in the universe." Additional visual currency is given the metaphor when the impersonators of the Mountain Gods dance during each of the four nights of the ceremony. Their appearance is awesome, a feast for the senses, with their vividly painted bodies, tall headdresses, and costumes whose metal jingles sustain an impelling rhythm (see Farrer 1978, 1980); for the present, however, our interest is centered on the designs painted on the dancers' torsos.

Figure 13.4 is a representation of a common motif painted on the dancers. The upper torso is divided into four fields: two each on the back and on the front of the dancer. The back presents a mirror image to the front in that dividing the torso on either a north-south or an east-west axis produces the complete design. One night the dancers appear with the colors indicated in Figure 13.3; the next night the same motif will be used but with the colors reversed, so that there is a four-pointed green star on a yellow field and so on.

> CRF: Is there any meaning to the design they paint on those dancers?
>
> A: Oh, sure.------------------------------It is good to live up here. You wake up in the morning. Maybe the crescent moon is still

---

[4]Note that balance does not always rest on a 1:1 correspondence. Other factors, such as the greater power of the Grandfathers here, often enter into perceived and ascribed balance.

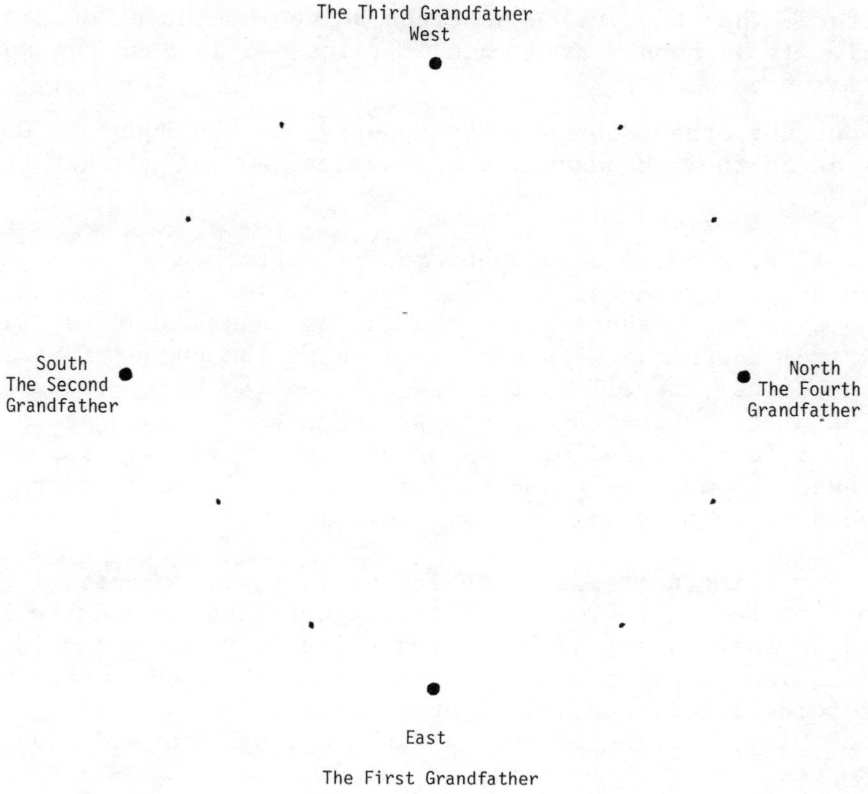

Fig. 13.3 The Four Grandfathers

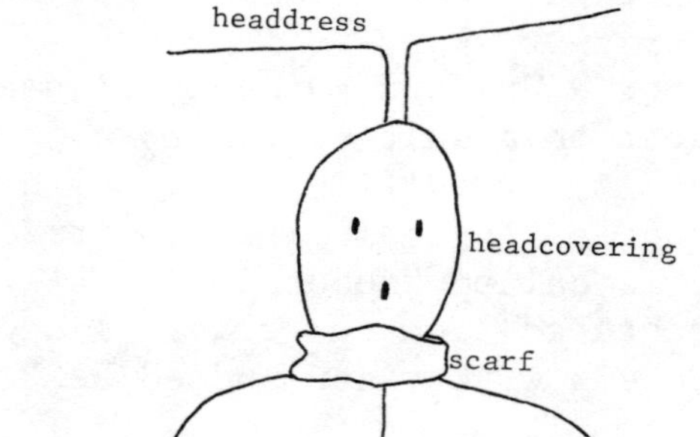

Fig. 13.4 Artistic motif painted on dancer's body

out and a star. Then the yellow sunlight shines on the green grass. It is beautiful. It is good. Even when it rains, it is good. A man lives a longer time up here.[5]

When queried about the other designs that appeared on the Mountain God dancers as well as on their headdresses and sashes, A replied that they were "the same thing."

Figure 13.5, showing a 1975 ceremonial performance, pictures the Mountain God dancer in more detail. This back view includes the distinctive sash that, along with the headdress and the use of painted motifs, distinguishes one dance group from another. Note also the use of the number four in the headdress; the headdress, as well as the dancer's entire body, is bilaterally symmetrical. The night before this photograph was taken, this group of dancers had appeared with a four-pointed black star on a white field. The motifs used on the dancer's body as well as those that appear on his sash and on his headdress are all said to be "the same thing."

"The same thing" extends to other motifs that, to Western European eyes, appear to be different. The mannekin in the tribal museum in Figure 13.6 illustrates another aspect of "the same thing." Here it should be noted that the reduplication of points on the star does not produce a different motif; the four-pointed star and the eight-pointed star are still pronounced to be "the same thing." It is balance and the use of four and its aspects that produce the feeling of isomorphism in the designs. The eight-pointed star motif does not "belong" to anyone, but people agreed that it looks Mescalero; therefore, it is appropriate for representation in the museum. The mannekin is in a display with a photomural background. Note that the design on the torso of the dancer behind the mannekin is said to be "the same thing" as the star and the "triangles" on the headdress. Furthermore, the lightning painted on the mannekin's arms is also "the same thing."

While it is easy for a Western European to see a positive/negative shift (e.g., from a black star on a white field to a white star on a black field) as being the same, it is more difficult to see how a four-pointed star is "the same" as a triangle or a diagonal line with triangles. [The diagonal with triangles, by the way, is called "mountains."] Figure 13.7 illustrates how each of these can be generated from the original depiction of the universe, the visual metaphor, and hence how they can indeed be "the same thing."

In Figure 13.7 we again begin in the east with the First Grandfather; heavy black dots indicate the cardinal positions of the Grandfathers. Dashes describe the circle, the sunwise movement. By following the dot-dash-dot-dash line, the four-pointed star is generated--and of course, a slight shift will generate the eight-pointed star of the mannekin. By following the dot-dash-dash-dot line in the upper (western) portion of the drawing, we can see the crescent moon. All that remains is to follow the dot-dot-dot-dot line that joins the north and south Grandfathers; by imaginarily cutting through this line and spreading out the figure, we have duplicated the four-peaked mountains, as Figure 13.8 illustrates. Other transformations generated appear

---

[5]The speaker (A) remains anonymous at his request. In 1964 when this material was recorded, he was in his late 50s. In his youth he had been a Mountain God dancer and was related to those who owned sets of the dancers. For data on ownership, see Farrer 1978 and 1980.

Fig. 13.5 Mountain God dancer and four-pointed star

Fig. 13.6 Mescalero Apache museum figure

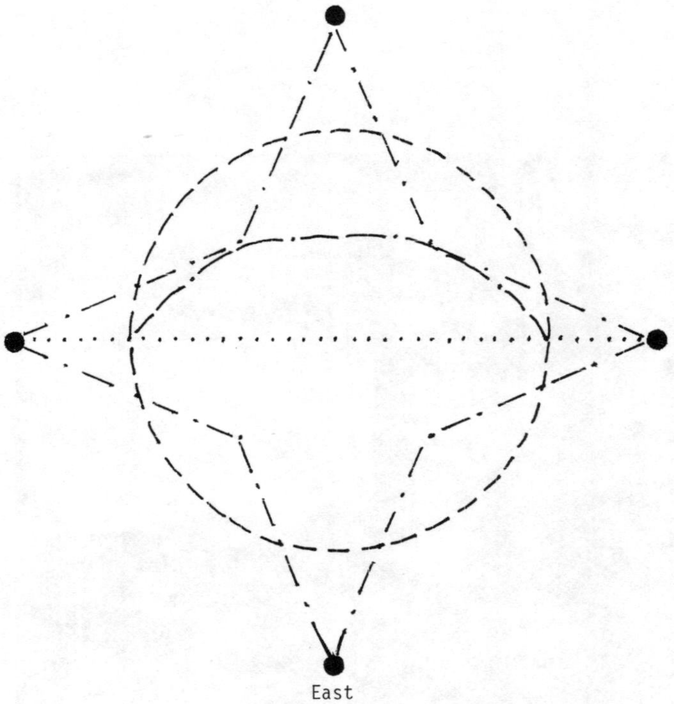

Fig. 13.7 Circle, star, and crescent, superimposed

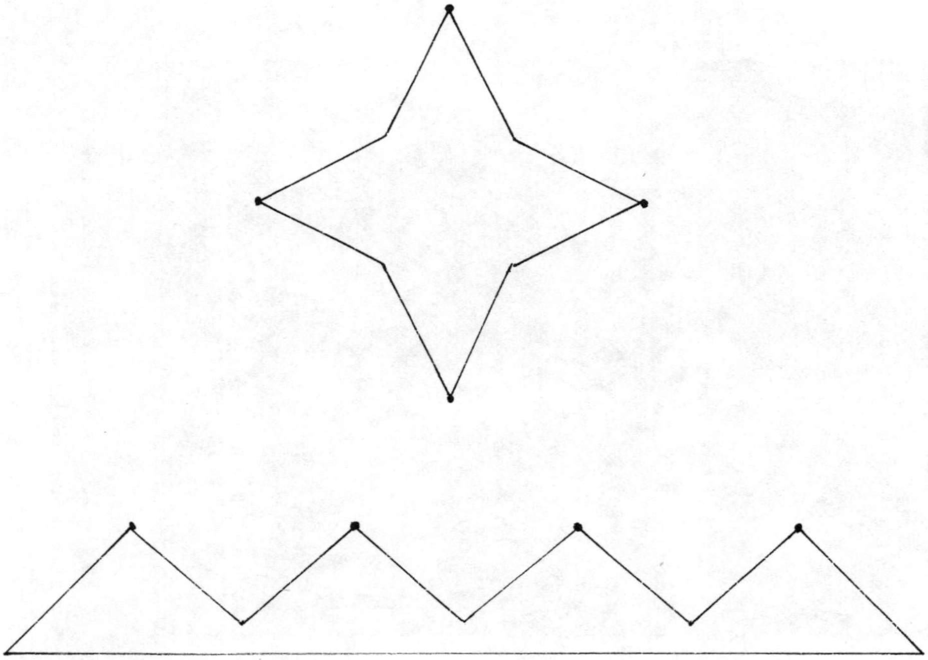

Fig. 13.8 Transformation of the star to mountain motif

in Figure 13.9. It is worth stressing that reduplication of motifs is not considered to destroy their inherent identity.

It should also be noted that the designs that are painted on the Mountain God dancers are given names and represent elements created on the first two days of the beginning time. There are no animate representations in the motifs.

While this essay will not be concerned specifically with the girls for whom the ceremony is being held, we wish to note in passing that the girls' attire also expresses the base metaphor. Their clothing also has balanced design fields, and many of the designs that are on the Mountain God impersonators also appear in beaded form on the girls' costumes. During the ceremonial time the girls are especially holy and have the ability to bless; and their blessing sequence, too, forms the familiar visual metaphor. Pollen is sprinkled to the east, south, west, and north to form the circle and cardinal points. Then pollen is sprinkled from the crown of the head to the forehead (from west to east), and from one shoulder to the other (from south to north), thus connecting the cardinal directions (see Fig. 13.1). Finally, in approximation of a crescent moon, pollen is dusted over the bridge of the nose. Thus the Powers of the universe are invoked, while creation is memorialized.

METAPHORS IN EVERYDAY LIFE

The base metaphor has its genesis in Mescalero cosmology, in observations of the natural order of the universe. However, expressions of that metaphor are not limited to visual or narrative transformations, or to aesthetic concerns alone; there are many other correlates as well. In formal speech events, such as speaking during a tribal meeting, a speaker's talk has four parts: (1) an introductory disclaimer denying expertise on the subject, (2) a brief précis of what is to be discussed, (3) the argument, and (4) a closure that includes statements to the effect that the speaker may not know very much but the preceding opinion is how the situation appears to him.

Mescalero people do not speak of the seven ages of man, as do Western Europeans. Rather, they adhere to a four-stage theory: infancy, childhood, adulthood, and old age. It is interesting to note that the first and last are seen in structural opposition to the middle pair; that is, it is assumed that individuals will be dependent upon others in infancy and old age, while they will be independent during childhood and adulthood.[6]

The girls who are having their ceremony must complete two sets of runs during the event. One set is on the first day and the other set on the

---

[6]This causes havoc with school administrators, most of whom are either Anglo or of the American mainstream establishment. When a child is not in school the parents are notified, in keeping with Anglo tradition. However, there is a prevailing notion among Indian parents that if a child is not in school, he/she has a reason for not being there. Children have much greater autonomy and responsibility than their age-mates in Anglo-American society. This trend, at least as regards school, appears to be shifting toward the Anglo norm, however.

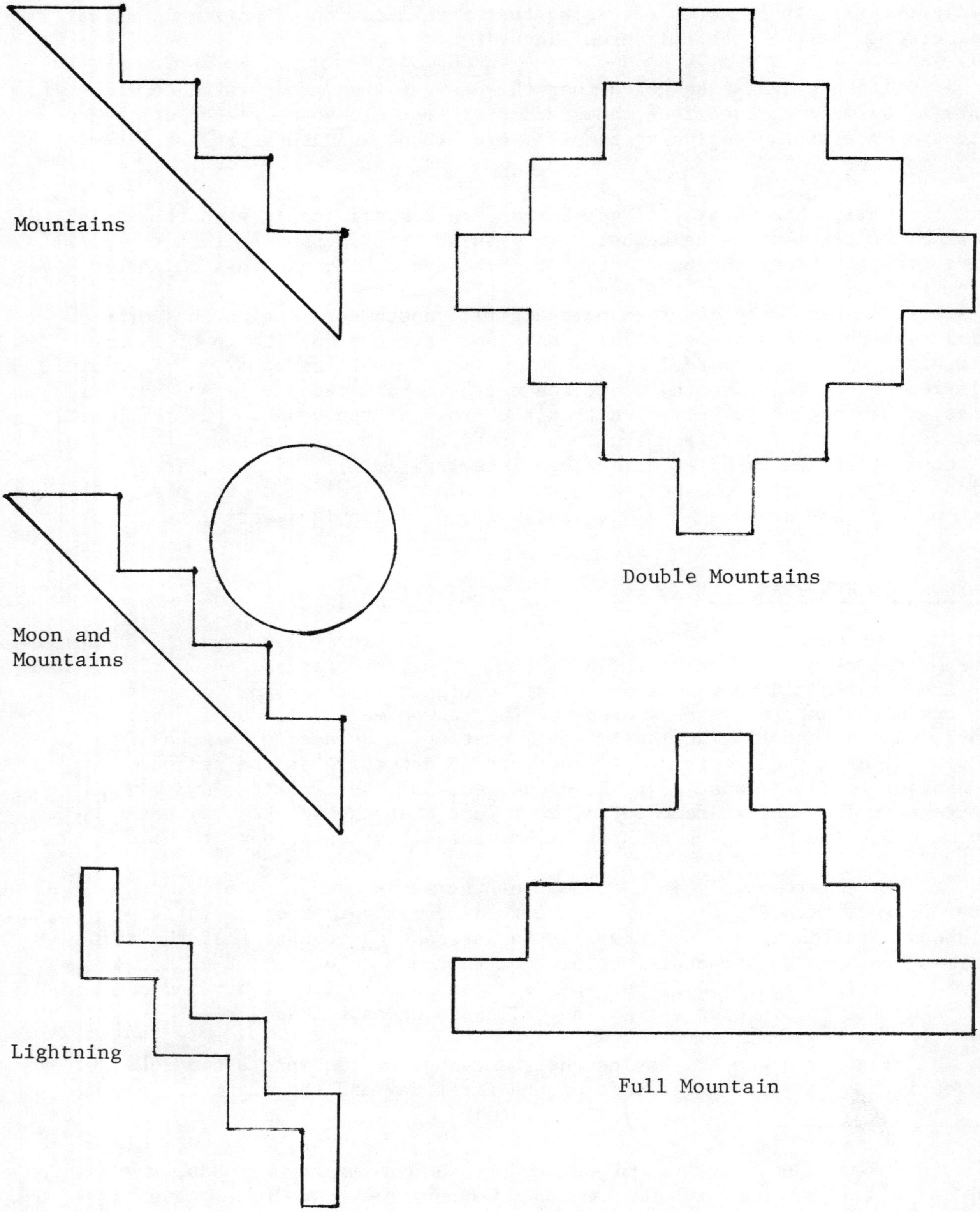

Fig. 13.9 Other design transformations

last day of the public portion of the ritual. In each set, they circle (sunwise, of course) a basket four times: once each for infancy, childhood, adulthood, and old age. The two sets of runs balance each other, especially since one is held at the beginning of the event and one at the end of it. The girls not only run through their life cycle symbolically, they also duplicate the cycle of the tribe while metaphorically announcing its perpetuation through its women.[7]

Traditional homes were circular, whether they were tipis or "arbors" (a wickiup kind of structure). When entering those homes, even today, a polite person always follows the "proper" route from the entry point around to the left; that is, sunwise from the real or presumed east entrance to the south, west, and north. While almost all Mescalero families today live in contemporary, ranch-style homes, many families erect a tipi or arbor near their home during the summer for guests, for coolness, or for special occasions.

Round corrals are still commonplace at Mescalero. The circle shape is felt to be the natural one in the universe and, therefore, the one most worthy of imitation by humankind. Dance patterns also proceed in a circle (Farrer 1978), and those circles follow a sunwise circuit. Play, too, has many aspects of circularity (Farrer 1977).

The tribal museum and cultural center is a structure with a clover-like, four-circle floor plan. Its exhibits were set up, and are maintained by, Mescalero residents. Anglo tourists enter the museum and almost invariably turn right to view the exhibits. However, the museum is arranged so that a turn to the left in a sunwise progression makes chronological sense. One portion of the circular exhibit space is devoted to a slide-tape presentation that is triggered by passing before an electric eye. If one enters and proceeds clockwise, the slide-tape show begins when one enters the exhibit room; however, going counterclockwise, as do most mainstream Americans, results in the slide-tape presentation being triggered as one *leaves* the exhibit space, thus requiring a rather quick about-face to catch the proceedings.

Such examples can be multiplied many times over. Suffice it to say that the verbal/visual metaphor has a myriad of expressions in everyday life. The base metaphor with its verbal/visual components has its genesis in cosmology and is most readily visible to the untrained eye in the public ritual of the girls' puberty ceremony. It is also in its most complete form during the ceremony. However, it is present each day, every day, all day--and night. Ask four times and it must be granted you. Request a favor with four gifts; when accepted, they assure your being accommodated. Keep your life and thoughts and actions in balance or risk jeopardizing the inherent balance of creation.

Whether one talks to a lay person or to a ritual specialist, the ideas are the same: circularity, balance, directionality, and the number four permeate the world. Such "rightness" of the world is expressed every day in the rising of the sun and in its daily circuit through the sky. Such rightness leads inexorably to personal adornment that carries out the theme of the base metaphor. Expressions of the base metaphor are good, pleasing, right, beautiful, and (most importantly) grounded in the cosmos at the time of creation.

The symbols generated by the depiction of the universe (which, of course, is itself a symbol) refer to the base metaphor upon which Mescalero

---

[7]The Mescalero are a matrilineal, matrilocal, matrifocal people.

Apache society is predicated. The Mescalero Apache maintain both visually and verbally that a dually-bisected circle is the foundation of life and existence, and that its expression in many transformations is beautiful, pleasing, right, good, and so on. The ethnoaesthetic canon is inextricably tied to the ethnoastronomical canon. The base metaphor generates symbols that tie together various systems of Mescalero belief and life. Those symbols, in their manifold realizations, have become so entrenched as to be signals.[8] Thus, a crescent moon cut into the fabric of a cradle(board) is a signal that the child inside is a girl, since the crescent is associated with females.

> Remember always . . . the reason for the woman being moon here is that the moon is feminine and also it is because of its ever-changing shape. It is like the woman with her changing ways.

The crescent shape recalls events in the girls' puberty ceremony and *that* recalls the Mountain Gods[9] who, in turn, recall beginning-time narratives, and so it goes. The crescent symbol invokes a series of symbolic representations of the base metaphor, and the base metaphor generates symbols--and so the circle is complete once again.

The coming together of the symbols and signals is seen most completely in the girls' puberty ceremony that is a self-conscious re-enactment of creation: "When we were created as a people and we became a people . . . in order for us to return respect for our Mother, we organized this ritual. . . . This is a re-enactment of our creation as a people." A creation, it will be remembered, that began with a circle of life supported at the four cardinal directions and following the path of the sun. What better place is there to look for organizing principles of society than to the natural world? And what can be more basic, yet put humankind more in touch with the creative force, than an astronomical replication?

Leach (1976:45) states, "We engage in rituals in order to transmit collective messages to ourselves." We engage in art and aesthetics, it seems to us, in order to comment upon those messages. And we make metacomments upon our comments throughout our lifetime, as we persist in living the sky.

---

[8]We use symbol and signal as does Leach (1976).

[9]There are genesis stories for the Mountain Gods as well, most of which attribute their becoming a part of Mescalero ceremonialism to a time after they had "come from the north country." Furthermore, most Mescaleros state that at least some aspects of the Mountain Gods were "borrowed" from the Pueblos. The base metaphor is readily apparent in the case of the Mountain Gods; however, the scope of this essay does not allow us to consider similarities and differences between the coming of the Mountain Gods and the genesis narratives of the beginning time.

14

Winter Solstice Observatory Sites in Kumeyaay Territory, San Diego County, California

Ken Hedges

In the American Southwest, the importance of calendrical systems in general and the winter solstice in particular was documented at an early date for agricultural societies like the Hopi and Zuni, where the calendar is important for the regulation of ritual and agricultural cycles (Ellis 1975). For the rest of the American west, occupied primarily by hunting and gathering societies, calendrical systems are not so well documented, and it has generally been assumed that such systems were not necessary for the subsistence activities practiced by these cultures.

For a century and a half, data have been accumulating in the ethnohistoric and ethnographic literature which demonstrate that solstice observations and calendrics among Native Californians were not as imprecise as some would have us believe. A recent survey (Hudson, Lee, and Hedges 1979) shows that no less than 40 California Indian groups are known to have made solstice observations or are presumed to have made them on the basis of the importance of the solstice in their calendrics. In addition, ten well-documented archaeological sites indicate that solstice observations were being made from specific observatory sites, or that solstice sunrise phenomena played an important part in Native California ritual. In southern California, the importance of the winter solstice event was recognized as early as 1826 by Father Geronimo Boscana, the missionary at Mission San Juan Capistrano, who wrote:

> They observed with greater attention and celebrated with more pomp the sun's arrival at the Tropic of Capricorn than they did his reaching the Tropic of Cancer for the reason that, as they were situated ten degrees from the latter, they were pleased at the sun's approach towards them. Its return meant much to them for it ripened their fruits and seeds, gave warmth to the atmosphere, and enlivened again the fields with beauty and productivity [Boscana 1933:65-66].

This paper focuses on data demonstrating the calendrical importance of the winter solstice among the Kumeyaay (Southern Diegueño) Indians of southern California and northern Baja California, for whom the method of solstice observation has been recorded:

> The solstices (hiLyatai) are observed. My informant knows at which points on the mesa east of his house the sun will appear to rise at the solstices. The winter solstice occurs during the month of xaLapisu [December] . . . [Spier 1923:358].

Unfortunately, calendrical data for the Kumeyaay are sparse. The months are clearly based on lunations, and each month begins with the rising of a particular star or constellation (Spier 1923:357; Kroeber 1925:718). The names of the months have been recorded (DuBois 1908:165; Gifford 1918:169; Spier 1923:357; Kroeber 1925:718), and it is evident that a system of two six-month periods, with the winter names repeated during the summer, was in effect. This system is also recorded for the Zuni, who begin the halves of the year at the winter and summer solstices (Kroeber 1925:718). In the recorded data for the Kumeyaay, there is no agreement on whether the year begins in September (Gifford 1918:169; Spier 1923:357), October (DuBois 1908:165), or November (Kroeber 1925:718), but it seems clear that it does not begin in December, and there are no data on methods of correlating the lunar and solar years. Kroeber noted that "the annual repetition of the Diegueño count very strongly suggests a primary recognition of two fixed points within the year; and these can hardly have been any events other than the solstices." He suggested that the failure of the calendar to begin at the solstice may be due to an unexplained imperfection either in the calendar or in our record of it (1925:718).

In terms of the importance of the winter solstice as a calendrical event, it matters little whether the year begins with it or not. As recorded by Boscana (1933:66), the return of the sun after the winter solstice "ripened their fruits and seeds, gave warmth to the atmosphere, and enlivened again the fields with beauty and productivity." Among Native Californians, the winter solstice was the major crisis point in the year, with ceremonies held to insure the return of the sun from its southward journey. Various references in the California data (Hudson, Lee, and Hedges 1979) suggest that a solstice period of a few days was recognized, as was the case in the Southwest (Ellis 1975:63).

The only known reference to a possible solstice ceremony of the Kumeyaay was recorded in 1914. Although somewhat romantically rendered, the pertinent portions are worth quoting.

> One of the ancient rites performed in by-gone days by the Indians dwelling in the village Helsh-ow' Na-wa' (Rabbit House) at the foot of the Cuyamaca Peak was the Kwut'ah Lu'e-ah (Song-Dance) given in honor of In'ya (Sun). The summit of Viejas mountain, just on the other side of their village, was chosen as the place for this celebration to be held, and they named the mountain Kwut'ah Lu'e-ah (Song-Dance) on account of it. . . .
>
> . . . The Indians made a pilgrimage once a year to its very top to watch In'ya (Sun) come out of En-yak' (East), and praise and honor him with song and dance. For In'ya (Sun) was the great Ruler of All Things. He governed the universe; he commanded the earth; nothing grew unless he caused it; he even dominated the bodies of men, some of whom he made energetic and strong, others weak and lazy. When he disappeared at night he cast a drowsiness o'er the world, so that everything slept until it was time for him to come again in the morning. Such a great ruler as he, received due reverence and worship.
>
> . . . The big circle on top of the mountain had been freshly prepared and cleared for the dancers and singers. . . .
>
> When all had reached the summit, the ritualistic ceremonies began. With song and dance in the blushing dawn, they watched for In'ya (Sun), Ruler of All. Opalescent streamers of golden radiance and flaming banners of crimson flaunting across the pearly tints of the receding night, heralded his arrival; while the people chanted songs of praise in honor of his wonderful light, and made obeisance in the dance in homage of his great

power over all things.

. . . The old trails are faintly discernable and traces of the ring where they danced and sung still remain . . . [Johnson 1914:24-25].

This ceremony "given in honor of In'ya" may well have been a solstice event. The mountain in question is Viejas Mountain, located just northeast of Alpine in San Diego County, California, overlooking the Viejas Indian Reservation, which encompasses prehistoric village sites. To test the recorded presence of a dance circle on the summit of the mountain, the site was examined by the author and Kre Easland in March 1977; it was found that the entire summit had been cleared and the rocks used to build windbreaks, sleeping shelters, and stone-lined paths. However, in a saddle at an elevation of 4100 ft on the ridge just southwest of the summit, a cross-shaped alignment was recorded. The age of such an alignment is difficult to determine, but heavy ground stain and dark weathered upper surfaces on the stones indicate that the alignment is not a recent feature. Using a hand-held lensatic compass, it was determined that the azimuth of the long axis of the cross was approximately 119.5°. At this latitude, the astronomical azimuth of winter solstice sunrise is slightly more than 118°. On a return visit to the site for further observations, measurements and mapping of the alignment were thwarted by vandals, who had apparently cleared the spot for bedrolls and built a fire ring of some of the stones.

A second probable solstice alignment was first recorded by archaeologist Janet Eidsness in 1978, during an archaeological survey of the Cowles-Fortuna Mountain Regional Park in San Diego, California. Several major village sites occur in the general area. The feature, at first not recognized as a possible solstice marker, was recorded as a circular rock alignment with a bisecting line, located at an elevation of 1250 ft on a flat-topped spur or bench on the south flank of Cowles Mountain. Now greatly disturbed, the alignment was originally in the form of a circle composed of small to medium stones stacked two or three high, with a wall or line of stones dividing the circle in half. As with the Viejas Mountain alignment, the age of the Cowles Mountain feature is difficult to determine, but good-sized black sage bushes have grown up among the stones, providing protection for the best-preserved portions of the circle and demonstrating that the alignment is not a recent camper's product. Within the errors imposed by the disturbed nature of the alignment and the hand-held compass, it was determined that the bisecting line of the Cowles Mountain feature exhibited approximately the same azimuth as the Viejas Mountain alignment.

With the discovery of two stone alignments exhibiting virtually identical azimuths, observations were conducted during winter solstice with the following results. On 21 December 1978, the winter solstice sunrise was observed from the site of the Viejas Mountain alignment. The sun was observed to rise directly behind a peak on the distant horizon (Fig. 14.1). Subsequent mapping identified this peak as Buckman Peak, elevation 4641 ft, at a distance of 16.6 miles and an azimuth of approximately 118.5°. On 20 and 22 December 1978, sunrise observations were conducted at the Cowles Mountain alignment site. The first gleam of winter solstice sunrise was bisected by a small peak just south of Lyons Peak on the horizon (Fig. 14.2). Subsequent fieldwork and mapping identified the horizon marker as a prominent rocky outcrop, elevation 2840 ft, located at a distance of 14 miles and an azimuth of approximately 119°.

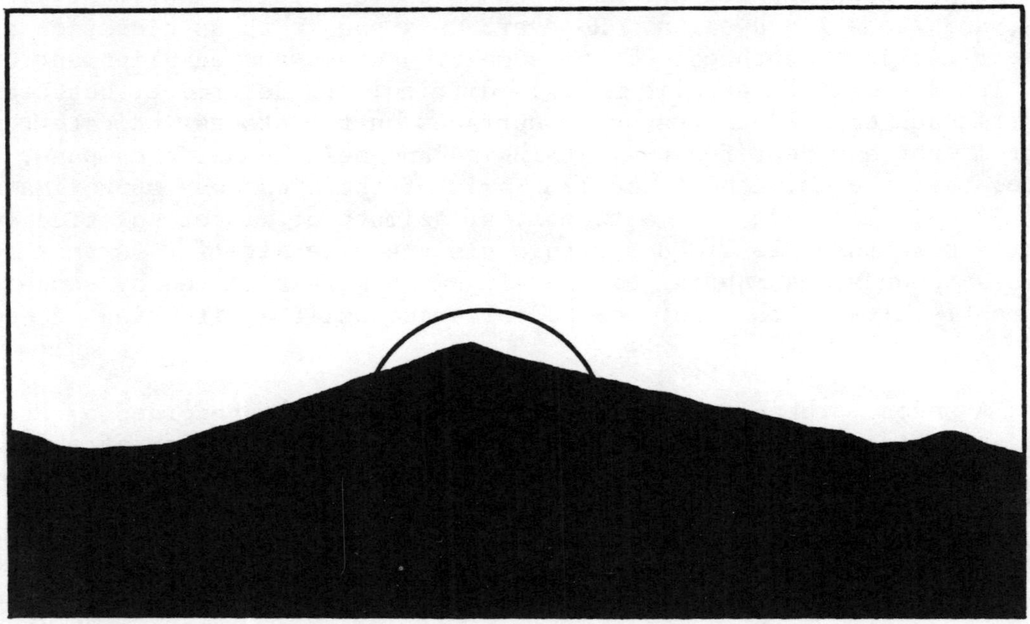

Fig. 14.1 Winter solstice sunrise viewed from Viejas Mountain

Fig. 14.2 Winter solstice sunrise viewed from Cowles Mountain

From the observational data at hand, it is clear that both alignment sites have horizon markers which function as indicators of winter solstice sunrise. Whether these sites were actually used by the Kumeyaay as winter solstice observatories is more difficult to demonstrate, but the identification of the Viejas Mountain site from ethnographic data concerning solar ritual, in conjunction with the existence of two sites with virtually identical azimuths and corresponding horizon markers, strengthens the conclusion that these are winter solstice observatories.

From archaeoastronomical studies conducted on these two sites, several general conclusions are suggested:

(1) The precise azimuth of a rock alignment by itself is not of primary significance. With alignments no more than 3 m in length, composed of stones up to 30 cm in diameter, great precision is not possible within the alignment itself. The alignments serve to identify the exact observatory location, and the general direction of winter solstice sunrise.

(2) With the horizon markers located at a distance of 14 miles or greater, great precision in determining the time of winter solstice is possible. With the Cowles Mountain site in particular, the bisection of the sun's first gleam allows for only a very small movement along the horizon before the marker ceases to function.

(3) With the postulated use of observatory sites placed within the protohistoric period, the sunrise phenomena observed by Kumeyaay sunwatchers should be essentially the same as those observed today. For sites assigned to relatively recent periods, direct observational data are at least as important as comparisons of computed sunrise azimuths with measured alignment azimuths in determining the significance of a suspected solstice site.

(4) For the Kumeyaay, the winter solstice sunrise was defined as the first gleam of the sun. At mid-point or lower limb observations, the horizon markers would not function. Furthermore, observation has demonstrated that the glare of the rising sun is so great immediately following the first gleam that accurate observation is impossible.

It appears, then, that the ritual in honor of In'ya the Sun on Viejas Mountain was a winter solstice ceremony, and the observational data reported here lends confirmation to the ethnographically-recorded technique of watching the sunrise in relation to horizon markers. The association of unequivocal horizon markers with winter solstice sunrise provides strong evidence that these sites functioned as observatories for the solstice event.

15

Possible Astronomical Alignments in a Fort Ancient Settlement at the Incinerator Site in Dayton, Ohio

James M. Heilman and Roger R. Hoefer

INTRODUCTION

The Incinerator site--an Anderson Focus-Fort Ancient site--is located on the flood plain of the Great Miami River in Dayton, Ohio. Excavation within the circular stockade has revealed a rigid patterning of concentric rings of houses, storage/trash pits, and burials around a central plaza (Figs. 15.1, 15.2). This plaza contains an unusual pole structure built around a large red-cedar center post. The hypotheses advanced in this paper are based upon the apparent alignment of the rising and setting sun as viewed from the dwellings, and pole structures found on the west side of the village. The combination of these features is thought to have played a significant part in the village layout and in the scheduling of the planting season, "Green Corn Rituals," and other cyclical celebrations.

The Incinerator site is located within a pocket of Wea soil, adjacent to the flood plain of the Great Miami River. In prehistoric times, this Wea soil would have been a prairie opening (Meeker et al. 1973), which was well suited for corn agriculture. Wea soil is the most productive soil in the country for corn agriculture that is not frequently inundated by flooding. Seven C^{14} dates for this village indicate that it was established near the end of the twelfth century. It was probably occupied for about a quarter of a century and then abandoned after fire consumed a large portion of the village.

The importance of this site was first realized by Kenneth McNeal, who encouraged John Allman's initial excavations in the 1960s. The Dayton Museum of Natural History began its program as a salvage operation in 1971, since the site was scheduled to be destroyed by the construction of sludge treatment ponds. By 1974, when the site was nominated to the National Register of Historic Places, the rigid patterning had become apparent. By 1978 about 50% of the extant village area had been excavated.

ARCHAEOLOGICAL FEATURES OF THE SITE

Stockade

Thus far, excavations have revealed over 500 ft of the roughly circular stockade, which enclosed an area about 420 ft in diameter. The posts of this wall, which averaged about 0.6 ft in diameter and 0.63 ft in depth, were

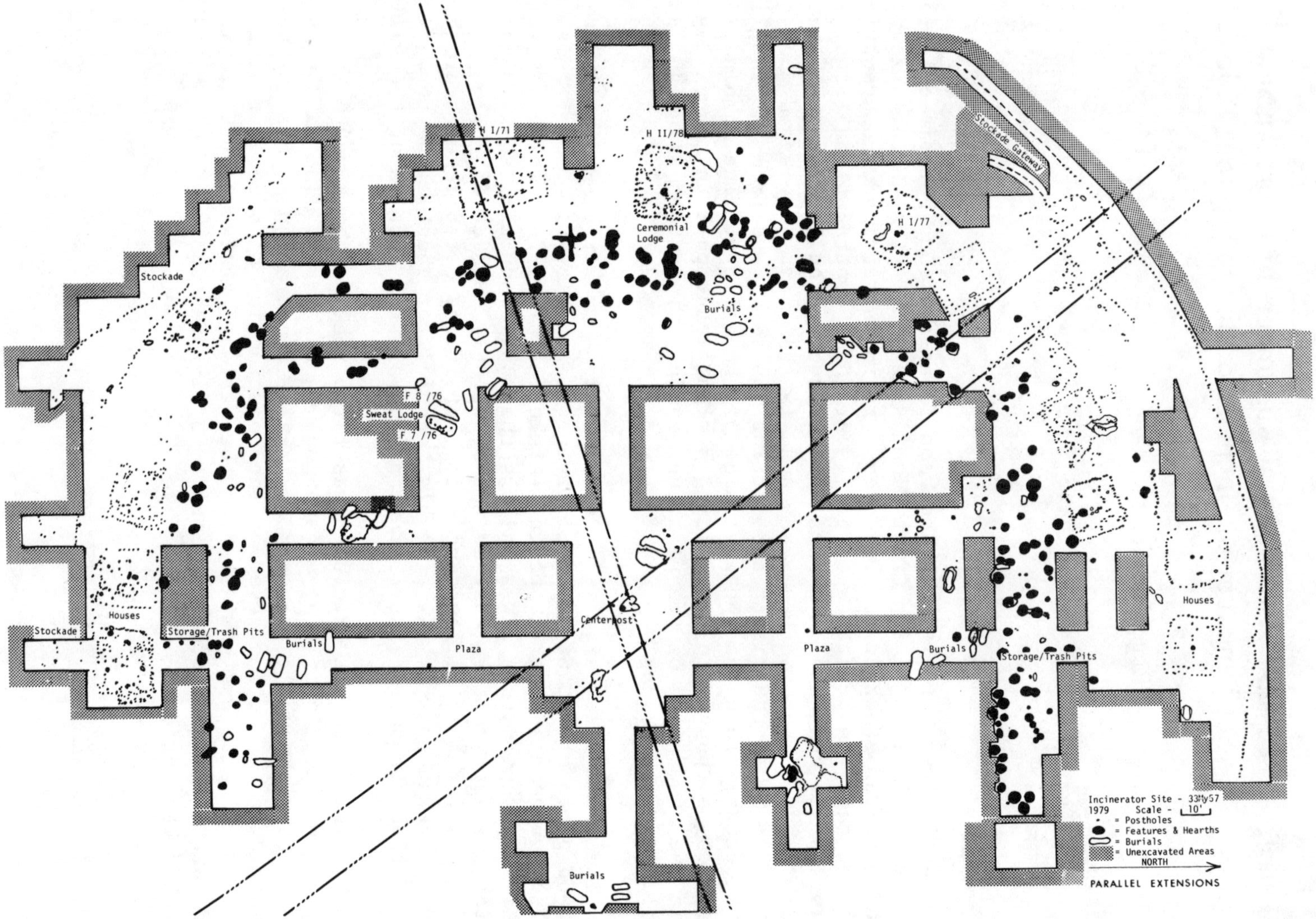

Fig. 15.1 Map of Incinerator site with parallel extensions

spaced approximately 1 ft apart. The northwest portion of the stockade was made up of overlapping walls (Fig. 15.1); built over the interior wall and constricting the opening of the overlapping walls was a gatehouse. Other posts indicate the presence of a series of gates which would have controlled the entrance to the village. Problems with neighboring groups are suggested by the very defensive nature of this stockade and its overlapping gateways and gatehouse.

House Structures

A concentric band about 40 ft within the stockade was reserved for housing. Fifteen wattle-and-daub structures have been excavated, and an additional fifteen are anticipated. Fragments of the clay which have been recovered show the imprint of small branches and grass. Other evidence of architectural details have been recovered from impressions on mud daubber wasp nests; the majority of these indicate that the nests were built against a smooth wall face, while a few show evidence of thatch roofing and roof beams.

The houses ranged in size from 18 x 18 ft to 22 x 30 ft for the largest dwelling (House I/81), which is situated on the west side of the village. With the exception of this house, all seemed to have hearths located within the geometric center of the dwelling.

The houses seem to be clustered in their distribution, and the clustering may well indicate distinct clan or lineage areas. In the house groupings on the north and south side of the village, the structures are 7 to 9 ft apart; there is a wider spacing between the different areas (Fig. 15.1). The west side of the village has wider spacing between the houses and larger structures.

Storage/Trash Pits

A concentric band, 25-35 ft wide, is dominated by storage/trash pits which are situated between the houses and a concentric ring of burials that surround the plaza. The majority of these are bell-shaped features about 3 ft in diameter and depth. During the excavation of a few of these pits, remnants of what appear to be grass and bark liners much like those used by the Hidatsa (Wilson 19717:87-93) have been found against the walls; these pits were originally used for the storage of corn, squash, etc. (Mills 1904:26). After the storage pits had been emptied of their contents, they were backfilled rapidly with trash consisting of broken pottery, animal bone, food refuse, and the by-products of various activities such as flint knapping and bone and antler working.

The contents of these pits have been a great source of faunal data. This material is being analyzed by Dr. Orrin Shane of the Science Museum of Minnesota in St. Paul. Dominating the meat consumed in the village are deer (75%), elk (9.2%), turkey (6.2%), and racoon (1.44%), which combined represent 92% of the usable meat selected by the villagers. The seasonality of the species in these pits indicates that the village was occupied perennially (Shane and Barber 1975).

During the excavation of these features, a quarter section of each pit is left intact as a profile block to give the stratigraphic series of the various lenses of fill. After these have been mapped, the profile blocks are sectioned out of the pits in arbitrary three-inch levels, and the soil is saved for flotation. Gail Wagner of Washington University has devised the soil flotation procedures for this site (Wagner 1977), and is doing the complete floral analysis of the material recovered. Thus far beans, 8-, 10-, and 12-row flint or flour corn, as well as popcorn, walnut, butternut, hazelnut, hickory nut, and the seeds of plum, cherry, grape, and sumac have been recovered. The 2500 flotation samples have not yet been analyzed, but sunflowers, squash, and wild legumes are anticipated (Wagner, personal communication).

Analysis of the 50,000 pieces of pottery from the site is being done by the Dayton Museum of Natural History staff. About 90% of the pottery is grit tempered and cord marked. Mica and shell tempering agents also occur. The majority of the ware has decorated necks and rim bands, with line-filled triangles (chevron) and rectilinear and curvilinear guillouche being the most popular motifs, although a number of other motifs occur as well; there are also several crude, negative painted vessels. Preliminary analysis of the pottery from this site (Heilman 1975) suggests that certain ceramic attributes are clustered in distinct regions of the village; these may well indicate matrilocal residence areas.

The large assemblage of bone and antler tools is being analyzed by the Museum's staff. Jim Robertson of the University of Illinois' Circle Campus has undertaken the analysis of the flint tools from the Incinerator site.

Burials

Along the margin of the central plaza, in a concentric band 20-40 ft wide, are located the preponderance of the 109 Incinerator site burials. These seem to occur in discrete family burial plots (Robbins 1975; Knick 1977). In general, the burials nearest to the plaza are of the highest ranking individuals, while those closer to the storage pits are of lower rank. The most common indicator of high social ranking in mortuary practices is the inclusion of a fossiliferous limestone slab vault over the burial. These slabs were brought to the site from an Ordovician limestone outcropping at a cliff 0.75 mi west of the village.

The most common items included with the 24% of the burials provided with grave goods were marginella shells imported from the Gulf of Mexico, which were strung as necklaces, bracelets, and earrings. Also common among the grave goods were pairs of shell discs with a single central perforation, which were found near the top of the manubrium. A few very high-ranking burials had pipes near the elbow or head and shell ear-discs, and several also had headgear made of marginella shells, lightning whelks, and cut animal jaws.

A few other burials have been found interred in the house floors and walls. In addition, occasional finds indicative of infanticide have been recorded (Robbins 1974) in the areas between the houses and stockades, or in trash pits. The vast majority of the Incinerator site burials are from the 0-6 year-old age group; very few burials are from the 7-15 year-old age group, although there is a high clustering of burials in their late teens and early twenties. An individual 35 years of age would have been an elder in this community, although burials as old as 50 years have been recovered from the site.

Preliminary analysis of the Incinerator site population supports the delineation of family groups, and our division of these groups by ceramic attributes is paralleled in the distribution of the burial groupings as well (Robbins, personal communication; Knick 1977:61). The burial population is also divided among Ilinid and Muskogid varieties at this site.

Plaza and Sweat Lodge

The plaza area, which is at least 150 ft in diameter, has been transected by two 10-ft wide north/south trenches and three east/west trenches. The squares opened in the plaza have consistently produced very little debris--a few flint chips and tiny fragments of pottery are the norm for a 10 x 10 ft square. This strongly suggests that the plaza was an open, cleanly swept area like those that Hanson (1975:40) noted at the Buffalo and Sloane sites in West Virginia. Only in the southwest portion of the plaza, in the vicinity of the sweat lodge, was any significant debris encountered; in this area, a large quantity of fire-cracked rock was recovered to the east and southeast of the structure.

The sweat lodge is composed of two parts--F#7/76 and F#8/76 (Fig. 15.1). F#8 is the western portion of the 13 x 8 ft structure; it appears to have been a crude wall trench in which posts were inserted along the north/south axis, and which was then backfilled to support the posts. F#7, the eastern portion of the lodge, was an open basin with posts along its eastern wall. This area, which contained a few pieces of fire-cracked rock in the lower levels, appears to have been open for a period of time before it slowly silted in. The area between the two features would have served well as a bench for the participants when F#7 was filled with hot rock. The fire-cracked rocks were primarily igneous and metamorphic in nature, with the granite family being the most popular; several bushels of fire-cracked rock were recovered from this area of the plaza.

WEST SIDE STORY

Thus far, this paper has emphasized the features that are common to the overall pattern of concentric bands, which are arranged by specific activity areas about the central plaza. At the same time, these areas are not truly homogeneous, since the pie-shaped wedges dominated by particular "social" groups each has its own eccentricities. The west side of the village is the most divergent--particularly that triangular area (Fig. 15.1) which lies between the solar alignments (April 24-August 20) and the center post, with House I/71 to the southwest and House I/77 to the northwest. House I/77 is the largest structure (22 x 30 ft) noted on the site and the only one which does not have a central hearth. Instead there appears to be a structural variation, with the hearth centered in the south end of the building, placing it in alignment with the central pole structure (see Fig. 15.4, discussed below). Adjacent to this house on the north is the partially excavated Ceremonial Lodge, House II/78. This 24 x 20 ft structure protrudes out of the normal arc of houses with its narrow side extended toward the central plaza. The wall posts for this structure, which averaged 0.78 ft in diameter and 1.94 ft in depth, were more than twice the depth of those found in any of the other structures on the site. Since the post molds are of such a large size, we suggest that this lodge was considerably taller than the surrounding houses and more monumental in appearance.

Although the normal burial pattern for the higher ranking burials was to place them in close proximity to the central plaza, here the relationship was inverted, with the high ranking burials sharing a common close proximity to the Ceremonial Lodge. The burials extend from the north side of the lodge northeastwardly through the band normally reserved for storage pits. Almost all of these burials had grave goods; several had large quantities of shell beads arranged over the head.

The storage pits on the west side of the village resemble those throughout the rest of the site. However, the back-filled garbage from these contained more exotic material than from the pits in the rest of the village combined, although everyday debris was encountered as well. F#19E, F#4/76, and F#5/78 contained lenses of burnt corn in tight association with a pipe and/or other exotics. These associations, which were all clustered around the Ceremonial Lodge, may represent the residue of Busk or Green Corn ceremonialism. The pits are similar to one that Shane postulates as a Green Corn Ritual Pit (Prufer and Shane 1970:27-29) at the Blain Village in Ross County, Ohio, although they are not identical.

It is interesting to note the disproportionate amount of burnt corn which has been recovered from the western part of the village, where burnt corn or beans are found in 69.2% of the storage/trash pits; in the south it occurs in 26.5% of the pits and at the north end in only 8.1% of the pits.

Having examined the depositional history of the material recovered from this site, we may assume that the distribution of the botanical materials is cultural rather than post-depositional in nature. One explanation for the great preponderance of burnt corn being located in this area could be that it is the residue of ritual garbage, where corn has been intentionally sacrificed in a fired offering and has subsequently been preserved because of its carbonized nature. These burnt offerings seem to be part of the core of "Green Corn Rituals" (Witthoft 1949:18, 32, 33, 54, 56).

In general, the west side of the village (particularly the pits surrounding the Ceremonial Lodge) have the greatest clustering of exotic material found on the site. Immediately in front of the lodge was a small pit in which a male and female dog appear to have been ritually interred. Other pits in this area have also produced pipes, a discoidal, and large bone hair pins, in addition to numerous bone, antler, and shell artifacts.

Because of the abundance of exotic debris, the significantly increased quantities of burnt corn and other plant remains, and the oversized structures, it is postulated that this was a ritually significant area within the site. A series of astronomical alignments with the center post, which dominates the central plaza, reinforces this ceremonial interpretation.

THE CENTRAL PLAZA AND ITS WESTERN ALIGNMENTS

The pole arrangement in the central plaza area (Fig. 15.2) is one of the most noticeable features of the site. A large (approximately 1.9 ft)

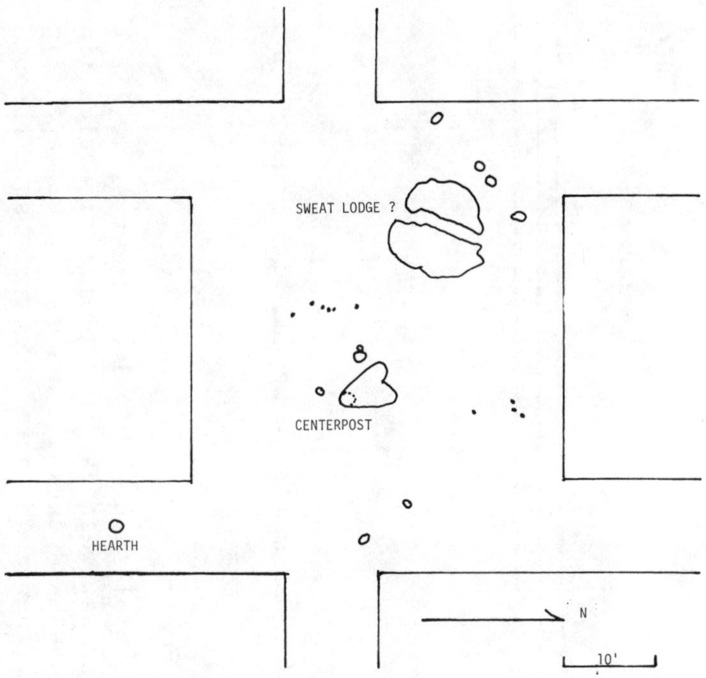

Fig. 15.2 Central plaza area, incinerator site

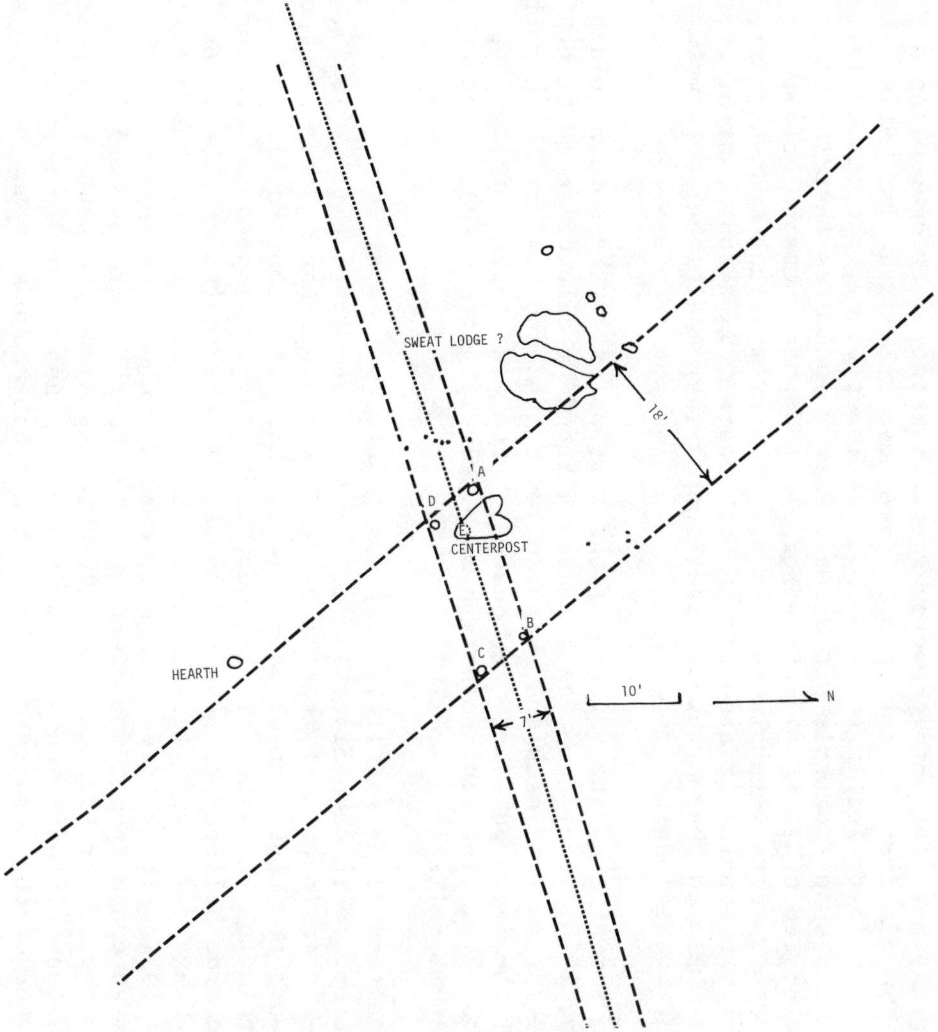

Fig. 15.3 Five poles in central plaza

diameter pole had been set into the ground approximately 4 ft deep using a slide ramp. Two distinct slide ramp pits have been excavated, indicating that the pole had been installed in the same spot at least twice. Recent analysis of the carbonized wood from this pit by Gail Wagner indicates that this large post was made of red cedar (*Juniperus virginiana*; Wagner, personal communication). Four other poles were placed in a parallelogram arrangement that framed the centerpost. A single isolated hearth was located 29 ft southeast of the central pole, while to the northwest were located four additional post molds and a possible sweat lodge.

The symmetry of the pattern of the five poles in the central plaza is striking (Fig. 15.3); they comprise a neatly aligned parallelogram 18 ft on each side and 7 ft across each end, with angles of 66° and 114° at the corners. The center of the large post hole is located on the center line of the long axis of the pattern. The five holes are so precisely aligned that a chance arrangement seems unlikely.

We extended the parallels along each outside line of the central plaza pole structure in each direction across the site plot map (Fig. 15.1). To the southeast and northeast, the lines lead across unexcavated areas. They also lead toward the east side of the village, where road construction has damaged a portion of the site. To the northwest, the lines pass through the center of the area believed to be a gatehouse. To the southwest, the lines pass through House I/71 (Fig. 15.4). A post hole somewhat larger than average (Fig. 15.4:F) is located within the house, exactly on the center line of the parallelogram extended from the centerpost. The hearth (Fig. 15.4:H) is located between the center line and the line extended along the southern parallel. This house is the only one excavated to date with a hearth not located in the geometric center of the structure. The distance between post E and post F is 142 ft.

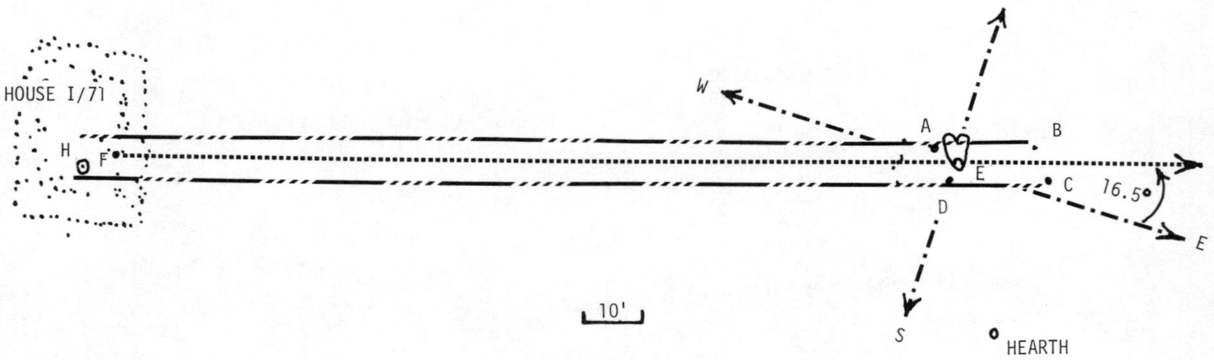

Fig. 15.4 Alignments connected with House I/71

CENTRAL PLAZA SUNRISE ALIGNMENTS

To check for possible astronomical alignments, we first considered the sun. The north/south and east/west cardinal points direction lines were superimposed upon the center post and extended across the village (Fig. 15.5). The center line of the parallelogram was extended from post F through post E (Fig.

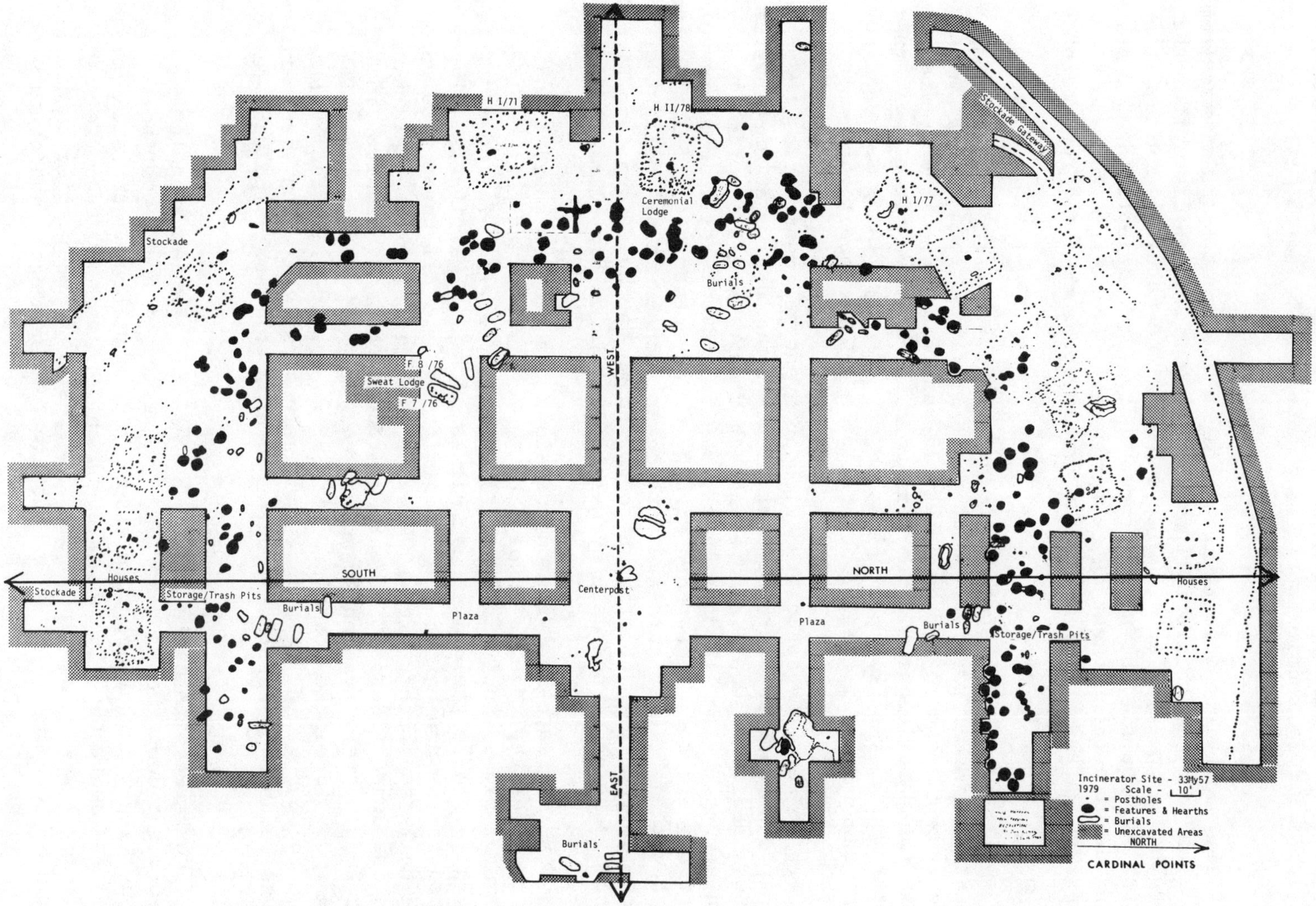

Fig. 15.5 Map of Incinerator site with cardinal points

15.4) at an angle of 16.5° north of the east/west directional line (equinox sunrise/sunset line; see Fig. 15.6). If this post structure had been created for the purpose of observing the sunrise at the time of the summer solstice, the angle of the intersection would have to be 31.25° north of the equinox line.

The rising sun will be in alignment with post E (as viewed from post F) twice each year. The first alignment time would be 34 days after the vernal equinox. Then, 118 days later, 32 days before the autumnal equinox, the second alignment time occurs. The two alignment dates are now April 24 and August 20. They coincide with the present dates for a safe, frost-free time to plant corn in the spring and the end of a 120-day growing season--the "green corn" stage in summer.

Is it possible that the occupant of House I/71 used post F as a backsight to align himself precisely with post E? If so, then 33 days after the vernal equinox the sun rose framed between post E and the line defined by posts D and C in the parallelogram. The next morning, 34 days after the vernal equinox, post E (as viewed from post F) would obscure the rising sun, and for a brief time the shadow of the tall center pole would connect the two.

For a ceremony associated with an event as important as the planting of the new corn crop, this arrangement of posts would give advance notice that the time was near, and would also establish a safe-from-frost time to plant the corn each spring. Some 118 days later, the filled but still immature corn would be ready for the "Green Corn Ritual" or any other type of late-summer festival observed. These data are particularly interesting in light of Witthoft's observation (1949:68) that a common theme in the "Green Corn Ritual" is the advanced setting of the date by male officials at a preliminary meeting.

Several features lying on and along the equinox sunrise and sunset lines, as well as their relationships to the central plaza pole configuration, are also suggestive of solar alignments (see Fig. 15.7). To the west, features 19E, 19W, and 19WW/74 yielded an unusual assortment of artifacts. These pits are located on the equinox sunset line (Fig. 15.8). All three pits were chronologically separated, with 19WW being the first filled. This large pit contained the densest deposit of garbage recovered from the site, including more than 2000 pottery fragments, and an equally high density of animal bone that appeared as a single deposit. Pit 19E/74 was the last and smallest pit in this group to be dug, and may represent the residue of "green corn ceremonialism": considerable quantities of burnt corn as well as a pipe were recovered from the fill.

The three pits begin 115 ft due west of the center pole and extend 10 ft west straight down the equinox line. A series of small post holes, forming an apparent lane 3.5 ft wide, then extends due west behind the pits for 48 ft to an unusual assortment of post holes which form an abrupt barrier line across the lane (Fig. 15.8). This arrangement suggests an interesting possibility. If any type of ceremony was held at the time of the vernal and/or autumnal equinox, this arrangement of poles could have been used to align the observer whose responsibility it was to note the occurrence of these two astronomical events each year. There were two possible methods of observation.

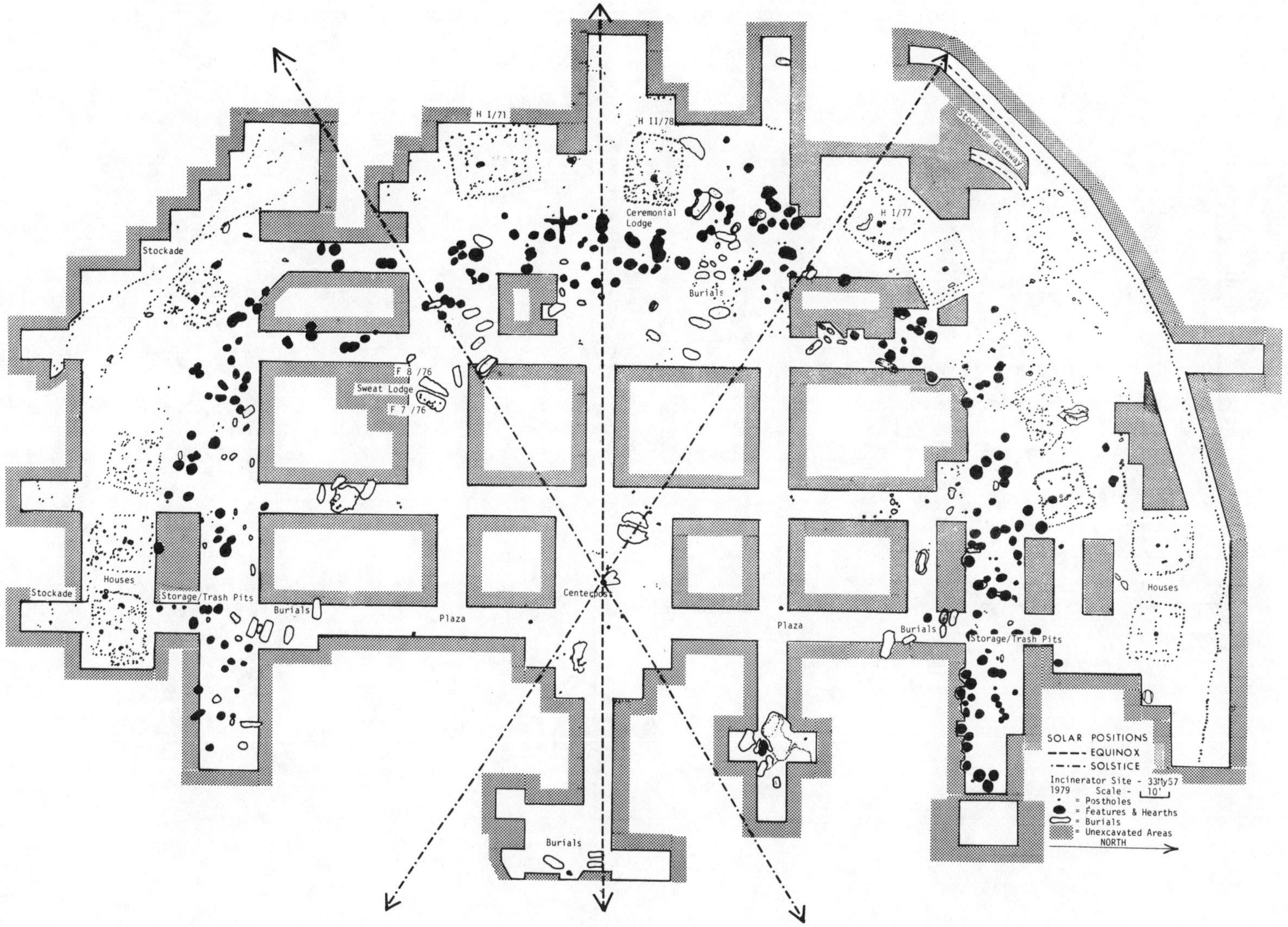

Fig. 15.6 Map of Incinerator site with solar positions

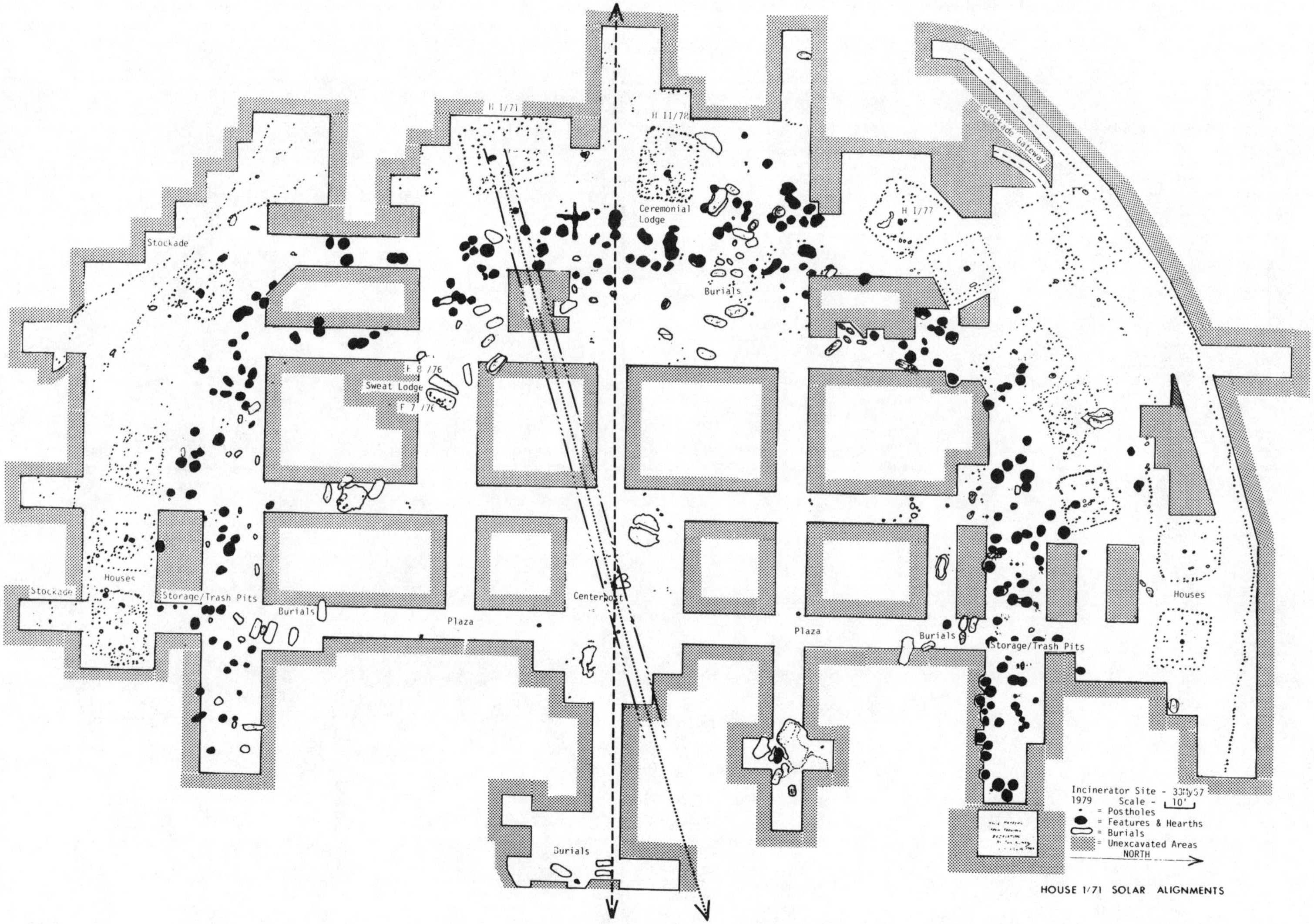

Fig. 15.7 Map of Incinerator site with solar alignments

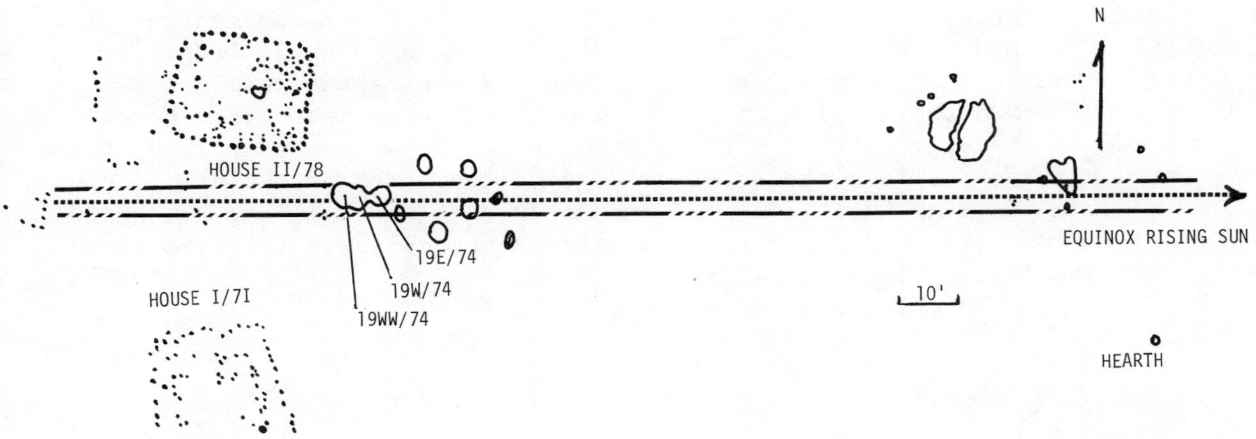

Fig. 15.8 Equinox alignment, Incinerator site

One method would involve observing the rising sun through the central plaza parallelogram of poles from near the point marked X (Fig. 15.8). The sun, on the equinoxes, would rise in the space between posts E and D. The lane of smaller posts would align the observer with the configuration in the plaza, and also with the ceremonial pits 19E, 19W, and 19WW. This alignment of the sun, center poles, pits, and lane of poles would occur at sunrise on only two days each year: the days of the two equinoxes.

The second method would involve observing the shadow of the tall center pole instead of the rising sun. On the two equinoctial days only, the shadow would fall across the pits and down the lane of poles to the structure that blocked the west end of the line.

CENTRAL PLAZA SOLSTICE ALIGNMENTS

Another possibility for alignments for astronomical purposes would be a system enabling observations of the rising and setting sun at the different solstice points (Fig. 15.6). At the summer solstice at the Incinerator site (at latitude 39°42'), the sunrise and sunset points would be 31°16' north of due east, while at the winter solstice they would be an equal distance to the south.

At the present time, the two solstice sunrise points and the areas of the site that lie along the sunrise solstice lines are either unexcavated or have been destroyed by historical construction projects. The winter solstice sunset line passes approximately 25 ft south of House I/71. Part of this area has been excavated, but except for F#8/75 (which has yielded massive quantities of burnt material, including corn and fabric) nothing unusual has been located in that area. However, a large percentage of the village area that lies along the winter sunline remains to be excavated, and here we may find features that could have been used as alignment points.

On and along the winter solstice sunrise line, unusual groups of post holes have been located in every area that has been excavated (see Fig. 15.9). Four large post holes located in a rough line and forming a near perpendicular to the sunline are only about 30 ft northwest of the central plaza parallelogram, and a large pit (possibly functioning as a sweat lodge) is located on the sunline between the four posts and the center pole. Approximately 100 ft down the sunline there is also a series of seven post holes that form a nearly circular pattern 8 ft in diameter. The southwest wall of House I/77 also nearly parallels the solstice sunset line. This structure is unusual because it is the only dwelling excavated that was built in the shape of a trapezoid instead of a nearly equal-sided rectangle, so that its southwest side parallels the solstice line.

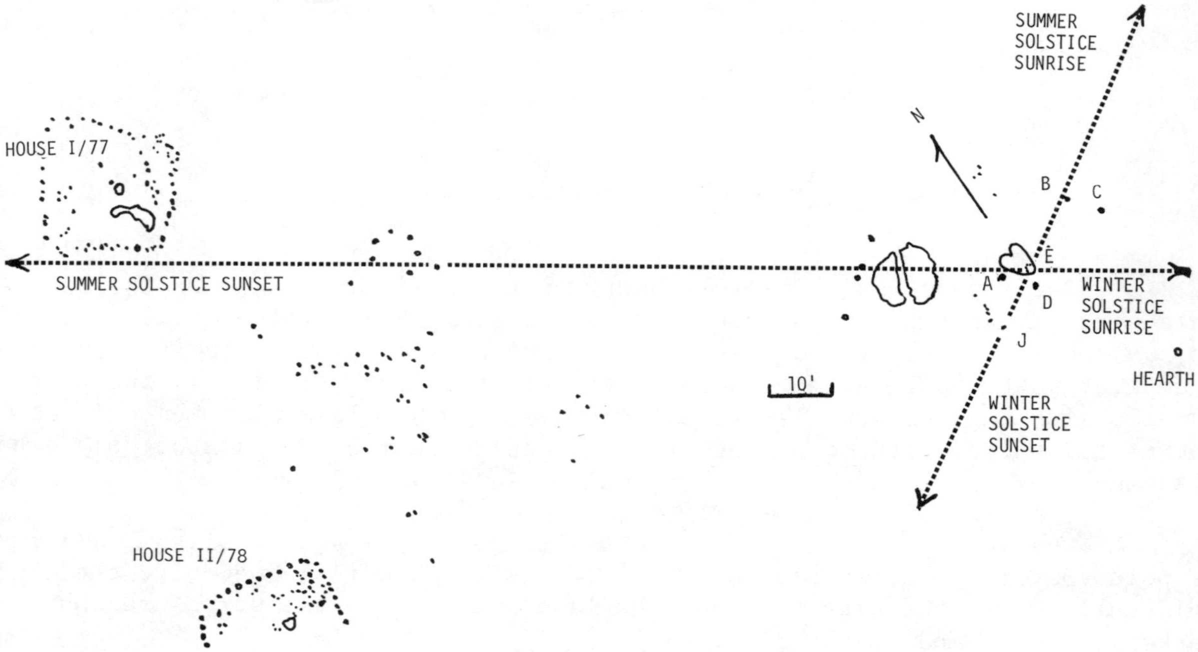

Fig. 15.9 Possible solstice alignments, Incinerator site

This alignment presents two possibilities. First, the setting sun at the time of the summer solstice would be in alignment with the features illustrated in Figure 15.9 and an observation point located near post A in the central plaza parallelogram. Second, at the time of the winter solstice, the rising sun would cast the shadow of the center pole down the alignments. Hence, two of the solar standstill points could be observed. Furthermore, a line drawn from post J through post E and touching the north side of post B (Fig. 15.9) very accurately defines the summer solstice sunrise and winter solstice sunset line. Thus, the central plaza pole structures and other features thus described accurately locate the horizon positions of the sun at the four solstice or standstill points.

The winter and summer solstice sunset lines also seem to define the boundaries for the majority of the trash pits that yielded ceremonial-type refuse and artifacts. This area also contains the two proposed observation points described earlier in this paper, as well as the largest house and the "Ceremonial Lodge" near which the majority of the ritual garbage on the site is buried.

OTHER POSSIBLE ALIGNMENTS

Another unusual group of posts makes up a roughly linear arrangement running north of the east/west line at an angle of 23°30'. Approximately 100 ft northwest of the center pole, a series of posts all lying within 1-2 ft north or south of this line and extending for about 25 ft to the northwest have been excavated. Within 25 ft north and south of the 23°30' line are 21 more post holes not associated with a dwelling structure. At this time enough unusual post holes and post-hole configurations have been found to indicate that specialized activities were carried out in this area, although no definite alignments can be clearly identified. However, only about 40% of the squares along the line have been excavated to date; with further excavation, more clearly defined patterns may be detectable.

CONCLUSIONS

While the continuing excavation of the Incinerator site may well clarify or expand our knowledge of Fort Ancient-period solar observations, the patterning of those posts already excavated strongly indicates that they were deliberately placed to enable monitoring of celestial processes and events. It is thought that the layout of this village was dominated by the relationship of the centerpost and its alignments to the equinoxes, summer solstice, and planting and harvesting times for corn. The alignment of the buildings and posts on the west side of the village with the central plaza posts must have been determined prior to the laying out of the site plan. Further excavation is necessary not only to clarify the known alignments, but also to check for other alignments, including summer solstice (rising sun), lunar standstills, and rising/setting positions of other bright celestial objects.

Although numerous Fort Ancient sites have been excavated, none has been thoroughly explored. We are aware of the plaza areas at the Anderson and Blaine sites and know that "some post molds with no recognizable pattern" (Hanson 1975:40) occurred at the Buffalo and Sloane sites. Although none of these sites appears to have astronomical alignments similar to the Incinerator site, there has been such a paucity of excavation in the plazas that the data are not truly comparable. Most of these sites were excavated under salvage conditions, which do not lend themselves to recognizing the correlations between widely-distributed features that make up solar alignments. Although the center post would be readily recognizable, at none of these sites--with the possible exception of Sloane--has the center of the plaza been excavated.

We hope future excavations at Fort Ancient sites will produce data indicating whether the Incinerator site is unique, or whether specialized activities in the plaza area were a common feature essential to the well-being and continuance of most major Fort Ancient agricultural villages. Without such seasonal indicators, the risk of crop loss in such a farming economy would surely have been considerably increased.

16 Transformations of the Hopi Calendar[1]

Stephen C. McCluskey

There is a note in one of the unpublished notebooks of Alexander M. Stephen (ca. 1883:60-61) that reveals much about the nature of Indian astronomy, and of Indian science in general:

> Intiwa [the Powamu chief at Walpi] says--"When my people had learned to build houses and men had grown accustomed to life in the 'Estufa,' Masau came and taught them many things concerning growth of plants and trees and instructed them about planting beans when the moon should be at a certain age and after the sun had come a certain distance on his way back to the north. Many, many days this has been the custom and we have no right to forsake the ways of our fathers."

This little passage mentions coordinated observations of the sun and the moon, the connection of these observations to the planting of crops (in this case a ritual planting for the Powamu festival), and two of the factors that lend importance to these observations: respect for the customs of their ancestors and the ultimate source of these customs in a divine revelation by Masau. The astronomical observations described did not provide new knowledge, but they were the stable empirical core of an agricultural and ceremonial calendar based upon a traditional body of knowledge and techniques. Such an unchanging science can exist in an isolated culture that is free from external influence, but there are usually disturbing influences. Of special interest is the case when such a traditional culture comes into contact with a powerful culture equipped with a calendar based on the predictive models of mathematical astronomy.

The Hopi Indians were faced with such a situation in the period beginning around 1870, when the United States government first seriously began to direct its attention to its recently acquired wards. The transformation of the Hopi calendar during this period of contact reflects the fragility of such

[1]I would like to express my thanks to numerous archivists throughout the Southwest, in particular Katharine Bartlett of the Museum of Northern Arizona, W. David Laird and Louis Hieb of the University of Arizona Library, the staff of the L. D. S. Church Historical Department in Salt Lake City, and John F. Schmidt of the Mennonite Library and Archives in North Newton, Kansas, for their assistance in my search for accounts of Hopi ceremonials. I am also grateful for the cooperation of numerous members of the Hopi Tribe who contributed to my understanding of Hopi culture. Finally, a generous grant from the Program for History and Philosophy of Science of the National Science Foundation (SOC 78-05768) made my research possible.

a traditional structure and the ways in which foreign emphases can disturb native elements.

In order to study the transformations of the Hopi calendar, I have analyzed an extensive body of festival dates from 1860 to the present collected from accounts in books, periodicals, manuscripts, and newspapers. The most extensive portion of these data concerns the Snake Dance, which gained early notoriety as a "weird" and "hideous" spectacle (to use the favorite adjectives of contemporary journalists) and drew a regular clientele of ethnographers, photographers, and journalists from the 1880s to the 1900s, followed by hordes of tourists from the 1920s to the present. These, and the industry of guides, Indian traders, and various boosters of the towns along the Santa Fe railroad and Route 66 that profited from tourism, provide abundant documentation for the Snake Dance. There is an almost continuous series of dated descriptions of the Snake Dance at Walpi Pueblo from 1877 to 1963; only 7 of the 44 occurrences are undescribed (Table 16.1). Similar bodies of data are available for the other pueblos, but that of Walpi begins fifteen years earlier than that of any other pueblo and offers the best representation of the original state of the Hopi calendar.

But what can such a body of festival dates tell us about the astronomical principles underlying the ceremonial calendar? If we consider the description of the bean planting for the Powamu festival in the passage quoted above, it is apparent that the festival is a luni-solar one, occurring when the moon is at a certain age, after sunrise had moved back north a certain distance along the horizon from the point of winter solstice sunrise. We can confirm this by analyzing the festival dates of Powamu using the relationships between the lunar month and the solar year.

The classical "Metonic" cycle requires the insertion of seven intercalary months every nineteen years. A more precise calculation indicates that one month is inserted every 2.7154212 tropical years. Employing a technique which I have discussed elsewhere (McCluskey 1977), we can perform a Fourier analysis with this 2.71-year period of the festival dates (or their calendar-independent equivalent, the longitude of the sun on those dates). The graph resulting from such an analysis of the Powamu festival displays a sawtooth pattern characteristic of luni-solar intercalation. A discontinuity in the graph will occur at the same phase angle for all festivals in the same simple luni-solar calendar, which provides a sensitive test for common lunar elements in a series of festival dates. Festivals employing different calendars, such as Easter, will have discontinuities at different phase angles, while purely solar festivities will display no single well-defined discontinuity.

Using this technique, I had proposed a set of relationships among the principal Hopi festivals, including a particularly complicated set of relationships concerning the Snake and Flute ceremonies (McCluskey 1977:186-190). Since I have now obtained a much more complete set of data, I can state that the apparent complexities of the Snake and Flute ceremonies resulted from the fact that the principles by which these festivals were regulated underwent significant changes during the period under analysis. During the early period (prior to 1900), the pattern that emerges is that of an almost purely solar festival, but a solar festival that displays characteristics related to the Hopi luni-solar calendar. The Fourier analysis of the Snake festival displays a luni-solar element in a slight discontinuity near the same phase angle as the Powamu festival, while the combined analysis of the Snake and Flute festivals displays no luni-solar pattern but merely an irregular variation of ±3.3 days around the mean value.[2]

The mean value, then, fits very neatly into the pattern of the Hopi calendar. For the period prior to 1900 the average longitude of the sun for the Snake ceremony was 143.0° and for the combined Snake and Flute ceremonies 143.9°. According to my earlier analysis of the Hopi calendar, the summer lunar month Pa-muya begins (at the latest) when the sun and moon are at 142.7° longitude, and ends (at the earliest) when the sun and moon are at 144.5°. Thus, prior to 1900, the Snake and Flute ceremonies, although occurring at a fixed time in the solar calendar, always occurred within the summer Pa-muya. At times there was a slight deviation from the average date to insure this, as is indicated by the slight luni-solar element in the analysis of the Snake festival. These relationships are confirmed by ethnographic reports concerning the Snake and Flute festivals.

Alexander Stephen (1936:713, 723) and Edward Curtis (1970:135, 138) both specify that the Snake festival occurs during the summer Pa-muya. Stephen, the principal early source on things Hopi, quotes several leaders of the Snake Society, who link a minor Snake prayer-stick-making ritual in the winter Pa-muya with the Snake Dance in the summer Pa-muya to form a complementary pair of festivals.

Stephen (1936:666-667) and Crow Wing (1925:101-106) both relate the Snake Dance to the immediately preceding Niman (or homegoing of the Kachinas) festival, which has been shown to be based on solar observations. Stephen reports that the preliminary "smoke assembly" of the chiefs, at which the date of the Snake festival is decided, takes place after a four-day waiting period following the end of the Niman Kachina, unless the crops are not well developed. The announcement is made on the following morning and the Snake Dance occurs sixteen days thereafter. Like Stephen, Crow Wing mentions a four or eight day delay if the crops are not ready. In addition, it is repeatedly noted that the Snake and Flute ceremonies follow identical patterns on alternate years.

These ethnographic data correspond closely to the data derived from the festival dates. According to Stephen's account, 22 days elapse between the ninth day of Niman Kachina, when the Home Dance is performed, and the ninth day of the Snake or Flute ceremonies, when the Snake or Flute dances are performed. According to the Snake and Flute data prior to 1900, the interval between average solar longitudes of Snake/Flute and Niman is some 22.5°, corresponding to a time interval of 23.4 days, with a variation in the data corresponding to ±4.0 days. This variation may, in part, be due to changes in the condition of the crops. The early pattern thus corresponds to the descriptions in the early ethnographic accounts.

---

[2]Due to the small magnitude of the luni-solar element in comparison to the irregular variations of the festival date, a precise critical phase angle for the discontinuity in the Snake festival analysis cannot be established. It can be estimated as lying between 200° and 260°, compared to the 240.8° for Powamu given in McCluskey (1977:184-185).

TABLE 16.1

DATES OF SNAKE DANCE AT WALPI

Date	Day	Solar Long.	Source
15 Aug. 1877	Wed	142.71	GAD
12 Aug. 1881	Fri	139.86	SHJ, BSD
12 Aug. 1883	Sun	139.39	MSD, MTK*
18 Aug. 1885	Tue	145.66	NAC9/5, SHJ, FMP, MIS**
15 Aug. 1887	Mon	142.31	SHJ
17 Aug. 1889	Sat	144.73	ESD, SHJ
21 Aug. 1891	Fri	148.11	FMP, FOF, DVM, SHJ, SCB, LSC
14 Aug. 1893	Mon	141.87	SHJ, FOF, FMP, DVM
18 Aug. 1895	Sun	145.26	FOF
21 Aug. 1897	Sat	148.64	WBP
24 Aug. 1901	Sat	150.60	NCS8/10, NHA8/10, DVM
18 Aug. 1903	Tue	144.36	NCS8/15
15 Aug. 1907	Th	141.50	NWM8/10, 17, FSD
17 Aug. 1909	Tue	143.92	CHN, NWM8/7
22 Aug. 1911	Tue	148.27	HMS
21 Aug. 1913	Th	147.81	RHS, CMS, NWM8/9
25 Aug. 1921	Th	151.73	CWJ, NWM8/19
24 Aug. 1923	Fri	150.30	NAR8/12, NWM8/17
19 Aug. 1925	Sat	145.98	HMS
23 Aug. 1927	Tue	149.36	NTS8/23, HMS
23 Aug. 1929	Fri	149.86	NWM8/9, 24, HMS***
20 Aug. 1931	Sun	146.50	NTC8/20, HMS
21 Aug. 1933	Mon	147.96	THI, NAS8/18
25 Aug. 1935	Sun	151.36	AHC
23 Aug. 1937	Mon	149.92	NWM8/27, AHC
26 Aug. 1939	Sat	152.35	NWM8/11, 9/1, HMS
27 Aug. 1941	Wed	153.82	NWM8/22, 29
22 Aug. 1943	Sun	148.53	NWM8/13
26 Aug. 1945	Sun	152.88	NWM8/17, HMS
21 Aug. 1949	Sun	148.09	NPG8/12
24 Aug. 1951	Fri	150.52	NWM8/10
23 Aug. 1953	Sun	150.05	NAS8/8
27 Aug. 1955	Sat	153.44	GMS, NAS8/30
24 Aug. 1957	Sat	151.05	GMS
24 Aug. 1959	Mon	150.58	GMS
26 Aug. 1961	Sat	153.01	NAS8/11
25 Aug. 1963	Sun	151.58	GMS

*Moran gives the date 12 Aug. while McIntire and Gordon's secondary account gives 13 Aug.

**Mindeleef gives the date as 17 Aug.

***Hubbell's ms forecast differs with the accounts in the Winslow *Daily Mail*, Hubbell giving the date as 22 Aug.

SOURCES FOR SNAKE DANCE

Newspapers (all newspaper entries are followed by month and day of issue):

NAR Arizona Republican (later Arizona Republic, Phoenix)
NAS Arizona Daily Star (Tucson)
NAC Arizona Champion (Flagstaff)
NCS Coconino Sun (Flagstaff)
NHA Holbrook Argus
NPG Phoenix Gazette
NTC Tucson Daily Citizen
NTS Tucson Daily Star
NWM Winslow Mail

Others

AHC Arizona State Historical Society, Tucson. Hopi Clipping File.
BSD J. G. Bourke, *The Snake Dance of the Moquis of Arizona*. New York, 1884.
CHN Kate Cory, Hopi Notes, 1909-1910, Museum of Northern Arizona, Flagstaff, MS 208-2-2.
CWJ Crow Wing, *Pueblo Indian Journal*.
CMS Letter from Leo Crane, Ariz. Hist. Soc., Tucson, Richard van Valkenburgh collection, box 12.
DVM G. A. Dorsey and H. R. Voth, "The Mishongnovi Ceremonies of the Snake and Antelope Fraternities, " Field Columbian Museum Papers, Anthropological series, vol. 3, no. 3, 1902.
ESD William M. Edwards, "Snake Dance of the Moqui Indians." *Harpers Weekly* 33 (1889):871-3.
FMP Earle R. Forrest, *Missions and Pueblos of the Old Southwest*. Cleveland, 1929.
FOF J. W. Fewkes, "The Oraibi Flute Altar." *Journal of American Folk-lore* 8 (1895):277.
FSD Earle R. Forrest, *The Snake Dance of the Hopi Indians*. Los Angeles, 1961.
GAD Andrew Smith Gibbon, Diary (typescript copy), Utah State Historical Society, Salt Lake City, MS 1533.
GMS Helen M. Greene, Hopi Notes, Museum of Northern Arizona, Flagstaff, MS 169-I-6.
HMS J. L. Hubbell MS collection, University of Arizona, Tucson.
LSC Charles F. Lummis, *Some Strange Corners of Our Country*. New York, 1908.
MIS K. Mendelieff, "An Indian Snake-Dance," *Science* 7(1886):507-14.
MSD Peter G. Moran, "The Estufas and So-Called Sacred Fires of the Moqui Snake Dance of 1883." in Thomas Donaldson, *Moqui Pueblo Indians of Arizona and Pueblo Indians of New Mexico: Extra Census Bulletin*. Washington, 1893.
MTK Elliot G. McIntire and Sandra G. Gordon, "Ten Kate's Account of the Walpi Snake Dance: 1883," *Plateau* 41(1968):27-33.
RHS Theodore Roosevelt, "The Hopi Snake Dance," *The Outlook* 105 (1913):365-73.
SHJ Stephen, *Hopi Journal*.
SCB Julian Scott, in Thomas Donaldson, *Extra Census Bulletin*. Washington, 1893.
WBP Ben Wittick, photo captions, Museum of Northern Arizona, Flagstaff, MS 186-7-28 to 186-7-31.

But later reports by Crow Wing (1925:101-102) and Waters and Fredericks (1969:257) insert a period of sun watching before the Snake ceremony. This change is also apparent in a systematic change in the times of the Snake and Flute ceremonies after 1900. Although there are frequent minor variations, an unmistakable trend indicates that these ceremonies ceased to maintain their old established relationship with other elements of the Hopi calendar. The Snake and Flute dances moved later and later and did not settle down to anything like a steady state until around 1940. For the period from 1940 to 1970, these ceremonies were being celebrated some 7.6 days later than during the period from 1870 to 1900. Similar changes can be perceived in the less extensive data from the other pueblos. Despite this change in festival dates, no corresponding shift in the intercalation of the luni-solar calendar can be perceived, although the data on this point are somewhat limited.

This shift of one major element of the Hopi calendar, in which it lost its careful synchronization with the other elements of that calendar, presents something of a problem (and a lesson) to archaeoastronomers. We tend to look at the astronomical systems of traditional cultures in a timeless, ahistorical manner, assuming them to be static and unchanging. Yet here we find a transformation of the Hopi calendar taking place in a brief period of some 50 years. This situation seems to pose an obvious question in the historical mode. What factors were present that could account for this change and how did they effect such a change?

The pattern of change of festival dates reflects closely the historical pattern of white influences on Hopi society. The earliest contacts were comparatively innocuous. The first Indian agents spent most of their time two days away from the Hopi in the more civilized surroundings of Fort Defiance. The few white men on Hopi lands in the 1880s were sympathetic observers of the Indians, both out of their own curiosity and for the economic motive that a good trader should know and defend the interests of his customers (McNitt 1962:124-191).

While the traders to a large extent wished to deal with and profit from the Indians and made few efforts to change their traditional ways, most of the Indian agents saw their mission as civilizing a barbarous and pagan people. Traditional religious ceremonials were actively discouraged, while agency boarding schools held regular Christian services at which attendance was expected. The whole tenor of the operation was to teach the Indian the white man's ways (James 1974:106-145, McCluskey 1980).

After several unsuccessful attempts in the 1870s and '80s, a permanent school for Hopi children was established at Keam's Canyon in 1887. Since it was a boarding school located a half day's ride from the more distant villages, parents understandably resisted the compulsory attendance rules. Particularly at the more distant villages, already existing factions became further polarized over the problem of school attendance. Tensions became so great that at Oraibi two totally independent ceremonial cycles, run by these factions, were established in 1897 (Titiev 1949:69-95). The two ceremonial cycles had to be held at different times, as two public dances could not be held in the village plaza simultaneously. The calendar was beginning to change at one pueblo and (presumably through the informal cross-checks between pueblos) this contributed to the related changes at Walpi pueblo.

This factionalism eventually led to a split at Oraibi in 1906 in which the members of the traditional faction were evicted and established a new village, Hotevilla, six miles further from Keam's Canyon.[3] Shortly after the split, the Snake festival was abandoned at Oraibi. As further results of this incident, both the traditional and progressive leaders at Oraibi and Hotevilla were arrested and forcibly sent to boarding schools to learn English, and compulsory school attendance for children became the norm throughout the reservation. Both these factors further disrupted the ceremonial cycle. The arrest of leaders forced men to take up ceremonial roles for which they were not fully prepared. More importantly, however, compulsory school attendance seriously interfered with the manner in which young boys were taught matters of tradition and ritual by their uncles, and led many of them to question the values of their elders.

Hopis of this generation displayed mixed attitudes toward tribal traditions. Many of them, in various ways, embraced elements of white culture. Some became Christian, others adopted white political models and became active in the Tribal Council, instituted under strong pressure from Washington in 1936 (James 1974:201-205, 223-235). Others adopted an entrepreneurial model and became advocates of establishing Hopi-owned businesses, while a few found a working relationship with scholars studying the Hopi.

Those adopting Western modes of thought tended in varying degrees to look down on elements of their culture as backward. To some Christians, the Hopi religion partook of devil worship. Talashonga, a Hopi Christian, said: "The redheaded ghost [Masau] is Satan" (Udall and Sekaquaptewa 1969:237). In some cases the leaders of ceremonials were converted to Christianity and abandoned their ceremonial roles without training successors (Thompson 1950:80).

The most important instance of such a rejection of tradition, which relates directly to this study, occurred in 1907 with the conversion to Christianity of several members of the Snake Society at Walpi. At the time of their conversions, these individuals stopped participating in the Snake Ceremony. Especially significant was the conversion of Sa'lako, the matriarch of the Cactus lineage of the Snake Clan. This lineage controlled or "owned" the Snake Ceremony[4] and traditionally provided the leaders for the Snake Society. Sa'lako was the only woman who knew how to prepare the Snake medicines taken by the society's members after the Snake Dance and mentions that her son, the Snake Priest, "got very angry because I would not help him any more with the snake dance" (Hopi Indians Collection, 1921-22:Exhibit B). The

---

[3]There is an extensive body of literature on the causes of the split at Oraibi. The existence of a faction opposed to cooperation with the Americans at Oraibi was noted as early as 1858 (Ives 1861:124). A thorough historical discussion of the major factors, including ceremonial cycles, in the split is given by Mischa Titiev (1949:69-95). Bradfield (1971:23) argues that excessive soil erosion in the period 1901-06 was the "immediate precipitating cause" of the split. Jonathan Reyman brought this last factor to my attention.

[4]On clan ownership of ceremonials see Eggan (1950:89-106); page 90 deals specifically with transmission within the lineage occupying the clan house.

prominence of her matriarchal position is further attested by the fact that the smoke assembly preceding the announcement of the dance was held at her house and the fetishes of the Snake Society were stored there (Stephen 1936: 103, 666, 710-711, 865).

Affidavits by Sa'lako, her sister's son John Lomavoya, and Hooker Hongeva, located in a dossier collected in 1920 by an "E. M. Sweet, Inspector,"[5] allege that widespread deception and immorality were connected with the Snake Ceremony. Although the shared phrasing of these documents suggests the influence of the investigator, they do provide strong evidence for a conscious rejection of important elements of their tradition by these Christianized Hopi. Such an abandonment of tradition by important members of the Snake Society could not but have had its impact on the transmission of the society's lore.

In addition to conscious rejection of tradition, there is also the problem of simple ignorance of its significance. A number of progressive Hopi with whom I have talked expressed surprise when told that the Hopi had an astronomy. They knew that there were Hopi sun-watchers, but they didn't connect this with the Western model of "science." This lack of respect for their own traditions undermined one of the strongest sanctions maintaining Hopi calendric techniques.

At the other extreme, we find Hopi trying to force a reconciliation of traditional Hopi lore with Western culture and science. Frank Water's recent account of Hopi traditions attributes to his Hopi informants such Western concepts as the earth revolving around the sun and rotating on its axis, the latter motion caused by the magnetic polarities of the Hopi war twins Pöqánghoya and Palöngawhoya, who are said to live at the North and South poles. Even some Ptolemaic elements crop up; to the traditional Hopi concept of the present world as a fourth stage of development is added the notion that there are seven physical worlds, followed by two more beyond man's reach, "and the ninth is the indefinable, incomprehensible domain of the one divine Creator of all" (Waters and Fredericks 1969:5, 32, 194-195, 202-203). This sounds very much like the seven planetary spheres, beyond which are the spheres of the stars, the crystalline sphere, and the habitation of God and the elect.[6]

Clearly, this ambivalence between white and Hopi concepts does not contribute to the accurate transmission of Hopi calendric techniques. Further inconsistencies exist between white and Hopi concepts of the four directions, with the Hopi defining the solstitial directions as cardinal, and of the

---

[5]I have been unable to determine whether Mr. Sweet was an official government inspector or an agent for one of the missionary bodies or the Indian Rights Association, both of which were engaged in an active campaign to suppress Indian religious observances as immoral. He is not mentioned in the annual reports of the Hopi Agency for 1920 and 1921, although it is apparent from the reports that the agent, Robert Daniel, was sympathetic with the movement to abolish the ceremonies.

[6]The nine worlds idea probably is a recent import from Western mysticism, although it may represent an earlier Spanish influence. Renee Opperman (1979) and Gary Urton (1980) mention similar layered cosmologies of nine worlds in North and South America, but in both cases the earth is the fifth world with four above and four below.

month, with the Hopi employing a true lunar month. These conflicts may play some role, but although I have not completed my investigations of this matter, they seem to have caused more difficulty to the white man than to the Hopi.

The one white concept that did seriously influence the Hopi calendar was that of the weekend. As young Hopi children began to attend day schools near home, and their parents began to take employment in town, the weekend began to influence the Hopi ceremonial calendar, as the following data attest:

Number of Snake Dances at Walpi Pueblo
on Various Days of the Week

	Sun.	Mon.	Tues.	Wed.	Thurs.	Fri.	Sat.
1870-1929	2	2	5	1	3	4	4
1930-1970	7	3	0	1	0	1	4

From 1870 to 1929, the distribution is essentially random; but from 1930 on, the concentration on the weekends is obvious. Since the last two days of the Snake festival involve public ceremonies of primary importance, the scheduling of the Snake Dance on Sunday places both these days on the weekend, allowing most Hopi to attend both public ceremonies.

A further contributing factor is that the 1930s saw the development of a mass tourist industry focused on the Snake Dance, which brought thousands of spectators to the Hopi villages to observe the dances. Since their purchases of artifacts contributed to the depressed Hopi economy, there were reasons for the Hopi to consider the convenience of tourists as well. There is an instance of the Snake Chief at one pueblo, a man who had lived in town for a time, providing a local newspaper with the dance dates at the various pueblos (Winslow *Mail* 1938); it is interesting to note that only his pueblo was holding the dance on a weekend that year.

The historical factors considered here that played the greatest role in changing Hopi calendric practices seem to have been the duality of the ceremonial cycle at Oraibi prior to the split in 1906, with the consequent increase in irregular performances; the failure of Oraibi to maintain a complete ceremonial cycle following the split; the breakdown of the traditional transmission of ceremonial lore between the generations; and finally, the introduction of the weekend, which added a new complicating factor to the ceremonial calendar. Given that these specific factors did play a role in the transformation of the Hopi calendar, the question remains whether such changes could only have been caused by the arrival of the white man. Clearly, the introduction of the weekend and the decline of respect for traditional Hopi culture were exclusive results of the arrival of an alien culture. However, factionalism and its ultimate result, the breakup of an entire village, is evidenced both by Hopi oral tradition and by the archaeological record as having occurred prior to the split at Oraibi (Titiev 1949:96-99). It is not unreasonable to assume that such historical incidents would have had similar effects on the ceremonial cycle and may account for some of the discrepancies between the cycles of the various Hopi villages. If this is the case, transformations of the ceremonial calendar may have been recurring events in Hopi history.

Of course, the extent to which such transformations would be tolerated depends upon the importance of the ceremonial calendar. Since the arrival of the white man and his calendar, the Hopi calendar no longer plays the primary role that it once did in regulating the Hopi agricultural cycle, so that changes to it do not damage the economic well-being of Hopi society. Nevertheless, it must be stressed that the most powerful causes operating here are specifically dependent upon contingent historical events.

This leads me to suggest that archaeoastronomers and ethnoastronomers consider at least two historical dimensions of the astronomies that we study. First, we obviously need to recognize that traditional astronomies, although they change much more slowly than modern science, are historical phenomena that do change with time. Second, we should not let our desire to establish theoretical relationships linking astronomical systems to other elements of culture lead us to neglect the specific historical individuals and contingent events which effect transformations in these relationships, at least when evidence for such details is available.

17 Ethnoastronomy: The Zuni Case[1]

M. Jane Young and Ray A. Williamson

The paucity of data on how the Pueblo Indians viewed and understood the stars has led many to conclude that they had relatively little interest in the stars or stellar groupings. During a recent search of the J. P. Harrington material in the National Anthropological Archives of the Smithsonian Institution, we uncovered new information about Zuni knowledge of the stars and other celestial objects. On the basis of this material, it seems likely that anthropologists have misjudged the extent of Pueblo interest in the stars.

Anthropological records indicate that stellar observation was an element in the ritual practice, legends, and art of the Pueblo Indians. The star groupings most often mentioned are the Pleiades, Orion, the morning and evening stars, and a star often called "Big Liar" (Curtis 1922:155, 121; Parsons 1939:Table 2; Stephen 1936:9, 233). However, with the exception of the identifications made by J. P. Harrington in his *Tewa Ethnogeography*, there are relatively few data on other constellations or stars recognized by the Pueblos.

The information which we have gathered concerning stellar observation at the Pueblo of Zuni clearly shows an interest in star groupings beyond those just mentioned--it is likely that this wider scope of stellar observation pertains to the other Pueblo groups as well. In the published ethnographies, Cushing (1896:392), Parsons (1939:182), and Stevenson (1904:130, 131) have reported that certain star groupings were very important for ceremonial and agricultural purposes at Zuni. Of primary importance for distinguishing divisions of the night were the positions of the moon, Orion, the Pleiades, and the Great Dipper--their rising or passage over the uncovered hatchway in the kiva roof often signalled the culmination of ceremonies or specific phases of those ceremonies. For example, shortly after the rise of the morning star above the horizon, the women dancers at the August rain ceremony in 1891 donned headdresses decorated with the morning and evening stars and began their dance (Stevenson 1904:194). Further, the rising of the morning star during the winter solstice ceremony marked the time for the kindling of the New Year fire (Bunzel

---

[1]This research was supported by a Visiting Student Research Assistantship at the Smithsonian Institution (MJY), and by a Smithsonian Post-Doctoral Fellowship (RW). The source material is from J. P. Harrington's unpublished manuscripts in the National Anthropological Archives, Smithsonian Institution. We gratefully acknowledge the patience and enthusiasm of Elaine Mills at the Archives, and the help of Von Del Chamberlain of the National Air and Space Museum, who assisted us in identifying some of the constellations.

1932:536; Stevenson 1904:130, 131). Cushing explains that the Zunis first planted corn in the spring "by the light of the seven great stars which were at that time rising bright above them . . . looking at the stars they saw how they were set, four of them as though around a gourd like their own, and three others as though along its handle!" (1896:392).

A study of the collections of Zuni folktales has revealed some interesting stellar references in the stories. For example, Benedict relates a tale concerning the placement of the stars in the sky. According to this story, a monster called Cloud Swallower once existed who was a great menace because he took away all the rain--so the Twin War gods, after receiving magic help from Grandfather Gopher, killed Cloud Swallower:

> They took his heart and threw it to the east and it became the morning star. They took his liver and threw it and it became the evening star. They took his lungs and threw them and they became the Seven Stars. They took his entrails and threw them and they became the Milky Way [Benedict 1935:51].

The Twin War Gods mentioned here[2] are considered to be sons of the Sun--the one going before and the other following after the Sun in his journey across the sky (Cushing 1896:381; Cushing 1901:378-379). According to one Zuni legend these twins were born of the union of the Sun and Water (Benedict 1937:I:1; Cushing 1896:381; Stevenson 1904:24; Tedlock 1972:226), and hence they remind us of two of the most important aspects of Zuni ritual--prayers to the sun father asking that he allow the crops to thrive, and prayers to the ancestors that they may send rain. These twins are also significant because, according to several versions of the Zuni origin myth, they led the people from the four underworlds to the surface of the earth (Benedict 1935:I:3; Bunzel 1932:547-602; Cushing 1896:382; Parsons 1923:136-137; Stevenson 1904:26; Tedlock 1972:256-257).

Further, there is an abundance of stellar motifs on Zuni altars, murals, and sand paintings, and on the masks, headdresses, and costumes of the kachinas (Stevenson 1904:Plate CIV, 432, Plate CII, 428, Plate XXXVIII). For the most part, these seem to represent single stars rather than constellations.

Because of this attention paid to the stars, it is rather surprising that there are not more extensive references to Zuni constellations and observations of celestial phenomena in the published ethnographic material. Of course, one explanation is that most of the anthropologists who worked at Zuni did not have an adequate background in astronomy--indeed, the astronomical information they did record is often quite confused and contradictory. Bunzel states that although "some of the stars and constellations are named and recognized . . . no observations are made of the positions of the stars and movements of the planets" (1932:487); however, the Zuni were clearly aware of stellar positions

---

[2]Cushing names the Twins Áhaiyúta and Mátsailéma (1901:378-379). In this tale, Benedict's informant refers to the Twins as the Áhaiyute, using the elder's name to signify the pair. At times, as in Cushing's accounts, they are the morning and evening star. In this tale, however, they create the morning and evening star, a seeming contradiction. Typical of folklore, the conflict has no direct resolution. There are sometimes two different pairs of twins, one primary, one secondary. In other texts, they are both assimilated into a single pair.

and movements because they used them for timing ceremonies.

J. P. Harrington's notes about Zuni constellations and other celestial objects provide us with a new perspective on Zuni celestial lore. Harrington, an ethnographer with the Bureau of American Ethnology, collected information from a Zuni member of a Chaco Canyon work crew in mid-June through early July 1929. It is clear from his field notes that Harrington had some astronomical knowledge, but he does not articulate his methodology and his sketchy descriptions make only tentative identifications possible. The interview was clearly conducted outside beneath the stars during the course of one night. The Zuni informant identified several constellations whose stellar groupings are unlike familiar Western constellations. It is striking that often the Zuni name for the constellation is a descriptive one, containing information about its shape or placement in the sky.

Before describing these constellations, it is necessary to list and explain some of Harrington's celestial vocabulary (see Table 17.1). The Zuni words are written exactly as they appear in Harrington's field notes, but his orthography is quite similar to that used by Stanley Newman, author of the only published Zuni dictionary (1958). Although there is a basis for some linguistic comparison, Newman has few words in his dictionary which describe celestial phenomena.

TABLE 17.1

HARRINGTON'S CELESTIAL VOCABULARY

ya'tokkh^yah (sun)	"move above"
yätonne (day)	"one moving above"
yátchu	"month," "crescent moon"
móyatchunneh (star)	"spherical object, crescent moon"
mókhwa:n'ősenah	"big lying star"
móyatchunłánah (morning star or great star)	"spherical object, crescent moon, big"
yátokk^yayálonah (the evening star)	"the one following the sun"

The word Harrington lists for the "sun" is *ya'tokkh^yah*, which he says literally means "the one who passes by." Newman cites the same word for the "sun" (*yato-kka*), which he indicates as coming from *yato*, meaning "go over" or "move above." From this word comes the one which Harrington defines as "day," *yätonne*, which with the singular ending literally means "one moving above." Newman simply gives *yato* for day, without the singular ending. The word which Harrington lists for crescent moon, *yátchu*, is cited by Newman (*yachu*) as meaning both "crescent moon" and "month." This is significant, for Ellis (1975: 64) has suggested that according to the old Zuni calendar, the month began with the crescent moon, the new moon being the day that "doesn't count." Harrington's word for "star," *móyatchunneh*, is the same as Newman's, with the exception that Newman omits the singular ending. This word is an interesting combination of the stem meaning "spherical object" (*mo-*) and the word meaning "crescent moon" or "month" (*yachu*). Harrington names a puzzling star, described also by Cushing (1901:380), Dutton (1963:79, 109), and Parsons (1930b: 11, 33), as "big lying star"--*mókhwa:n'ő̋senah* (*mo*, "spherical object"; *khwa*, intangible particle; and *yosena*, "lying"). According to Harrington, this is "a bright star preceding the morning star, coming out at three o'clock in the morning, not as bright as the real morning star." Cushing (1901:380) gives the following rather different description: ". . . It is a great red star. It rises in summer-time and tells of the coming morning when it is only midnight; hence it is called *Mokwanosana* (Great Lying Star)." [We suggest that the "lying star" is a planet which rises too early to be the morning star.] For "morning star," Harrington lists the word for "star" plus the word for "large" or "big," *móyatchunłánah*. Bunzel gives the same name, translating it as "great star" and identifying it as Venus or Jupiter (1932:487). It is interesting to note that the word which Harrington gives for "evening star" does not even contain the word for "star," but rather the word for "sun" (*yátokka*) and the word *yálu*, meaning "to be behind or follow"--hence, *yátokkʸayálonah*, "the one following the sun."

With this background information, we will pursue Harrington's descriptions of the Zuni constellations, comparing them with the identifications made by Bunzel, Newman, Parsons, and Stevenson where possible (see Table 17.2).

TABLE 17.2

ZUNI CONSTELLATIONS

téhłiyakkhʸacíwanih	"Chief of the Night"
ha''im'on'úhanankhwátélen'onah	"three running down to the horizon"
há''ilekkʸa:khweh	"eight ones"
téhłiyakkʸan'ikʸe:nanneh	"heart of the night"
ipi'ⁱlakkhʸah (Orion)	"be in a row"
moyatchu:a̋łununneh (Delphinus)	"sling constellation"
méla:khweh (Cassiopeia)	"star zigzag"
khwílelekkʸa:khweh (The Great Dipper)	"seven ones"
píclankhwin'úkkhwatelen'onah (The Little Dipper)	"those to the north"
kúpa:khweh (Pleiades)	"seeds"
'áwi:łaknanneh (Pegasus)	"four big one," "square"
móyatchu:c̋llowah (The Northern Crown)	"red star"
yúpyałanneh (Milky Way)	"ashes stretched across"
cóli:khweh	"stick game ones"

Also for comparative purposes, we will cite Harrington's identifications of similar Tewa star groups. The latter give some indication of which constellations Harrington knew about or paid attention to. Although all of the Zuni material comes from Harrington's unpublished field notes, the Tewa material is from Harrington's published *Ethnogeography of the Tewa Indians* (1916).

The most striking of the Zuni constellations is the Chief of the Night, *téhłiyakkh^{y}aciwanih*. According to Harrington:

> This is the most majestic constellation ever reported from any tribe of Indians, as far as we know. It is a gigantic human figure, even bigger than the whole visible sky, for its head had already set in the west or had been obscured by clouds lying near the horizon, while its heart was pointed out in mid heaven, lying in the milky way, and its legs were said to extend beyond the horizon in the east.

Harrington then goes on to discuss particulars of the figure, only some of which are completely described. Again, quoting from Harrington:

> Attention was first called to its left arm, which is traced as three stars of considerable magnitude, one at the shoulder, one evidently at the elbow and one at the tip of the extremity. The star at the shoulder can be found by following the line of the tip section of the handle of the great dipper to the east for a distance about equal to the length of the handle of the dipper; the brightest star in that region of the heavens is this shoulder star.

We have tentatively identified the figure's left arm as consisting of the bright stars Arcturus (shoulder), Spica (elbow), and Antares (forearm). The Zuni name for this part of the constellation, *ha''im'on'úhananhkhwâtélen'onah*, means "three running down to the horizon."

This left arm apparently terminates in another constellation which Harrington describes as shaped like a hand. The Zuni name, *há'ilekk^{y}a:khweh*, is again descriptive and means merely "the eight ones." We have tentatively identified this as eight stars in Sagittarius. Harrington mentions a hand-shaped constellation for the Tewa also, but gives no indication of its placement in the sky.

The heart of the Chief of the Night, described by Harrington as a "V" consisting of five stars, is called *téhłiyakk^{y}an'ik^{y}e:nanneh*. This name means "heart of the night," a term which apparently refers to the astronomical fact that the constellation appears in the zenith around midnight in the summer (see Fig. 17.1--this constellation consists of Vega, Deneb, and part of the Northern Cross). In assigning the heart to the Northern Cross, we have chosen an identification consistent with Harrington's description of a V-shaped pattern in the Milky Way. It must be noted, however, that Harrington designates this "V" as the constellation which the Spanish-speaking people of New Mexico call *la campana*, "the bell." But in his account of Tewa constellations, Harrington claims that *la campana* is the Spanish name of the V-shaped group of stars astronomers call the Hyades, a grouping which is part of Taurus. However, the Hyades does not lie in the Milky Way, so it cannot be the heart of the Chief of the Night. It must be remembered that Harrington and his Zuni informant were viewing the sky directly. The resolution of this discrepancy in Harrington's description awaits field work in this area with the Zuni and

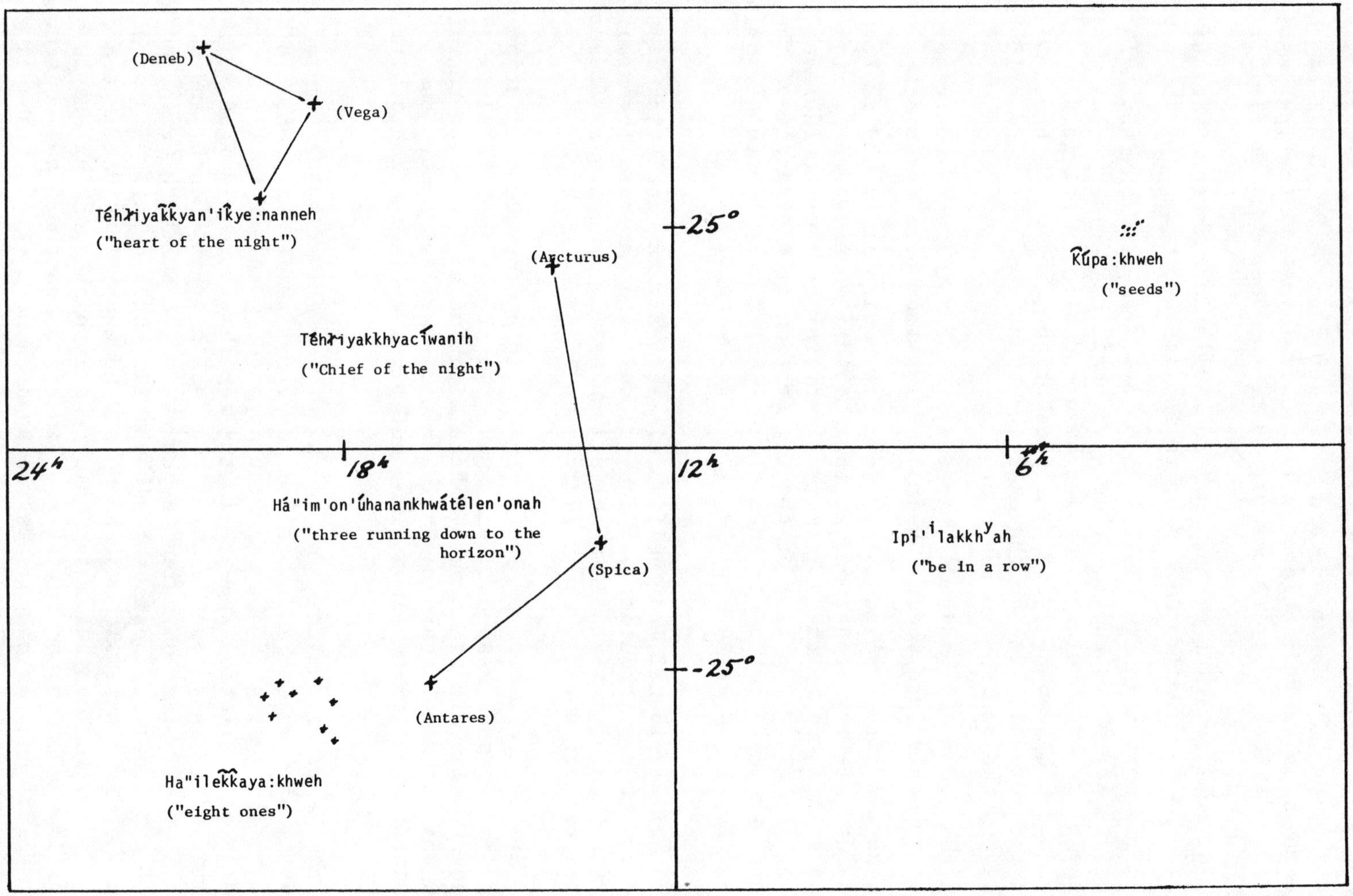

Fig. 17.1 Zuni constellation, *Téhλiyakkhʸacíwanih* ("Chief of the Night"). Zuni constellation *Kúpa:khweh* ("Seeds") corresponding to the Western constellation Pleiades is also included.

perhaps the Tewa. Either Harrington is incorrect in his identification of the heart as *la campana*, or there are two star groupings named *la campana*.

The right arm of the figure, *ipi'*^{i}*lakkhyah*, is "said to refer to four stars, one of them smaller than the other three, chasing each other towards the horizon." Although Harrington does not translate this word, it seems to come from the Zuni word *piʔla*, "be in a row." We have identified this right arm as Orion's belt. In one of Tedlock's translations of Zuni narrative poetry, the setting of "Stars-in-a-Row" is used as a way of measuring the time of night. Tedlock suggests that this stellar group "is Orion's belt, which sets in the west after midnight in midwinter" (1975:117, 150). Bunzel gives the same word for this group as Harrington (*ipi'laka*), describing it as Orion's belt, with the translation "the row" (1932:487). "In a row" is also the translation of the Tewa word Harrington identifies as Orion's belt. Parsons, however, translates the same word as the "Seven Stars" (1930:11). Although it can be claimed that there are seven stars in this part of Orion (belt and sword), the name "seven stars" is ordinarily used by the Zuni for the Western constellation the Big Dipper.

It is important to note that we are not suggesting that all of these star groupings of the Chief of the Night were seen at the same time of night; instead, they were probably pointed out by the informant as the night progressed. Some parts of the body were not seen at all that night, for Harrington states that the informant "did not trace the gigantic legs, which he says extend far to the northeast and southeast." Harrington goes on to mention that "some part of the body of the Chief of the Night is perpetually visible, guarding the earth throughout the year." It is evident that some one of these stellar groupings described as his body parts will always appear in the night sky.

Harrington describes one final constellation whose star grouping is unlike any Western pattern. Near Libra (but as yet unidentified to our satisfaction) is the Zuni constellation *côli:khweh*, or the "stick game ones." This name refers to players of the Zuni ritual game "sho'liwe," which is said to have been originally introduced by the Zuni Gods of War,[3] who are represented in the sky by the morning and evening stars (Cushing 1901:378-379; Risser 1941: 216).

> But we shall depart for our everlasting home in the mountains--the one to the Mountain of Thunder, the other to the Mount of the Beloved--to guard from sunrise to sunset the land of the Corn-priests of Earth. . . . Henceforth two stars at morning and evening will be seen, the one going before, the other following the Sun-father--the one Ahaiyuta, his herald; the other Matsailema, his guardian; warriors both, and fathers of men [Cushing 1901:378-379].

According to Zuni legend, the Gods of War and the rain priests played "sho'liwe" to call the rain. It is of considerable interest here that

---

[3]Anna Risser recorded her versions of Zuni folktales from Nachapani, the Zuni informant Harrington worked with while he was in Chaco Canyon. Thus the point of view about the Zuni War Gods and the morning and evening stars is not independent of Harrington's own data, which derive from the same person.

the sticks which are used in this game have markings engraved on them which represent the Zuni divisions of the day--morning, noon, and sunset. Stevenson gives the following description of the sticks:

> Each player takes the side of one of the Gods of War. There are four pieces of split reeds, two representing the side of the elder and two the side of the younger God of War. The reeds shown in the figure may be described as follows: *a* and *c* belong to the elder and *b* and *d* to the younger God of War; *a*, *kwin'na* (black), has the concave side of the reed colored black, indicating the whole day, and three sets of markings on the convex side denoting the three periods of the day, morning, noon, and sunset; *b*, *a'thluwa* (center), has a daub of black markings on the convex side also denoting noon; *c*, *ko·'hakwa* (white medicine), has a daub of black paint at each end of the concave side, indicating morning and evening or sunrise and sunset, markings on the convex side denoting the same; *d*, *pahl'to* (mark on the end), has a daub of black paint on the concave side of the joint end denoting sunrise, which to the Zunis is the first light of day, or the white light which comes first, and markings on the convex side indicate the same, while three dots sometimes found on the joint of the reed, not shown here, indicate eyes and mouth of the face [1904:330-331].

In addition to these constellations, which are so different from our Western ones, Harrington gives the Zuni names for constellations whose star groupings are familiar to us. *Móyatchu:a̋łununneh*, the group translated as "star sling," according to Harrington is identical to the Western constellation Job's Coffin or Delphinus. The Zuni name appears to be descriptive of the shape of this group of stars. Harrington translates the Tewa word for this constellation as "belly of a sling." *Kúpa:khweh*, meaning "seeds," is identified by Harrington as the Pleiades. Bunzel (1932:487) gives the same word for the Pleiades, but Newman lists the Zuni word *mela* as meaning the Pleiades--this, however, is the word which Harrington identifies with Cassiopeia. The very descriptive Tewa word which Harrington gives for the Pleiades translates "in a bunch." In late June, when Harrington was doing his field work, the Pleiades would have been rising just before sunrise--perhaps the designation as seed stars has to do with the sprouting of plants at this time. The Zuni word given for Pegasus is *'áwi:łaknanneh*, meaning "four big one" or "square." Harrington's identification of the Northern Crown as *móyatchu:ci̋lowah* is puzzling--he does not translate this word, but it literally means "red star." It is difficult to see how this applies to the Northern Crown, which contains no obvious red star. Harrington's translation of the Tewa word for the Northern Crown is "meal drying bowl"--a designation which seems to fit the shape very well.

The word Harrington gives for the Milky Way is *yúpyałanneh*, which translates "ashes placed or stretched across." In one Zuni narrative translated by Tedlock, the Milky Way represents the trail left by Nepayatamu when he travelled across the sky at the time when the Clown Society (*Newe·*k^{w}*e*) was created (1972:130, 132). According to Stevenson, a bar representing the Milky Way extends across the room above the altar of the *Newe·*k^{w}*e* Society (1904:432, Plate CIV). Further, Benedict suggests that the Twin War Gods made the Milky Way during their search for the Corn Maidens by throwing ashes across the sky (1935:I:38). Cushing indicates that the Zuni call the Milky Way "The Great Snow Drift of the Skies" (1901:381). The Tewa have two words for the Milky Way, one meaning "backbone of the universe," and the other meaning "whitishness."

For Cassiopeia, Harrington gives the Zuni word *méla:khweh*, which he translates as "star zigzag." The shape of Cassiopeia would make Harrington's rendering of *méla:khweh* a plausible one, but we have no other evidence to support this translation. As mentioned earlier, *mela* is the word which Newman lists for the Pleiades; however, he gives no other meaning for the word. Harrington states that Cassiopeia is not known to the Tewa.

Bunzel (1932:487), Cushing (1896:392), and Harrington all agree on the word for the Great Dipper, *khwílelekkʸa:khweh*, which translates "seven ones" or "the seven." The Tewa word for this constellation is translated by Harrington as "seven corner."

Finally, the word for the Little Dipper is given as *píclankhwin'úkhwatélen'onah*, which we translate as "those to the north." It is significant that this word contains within it the stem for the direction north (*pič̌le*), for the star at the tip section of the handle of the Little Dipper is the North Star; hence, this constellation always appears in the north. The Tewa translation given by Harrington for this constellation is "the yoke."

In summary, the existence of known Zuni and Tewa constellations seems to belie the usual notions that Puebloans are little interested in the stars, save for a few essential stars or star groupings primarily used for timing ceremonies. The numerous examples of stellar representations in the kiva murals at Awatovi (Smith 1952:243), Pottery Mound (Hibben 1975:48, 49, 82), and Kuaua (Dutton 1963:108-110), as well as those in prehistoric rock art (summarized in Brandt and Williamson 1979 and Koenig 1979), suggest that a stellar tradition is centuries old. Because of these examples, Harrington's data, and the obvious utility of the stars for timing ceremonies and for setting the dates for planting or ritual celebrations, it seems plausible that stars were observed more in historic and prehistoric times by the Pueblo people than has been previously thought. Further attention to the archival and archaeological record for Pueblo interest in the stars seems warranted. It will also be important to interview Pueblo elders to collect their non-sacred star lore.

18 Mesoamerican Archaeoastronomy: Parallax, Perspective, and Focus

Judith Ann Remington

It is not at all surprising that what we now call archaeoastronomy had its true beginning many years ago in Mesoamerica. In the New World it is only in Mesoamerica that we have inherited an independent source of verification of our present-day interpretations of the science of the past. These verifications come to us through the codices and the inscriptions which have survived to the present.

PARALLAX

Parallax, according to Webster's *Unabridged Dictionary*, is "the apparent change in the position of an object resulting from the change in the direction or position from which it is viewed." Using the term "object" rather loosely, parallax describes a situation that we encounter rather frequently in anthropology and, nearly always, in archaeology.

When we study another culture in anthropology, we have to accept as our first operating assumption the notion that we (as representatives of our own culture) are not in the same ontological place or moving in the same epistemological direction as the members of the culture we are studying. If our introduction to that other culture is through an interpreter, especially one from a culture different than our own and from a culture different than the one being studied, we can be sure that we are in a situation where considerations of parallax are appropriate. To the extent that we understand the culture of the interpreter and to the extent that the interpreter understands our culture, we can begin to assess the apparent displacement of the people under study relative to us, and differences in the cultural direction of that people relative to us. Without the two points of view represented by our hypothetical interpreter and ourselves, we are on even shakier ground. In that case, the anthropologist attempts to assume the second point of view while retaining his or her reified cultural outlook as an original comparison point. This is essentially a problem in relativism, and relativistic problems can only be solved in context.

In archaeology the problem is still greater, since only material remains are available for study in most archaeological work. Physical constraints such as terrain, subsistence possibilities, and available materials provide one set of parameters for assessment. Another set of parameters, consisting of factors such as type of social organization, motivation, and ethos, is not represented in middens, articulated stones, or polished bones. Much may indeed be inferred, but it is clearly inference. In anthropology we are able to learn context, to be taught context by the people we study; in archaeology the causal part of context must be inferred. At this point you may say, "Wait a minute. What about cultures that still survive? They may be different from the ancestral form, but they are still the same cultures in an evolved form." Perhaps they are. But what is our fixed point? How do we assess how different they are and how they are different?

In modern anthropology it is at least theoretically possible to learn what other cultures are but, if Feyerabend (1975) is correct, it may be impossible to translate them accurately and meaningfully into the terms of another cultural theory, our own included. If one is bilingual, or even conversationally deft in a language other than one's own, one can testify to the cognitive difference between understanding what is being said in both languages and translating accurately and meaningfully from one into the other. When we talk about other traditions and other formulations in the language of the reality that we assume to lie behind the physical constraints on our survival, we begin to sink into a mire of many possibilities.

In archaeology, we cannot check our inferences about cognitive parameters with the people who operated within those parameters. If there are living cultural descendants we can obtain clues, but we cannot get evidence. If they support our inferences, we must suspect common influences at work; if they deny our inferences, it may mean that concepts present in the past have been lost or replaced.

As in mathematics, the critical step in an anthropological "proof" is finding out how to put the question which makes the sequence of the proof fall into place. Unlike mathematics, in which the context of the problem is defined, research into present and past cultures different from our own occurs in an undefined context. In doing research on the cognitive past, we have three options. First, we may assume a context. In this case, the questions we ask are sure to be comprehensible to us and the answers will be as well. Unfortunately, given an assumed context, we have no way of knowing whether our questions would have been comprehensible to the people whose past we are studying or whether the answers we get would have been at all relevant. We will get answers, but answers of totally uncertain epistemological status.

Second, we may deduce a context based on the available evidence. In this case, the questions we ask will be good sensible questions in terms of our deduced premises. We will have advanced one step further in the pursuit of knowledge, in that we recognize our premises for what they are: our deductions. We are slightly more open to the possibility that our deductions may be wrong and that the answers we get to our questions may be absurdly biased.

Third, we may investigate context. This is by far the most difficult course, for it requires an examination and awareness of our methods, as well as an examination of the materials or the data that we find. We necessarily begin with an assumed context, which we refine deductively through the results of our investigations, and we continue the process until all the evidence (including that which has become evidential as a result of our work) has been explained consistently and coherently.

This, of course, is science. However, it is equally obvious that it is impossible to achieve in a discipline where we will always lack most of the evidence. This is the case in any discipline dealing with the past. The most we can hope for is the best possible synthesis of what is presently available.

After this rather negative discussion of the possibilities in disciplines which study the past, we should be better able to recognize the advantages which archaeoastronomy uniquely enjoys. There are two such primary advantages, one of which until now has been found in the Americas only in Mesoamerica. The first advantage is the regularity of celestial mechanics. We cannot be off by very much in predicting the occurrence of an eclipse, a transit, an occultation, or any other such event in the relatively short time-span within which we deal. The weather can affect the visibility of an event but nothing can interfere with its occurrence. This provides us with one of the two different points we need to measure parallax. The second advantage (found specifically in Mesoamerica) is the continued use of the 260-day calendar, which links us with over 2000 years of days, and provides us with the other point we need. The objects we can thus place in time and space are the astronomical events noted in the codices and inscriptions. The knowledge of how these events were recorded and predicted provides us with hard evidence on the cognitive context in which they were embedded.

Hypotheses based on this evidence may not lead to tight theory of the sort enjoyed by the strictly physical sciences, but they do allow for the elimination of hypotheses stemming from wishful thinking and fantasy. By and large, the study of Mesoamerican astronomy has been free of the worst abuses of speculative fiction, and has generally been exciting and worthy of respect and praise.

PERSPECTIVE

The earliest students of the astronomical knowledge of the ancient peoples of Mesoamerica were the Spanish priests and friars, such as Bishop Diego de Landa and Friar Bernardino de Sahagún. It is ironic that we must begin our attempt to see the history of archaeoastronomy in Mesoamerica in perspective with those names. Both were Spanish men of religion at a time when Spanish Catholicism was particularly parochial. During the age of the Copernican revolution, but within the miasma of the Inquisition and the self-righteous aftermath following the expulsion of the Moors (who at that time were the single major European/Near Eastern carriers of scientific knowledge), the status of scientific knowledge was at a low point in Spain. The only tolerable goals of knowledge were God and gold. For the religious contingent during the Conquest, knowledge was valued only insofar as it aided in the conversion of the native peoples. Sahagún had the grace to recognize complexity and sophistication when he saw it; unfortunately (and understandably), he did not have either the education or the informed scientific curiosity that would have enabled him to compile as rich a record of the scientific knowledge of highland Mexico as he was able to compile of the social organization, economics, and religion of the Nahuatl-speaking people.

Landa also compiled invaluable records of the Yucatecan Maya; he also tantalizes us with half-garbled and heavily-interpreted bits of scientific information. For the modern student, it is nearly intolerable to read:

> These people also made use of certain characters or letters, with which they wrote in their books their ancient affairs and their sciences, and with these and drawings and with certain signs in these drawings, they understood their affairs and made others understand them and taught them. We found a great number of books in these characters, and, as they contained nothing in which there was not to be seen superstition and lies of the devil, we burned them all, which they regretted to an amazing degree and caused them affliction [Tozzer 1941].

Of the seventeen existing codices which are surely prehispanic, ten are directly concerned with auguries, the count of the days, and other astrological/astronomical matters. Although this proportion may be biased, since such codices may have been more carefully protected by their keepers and the elaborately decorative hieroglyphics may have made them more valuable to the Spanish, it still seems significantly large. One of these books, the Mayan Dresden Codex, has been the single greatest source of information about Mesoamerican astronomy to date. Other books, such as the books of Chilam Balam, were literal translations of Mayan historical codices, written in an adapted Spanish alphabet in the century after the Conquest. Much of what is known about the cycles of the calendrical periods was learned from the chronicles in these works.

The Dresden Codex has been known to European scholars since 1739, when it was rediscovered in Vienna and sent to the Dresden Royal Library. It was published for the first time in the 1830s in Lord Kingsborough's *Antiquities of Mexico* (Volume 3). In 1864, Brasseur de Bourbourg published Landa's *Relación de las Cosas de Yucatan*, appended to which was the *Codex Perez*, which in turn contained the *Chilam Balam of Maní*. Many explorers, adventurers and archaeologists, among them Stephens and Catherwood, had published descriptions and drawings of Mayan inscriptions by 1870. By that time, both the Mayan Madrid Codex and the Mayan Paris Codex had been published.

In 1880 Ernst Förstemann, a German archaeologist, began the first serious attempt to decipher the codices; he also published a more accurate reproduction of the Dresden Codex. By 1887 Förstemann had identified the month signs, the signs for 0 and 20, and the 260-day calendar. He demonstrated the modified vigesimal (base 20) numerical system and showed that it was positional. He had figured out the pattern of the Venus table and had recognized it as an ephemeris for Venus. Finally, he showed that the Long Count began from the date 4 Ahau 8 Cumku and explained the ring numbers, which indicate a starting point prior to the base date. At approximately the same time an English archaeologist, A. P. Maudslay, named the set of long dating inscriptions the Initial Series dates, and the name stuck. Between 1889 and 1902, Maudslay published his five-volume work, *Archaeology*, which made good reproductions of the inscriptions available for the first time.

During the 1890s, J. T. Goodman, an American archaeologist, used copies of Maudslay's reproductions to decipher the time count. Both Förstemann's and Goodman's results agreed. Brinton's *Primer of Maya Hieroglyphs*, published in 1895, contained an early summary of Förstemann's work as well as reports of his own researches. Goodman published *The Archaic Mayan Inscriptions* in 1897, and he also summarized his own and Förstemann's work.

(Förstemann's success continued until his death in 1906. In the years after his explication of the Long Count, he identified the major time-period glyphs, the lunar ephemeris represented by the eclipse table, and the head variant forms of the time periods.) Among Goodman's other contributions were the identification of the numerical head variant glyphs and two partial time glyphs. Finally, in 1905 Goodman suggested the 11.16 correlation, which is presently the most commonly accepted; today it is called the Goodman-Martinez-Thompson (or GMT) correlation in honor of the three scholars who contributed most to its development. Although many other scholars were active in this first phase of the study of Mesoamerican astronomy, the completion of this aspect of the work was signalled by Charles Bowditch's publication of *The Numeration, Calendar Systems and Astronomical Knowledge of the Mayas* in 1910.

This early phase was exploratory. It was assumed only that the books would not have been written nor the elaborate inscriptions carved had they not contained information of some importance. Context was completely unknown when the work began. Once the numbers were identified as such, it was a relatively straightforward task to identify their values, the base, and the positional system employed. Context here was independent of culture because, regardless of base or notational system, numeration is tied to quantities. In the same way the number 584, no matter how written, is recognizable to one familiar with astronomy as the synodic period of the planet Venus (i.e., the time Venus takes to return to the same position in the sky relative to the sun). The values 177 and 148 are similarly seen as good estimates for six and five synodic periods of the moon, respectively, and their co-occurrence provides contextual support for the identification. In other words, some numbers carry with them their own contexts for inhabitants of this planet. The base unit, the day, is common to all of us. The synodic periods of the moon and the planets, expressed in days, are part of our shared human experience. Changes in these periods, even over millenia, are insignificant.

The early students of Mesoamerican writings had been able to assume context because of the uniquely-identifiable physical constants of the solar system. From these constants, it was possible to deduce textual content to some extent, and to generalize from clearly astronomical texts to less obviously astronomical texts. But in the first two-thirds of this century, the emphasis in Mesoamerican astronomical studies has shifted in focus from the identification of the kinds of things that had been known to the interpretation of their cultural meaning. The implications of this shift for archaeoastronomy are great.

By 1920, the absolute quantity of information available was considerable. Although anthropologists, archaeologists, and art historians did not have an actual monopoly on the information, they did have some control over what information was made available to scholars outside of the field. At this point, the deduction of context produced different theories, depending on the point of view of the scholar. Within the field, the theories were defined by at least three dimensions, which can be characterized as ultimate origin, local origin, and system.

Few Mesoamericanists seriously considered the possibility of major civilizing influences coming from outside of the Americas; instead, intense controversies arose over the allowable degree of minor external influence and the amount of diffusion permissible within the Americas. The Spinden hypothesis, which suggests a common archaic Mesoamerican origin for civilization in the

Americas, was fiercely debated, rejected, and reconsidered several times. A large amount of semantic energy was also spent on the word "civilization," which Old Worldites defined one way and New Worldites defined another.

Within Mesoamerica, the major argument raged over the lowland or highland origins of civilization. At the time, the arguments were based on over-interpretations of inadequate data and served ultimately only to create a rather sharp division between Mayanists and Mexicanists. This argument fed into arguments about the type of social organization characteristic of the ancient Mesoamericans. The most extreme form of this argument contrasted benevolent, peaceful, and scientific theocrats with dictatorial, militaristic, and elitist users of supernatural mystification. Obviously, semantics again played a larger role than available evidence.

During the period in which these dichotomies were taken seriously, Sylvanus Griswold Morley, Herbert Joseph Spinden, and J. Eric S. Thompson dominated the visible field of the investigation of the significance of astronomy in Mayan Mesoamerica. Morley was a field man; his forte was dates, and he was responsible for the discovery and decipherment of the bulk of the hieroglyphic texts available at the time. His interest in the correlation question led him to collaborate with R. K. Morley and Robert W. Willson in the determination of the function of the supplementary series (i.e., a lunar position counting system). Although he established many glyph identifications, the primary importance of his work lay in his use of current ethnographic data to interpret archaeoastronomical and historical records. Unfortunately, because of his friendship with and liking for the Yucatecan Maya with whom he worked, he became something of an apologist for their rather ethnocentric view of history.

The problem of the correlation of the Mesoamerican and European calendars resulted in theoretical positions that delayed the development of an objective archaeoastronomy. Primarily because of conflicts between the adherents of the GMT (the foremost of whom was of course Thompson himself) and the Spinden correlations, the degree of technical skill and conceptual mastery with which scholars in the European tradition were ready to credit the Mesoamerican astronomers of the past was consistently underestimated; however, this was also partly because of trends within Western science. As technical skills began to match conceptual skills, the values of logical positivism gained ground, and the testability of hypotheses came to be emphasized over the utility, plausibility, and consistency of hypotheses.

The GMT versus Spinden correlation controversy was literally a comedy of errors. Both men fudged their calculations, but Thompson justified his results by denigrating both the arithmetic skill and the scientific role of astronomy in ancient Mayan culture. Spinden was already unpopular with the academic community, in part because of the Spinden Hypothesis and in part because of his rather ungentlemanly defense of his correlation. Actually, the GMT does not fit the astronomical evidence very well. It usually requires a fairly major "amendment" of the text. For example, in discussing the Venus table, Thompson states:

> The figure 9360 is my own amendment, for the number is actually written as 9100 (1.5.5.0). However, there are good grounds for supposing that this is incorrect. . . . In two or three cases in the tables of Dresden there are apparent mistakes which can best be corrected by the addition or subtraction of 260 days [1960:225].

Even with this amendment, the correlation does not fit the data well.

The steady decay in the perception of scientific accuracy can be demonstrated by a few quotations. In 1906, Zelia Nuttall stated:

> Although the foregoing illustrations constitute merely a part of the material I have been collecting on the subject, they will suffice to establish beyond a question the hitherto undemonstrated fact that the ancient Mexicans not only employed their carefully oriented temples and ball-courts as astronomical observatories, but also invented ingenious devices for accurately registering the periodical appearances or disappearances of important celestial bodies [1906:295].

In 1946, Morley was a little less definite:

> It may be asked how the ancient Maya achieved such a high degree of astronomical accuracy without the instruments upon which modern astronomers depend. However, if the lines of sight are sufficiently long, accuracy to within less than a day's error may be achieved in fixing the synodical or apparent revolution . . . of many of the heavenly bodies. Maya temples are sufficiently high to obtain clear lines of sight from their summits to distant points on the horizon [1946:257].

Even Spinden, in 1940, was backing off from the requirements of purely scientific astronomy:

> . . . The Maya Indians had a high self-contained civilization, a truly primary civilization distinguished by strikingly original arts and sciences. They had their own special system of hieroglyphs for inscribing texts on blocks of stone and in paper books, their own notation of numbers in place values; their own calendar and long-range chronological machine. With clear means of record they evolved a true science of astronomy which, being only human, they combined with a pseudo-science of astrology [1977:162].

But it was left for Thompson to administer the final blow:

> The high hopes entertained for decisive results from a strictly astronomical approach have not been fulfilled. I think that is because the Maya priests did not use astronomy as an exact science. Instead, they fashioned it to their mystical and poetical approach. Associations of celestial phenomena with lucky or unlucky days, or connections deep in mythology were, I am sure, of more importance than an exact record of when they occurred. Since these are the methods of the astrologer rather than of the astronomer, the precise technique of the latter is often of little value in augmenting knowledge of the meaning of the inscriptions [1960:33].

This statement is misleading in that, until very recently, a strictly astronomical approach to the calendrical questions has not been taken. Both Thompson and Spinden reified a deduced context for Mesoamerican astronomy: Thompson's "mystical and poetical approach" and Spinden's "true science of astronomy . . . combined with a pseudo-science of astrology." Perhaps Thompson was too willing to accept a truly remarkable degree of inaccuracy from priests who dabbled in astronomy, and perhaps Spinden was too unwilling to consider the possibility that the priestly side of the astronomers' jobs may have demanded considerable mystification.

When outside scholars, particularly professional and amateur astronomers, were brought into the several battles that raged around the correlation, they were deployed strategically. The biases of the anthropological experts tended to determine what information was considered relevant and what information was considered reliable. The astronomers were dealt data from stacked decks--one hesitates to suggest conscious intent.

Robert W. Willson (the astronomer who, with the Morleys, identified the supplementary series as lunar) was one of the first, if not the first, to realize that the lunar table in the Dresden Codex was an eclipse table. Because of the advice of archaeologists, he limited his search for an eclipse series to a time range which may not have been sufficiently large. In addition, his search was predicated on the assumption that the table was based on a single observation series rather than a compilation of observations.

John E. Teeple, a chemical engineer and amateur astronomer, continued Willson's work on the supplementary series and the eclipse table. He was able to identify the node days of the table and to recognize the recession of the node days through the double 260-day calendar. However, he was not encouraged to investigate the possibility that the series of dates in the table may have indicated Mayan recognition of this recession in a highly sophisticated way.

Maud Makemson, another astronomer, was discouraged from considering the possible association of the Venus and the eclipse tables indicated by the tenth picture in the eclipse table. As Thompson says,

> The tenth picture, which ends the table, is that of a monster who plunges head first from a celestial band. . . . The head of the monster is hidden by a large glyph of the planet Venus. One is instantly reminded of the Aztec belief that during eclipses the monsters called Tzitzimime . . . plunged earthwards from the sky. . . . It has been supposed that the presence of the Venus monster in this picture indicates some relationship with the revolution of that planet. It is very much more probable, however, that . . . the synodical period of Venus in no way enters into the calculations [1960:233].

One more comment about the Spinden/Thompson correlation battle is relevant here. The GMT correlation was accepted less on its own merits than the Spinden correlation was rejected because of the unpopular ideas and attitudes of its major proponent. The first C^{14} dates tended to support the Spinden correlation, then well out of favor. As the C^{14} dating techniques were perfected, appropriate correction factors were developed. The appropriateness of the correction factors, however, was judged at first by how well they corresponded to the GMT. One wonders what would have happened had the possibility of an alternate correlation been more widely considered. As Coe somewhat ironically noted, when the latest run of C^{14} dates supported the GMT,

> Most Mayanists have given sighs of relief, for any other chronology would play havoc with what we now think we know about the development of Maya culture over two millenia. Even further, any displacement in the dating of the Maya Classic Period would disrupt the entire field of Mesoamerican research, for ultimately all archeological chronologies in this part of the world are cross-tied with the Maya Long Count [1966:29].

A determined and often defiant adherence to assumptions that were no longer tenable was characteristic of the second generation of Mesoamericanists. As the context of the knowledge of ancient populations became better known, it became possible to deduce what could have been possible; such deductions were often identified with the scholar who proposed them and the scholars who defended them--a case of "love me, love my theory." The evidence available was not sufficient (and still may not be) to raise any of the hypotheses to the status of theory. New discoveries, such as the Mayan frescoes at Cacaxtla, wreak havoc with old hypotheses. Nonetheless, the hypotheses were presented as theories and defended fiercely, to the detriment of the study of the scientific knowledge of the inhabitants of prehispanic Mesoamerica.

Quine's *From a Logical Point of View* (1953), Kuhn's *The Structure of Scientific Revolutions* (1962), and books by many other writers and thinkers led to a dramatic forward step in understanding the nature of Western science. It became possible for scientists and philosophers, historians, and sociologists of science to look critically at their own knowledge and to question its absolute validity. Ultimately, the very foundations of knowledge were questioned. As part of this intellectual climate, anthropologists began to reexamine their own work and saw that much of it had been based on unquestioned assumptions and epistemologies that were now becoming suspect.

In the post-Kuhn years, it became acceptable to question not only the methods of scientific investigation but also the assumptions underlying the methods and the very epistemological values and ontological definitions which gave them rise. It thus became acceptable to consider the possibility of truly alien methods of organizing the "same" data. With this, the third generation of Mesoamerican studies came into being. Enough information was available to begin to investigate assumptions, and scholars were willing to do so. Archaeoastronomy, as a discipline, became possible.

There have always been scholars who are willing to go against the perceived wisdom, but it is usually difficult to learn about their work. It may have been inevitable that, when the archaeoastronomical breakthrough came, it did not come about through the astronomical knowledge written in the codices nor through the astronomical events commemorated by the inscriptions, but rather through the investigation of the orientations of the buildings and sites themselves and through the informed consideration of alternative interpretations of the inscriptions. Proskouriakoff's recognition in 1960 of the historical rather than mechanistic nature of many of the inscriptions led to a resurgence of interest in Mesoamerican investigation. Not only was a fresh out-look possible, it was more than welcome after the decades of polemics. The Olmecs were suddently "discovered," nearly 20 years after Covarrubias (1942) had declared their existence a necessity. The exciting work at La Venta, in Tehuacan, in Teotihuacan, and in Oaxaca, as well as changes in both anthropological and archaeological field techniques, led to an introduction of scholars into the area who were not content merely to receive wisdom from their elders.

Many of us have been involved in this renaissance, either as active participants or as witnesses. Those of us who had already committed ourselves to Mesoamerican anthropology and archaeology had discussed Nuttall's theory (1928) of the zenith-passage origin of the 260-day calendar and, therefore, its connection with the latitude of Copan. We were also familiar with the apparent similarity in the orientations of many of the early sites, and we were suspicious of oddball buildings which were oriented quite differently from other buildings at a site. We had heard stories of worked magnetite found in suspicious places, and we wondered why the Olmec mirrors had been so carefully ground. When we talked about Morley's suggestion (1946:260) that the objects dangling from the celestial band in the Paris Codex represented the Mayan Zodiac, we wondered what constellations the ancient people of Mesoamerica recognized. We wanted to find out.

However, at that time (in anthropology at least) such questions were considered to be amusing problems, perhaps good for a term paper, but definitely not to be undertaken as serious research. Work on the correlation was anathematized; I was told that it led to madness (that may be true). Furthermore, it was unpopular. Speculations about technology (such as whether or not there were compasses or telescopic mirrors or abaci), about architecture (such as whether or not there were true astronomical observatories), about orientations (such as whether or not axes were aligned according to astronomical or geometric plans), and about survivals (such as whether or not living descendants of the ancient Mesoamericans knew any of the answers to these questions) were not too well received. Such speculations were considered to be dangerously close to talking about the mystical properties of pyramids, the advent of alien cosmonauts, or the search for the lost tribes of Israel.

Talking about the possibility of diffusion was not quite nice; Jomon pottery was discussed in hushed tones, and the suggestions of connections between the calendars and constellations of the Americas and Asia was considered to be in rather bad taste. As a friend said, "We can either count tortillas or set up non-semantic ratios based on French symbolism."

Fortunately, a few astronomers, historians, architects, and other such types who specialize in perspective decided to become cultural relativists. Their motto seemed to be, Europeans are Europeans, Mesoamericans are Mesoamericans, but astronomers are astronomers." Within a few years, many previously radical ideas had been considered seriously for the first time--such as the idea that Mound J could have been deliberately oriented for astronomical purposes (Aveni and Linsley 1972), that the Caracol was a measuring device (Aveni, Gibbs, and Hartung 1975), that a ruler of Palenque could have been buried in such a way that his tomb was given astronomical significance (Schele 1977), that the Castillo of Chichen Itzá could have been an equinox marker (Rivard 1970), that the pecked cross petroglyphs of Teotihuacan may have served as architects' bench marks (Aveni, Hartung, and Buckingham 1978), that Olmec city planners may have used compasses (Carlson 1975), that whole sites were surveyed and planned in advance for cosmological reasons (Hartung 1974), and that modern descendants of the holders of Mesoamerican high cultures may still have valuable knowledge (Neuenswander 1978). Furthermore, Michael Closs (1977) can now teach us a lot of important mathematics, David Kelley gets listened to when he talks about Asia (Mukerji and Kelley 1978), and Floyd Lounsbury can give us ideas about the cognitive significance of numbers and cycles (1973).

Even the correlation question seems to have opened up again. At the American Anthropological Association meetings in November of 1978, Neuenswander of the Summer Institute of Linguistics gave a paper in which she challenged Thompson's ascription of the meaning of completion to the variant forms of the lunar series Glyph D. She showed that on the basis of ethnographic evidence obtained from contemporary Achi Maya in Guatemala, lunar age is shown by hand positions which correspond to the hand positions found on the variant forms. She concluded:

> Since time is almost always counted by the period just begun, rather than the one just completed, it would appear that the array of diverse hand positions proposed by Thompson and others, all meaning "completion," is not realistic in the light of contemporary Achi structures [1978].

While this is very exciting news to those of us still working on the correlation problem, as it may shed light on the question of lunar position in the several competing correlations, it has a more general significance as well. Ethnographic evidence provided by our Indian contemporaries in Mesoamerica, and in the Americas in general, may help us solve problems with which we cannot otherwise deal. This is only in part fortuitous. Until recently, we did not know what questions to ask or how to ask them. Because of contributions made in the past hundred years, because we can validly assume that we are dealing with a limited context thanks to the astronomical constants of our solar system, and because we can learn from the deductive failures of the serious scholars of the past, we are now in a position to investigate context.

FOCUS

There are quite a few of us who want to investigate context; in fact, there are many more than the few I mentioned earlier. We are beginning to learn how to get at parallax, we are beginning to get some perspective, and we are beginning to find our focus. Perhaps we should say "our foci." Because the field is interdisciplinary, archaeoastronomers tend to waver back and forth between archaeology, anthropology, and astronomy. While this does provide the stereoscopic vision needed for measuring parallax and providing perspective, it occasionally leads to a rather fuzzy focus, a kind of diachronic aberration.

Although anthropologists have a tendency to sin theoretically when a pet bias gets rubbed the wrong way, they are useful creatures to have around when there is a question of cultural context. Although archaeologists have a tendency to get uncomfortable when asked to speculate about intangibles, they are useful creatures to have around when intangibles can be demonstrated through material context. Although astronomers have a tendency to take wild leaps into the great beyond of cosmological theory, they are useful creatures to have around when the practical possibility of an event, practice, or technique can be demonstrated in a technical context. Together, focus can be achieved; the work of the past decade has proven this.

In the near distance, archaeoastronomy faces problems. There is still a tendency to think of the field as first cousin to the occultist fringe. It is taken most seriously, perhaps, by astronomers and historians, because the scientific and philosophical importance of cosmology is clear to them. For anthropologists, the need to attain scientific respectability and funding tends to encourage studies which can be shown to be immediately practical. The combination of disciplines is our best defense against getting too wild-eyed, and a conscious openness to new ideas is our best defense against an overly cautious self-definition.

In the middle distance, archaeoastronomy faces many fascinating puzzles. We may deal with correlation questions, diffusion questions, transmission questions, technology questions, and chronology questions, among many others.

In the far distance, archaeoastronomy faces a single, major significant problem: that of establishing criteria for the development of a cross-cultural epistemology which is able to assess the science of the past and the science of the present, with the goal of enriching the science of the future. All of the pristine civilizations of which we have knowledge developed astronomy; perhaps looking out at the universe allows humans to see themselves as creatures capable of encompassing space through knowledge. Anthropologists learn to look more closely at contemporary others in order to see themselves more clearly as representatives of a single species. Archaeologists learn to look at the past in order to illuminate the present, to find temporal perspective.

Because of hieroglyphics, Mesoamerica became the first locus of archaeoastronomical study in the Americas. Because of the history of Mesoamerican investigations, with its attendant failures and achievements in understanding and method, Mesoamerican archaeoastronomy has also provided the first clear focus.

19 Numerology and the Astronomy of the Maya[1]

John B. Carlson

> Surely, to comprehend the aims, attitudes, and achievements of Maya astronomers who were first and last priests, an understanding of Maya mentality and outlook is essential. One must appreciate the impact of the divinatory aspects of the 260-day sacred almanac, and never lose sight of the fact that the ends of Maya astronomy were not scientific, but astrological. The Maya were interested in heliacal rising of Venus because then the world was in danger, and carefully recorded that happening; invisible conjunctions of planets with the Sun had no bearing on man's future, and therefore there was no call to try and calculate them. One must try and get in the skin of the Maya priest-astronomer. Also, some knowledge of Maya culture and history is essential. The findings of archaeology cannot be ignored. Had astronomers interested in Maya astronomy such a background, we would be spared assertions . . . which one can only designate as highly intemperate. . . . In light of such deductions . . . one is inclined to say that Maya astronomy is too important to be left to the astronomers [J. E. S. Thompson 1974:97].

The astronomy, mathematics, calendar, and recorded history of the Classic Maya are parts of a complex, numerologically-interrelated whole. These numerological relationships have their origin in (synodic) astronomical periods and other natural cycles and in Maya perceptions of the geometry of space and the human form. Most of our knowledge of the nature of Maya numerology comes from analyses of the surviving Post-Classic Maya books, such as the Dresden Codex (Thompson 1972), and from the Classic Period monumental inscriptions. The numerological content of the calendric portions of the Classic inscriptions has been largely understood since the middle of this century (see Thompson 1971), but the true nature and contents of the full texts are only now becoming clear.

As a result of the exploratory work of Proskouriakoff (1960, 1961), it is now known that the monumental inscriptions contain the recorded dynastic history of the Classic Maya elite. Following this pioneering work, which involved analyses of dates at Piedras Negras and Yaxchilan (Proskouriakoff 1963,

---

[1]I gratefully acknowledge the inspiration and guidance of Floyd G. Lounsbury who so generously has shared with all of us his unique feeling for Maya mathematical and calendrical astronomy. I first began to learn that "Maya astronomy is too important to be left to the astronomers" with the encouragement of Professors Lounsbury and Michael Coe of the Yale University Department of Anthropology. The generous support of a Research Fellowship from The Tinker Foundation made this study possible.

1964), a great many discoveries have been made regarding both the historical and mythological dynastic sequence at Palenque. Of particular relevance to the numerology of these inscriptions is Lounsbury's "A Rationale for the Initial Date of the Temple of the Cross at Palenque" (1976). The essence of this work is summarized in his major synthesis, "Maya Numeration, Computation, and Calendrical Astronomy" (1978). Lounsbury has succeeded in demonstrating that some of the "long reckonings" or "distance numbers" in Maya calendrical computations are "contrived numbers" composed of numerologically-important Maya factors. The historical events commemorated in an inscription may be related by these numbers to "like-in-kind" events in the mythological past or future. An example is cited from the Temple of the Cross at Palenque, where the birth date of the powerful ruler Lord "Shield" Pacal is related by a highly contrived distance number to an event just before the beginning of the current Maya era. The event is declared to be the birth of a female personage, an "ancestral mother" of the Palenque ruling dynastic lineage. The two events have the same position in an entire complex of significant Maya calendric cycles: 4 color-directions; 9 Lords of the Night; 5, 7, 13, and 20 days; and greater periods such as the 260-day Almanac, the 364-day Computing Year, the 780-day Triple Almanac, and the 819-day, 1820-day, 2340-day, 3276-day (see Berlin and Kelley 1961), and 16,380-day cycles. Table 19.1 contains a more complete description of these periods and their interrelationships.

Concerning this contrived number, Lounsbury concludes:

> It seems pretty clear what we are dealing with here. Surely it is a matter of mythic genealogical and numerological charter for the position of a King, and through him perhaps for his successor and the dynasty that he founded. If the mythical ancestor himself (or herself?) was not a fabrication for the purpose, then at least the ascription of his or her birth date was a bit of numerological manipulation for just that purpose--possibly tampering with an earlier tradition as to the day on which that mythical being came into existence [1976:218].

A similar situation exists in the Venus Tables of the Dresden Codex (Thompson 1972:62-71). On page 24, the Maya scribe has written a "ring number" which designates a special date 2200 days before the beginning of the current Maya era on 4 Ahau 8 Cumhu. This new "ring number base" date is then related by a highly-contrived distance number to the historical starting date of the Venus calendar itself.

Thus, in addition to the obvious possibilities for forecasting the most auspicious days for certain events to take place, such as the accession of a ruler, this system would have permitted the Maya priests/historiographers to tamper with or rewrite history after the fact. From an analysis of the Maya evidence, in conjunction with several cross-cultural comparisons, I have concluded (Carlson 1978) that we should be exceedingly careful in attempting to interpret the Maya monumental record as "history" as we might perceive it.

The evidence strongly indicates that the real purpose of Maya calendrical astronomy was to place man and historical events in harmonious context within the endless cycles of celestial bodies and of time itself. Astronomical knowledge and calendrical arithmetic were used for astrological divinations of the future. By understanding the patterns in planetary motions, the calendar priest-astronomers could predict future configurations and arrange for the propitious timing of rituals. They calculated eclipse-possible periods, tabulated the heliacal appearances and disappearances of Venus, computed the conception and birth-related cycles of the moon, and recorded them all in the context of human events on the monumental stelae.

There is also good evidence that transient phenomena such as comets, monstrous births, and unusual meteorological events were noted and interpreted as omens or portents. Maya cosmology embodied the mutual interaction of man and an animistic natural world, all within a cyclical conception of time; the celestial bodies and numbers themselves were deified anthropomorphic figures. The predictive "sciences" of astronomy, astrology, numerology, and divination must have been inseparable. A powerful, educated priesthood served a vital bureaucratic function of manipulation, mediating between the dynastic ruler, the people, and the personified forces of nature.

In short, the intent of Maya calendrical astronomy and numerology must surely have been to invoke an auspicious fortune, avoid the dire consequences of certain planetary aspects, and obtain cosmic sanction for human undertakings. In the present brief study, several examples of this Classic Maya numerological order--natural or contrived--are explored through an analysis of Maya celestial and calendric cycles.

THE NUMEROLOGICAL FOUNDATIONS OF MAYA CALENDRICAL ASTRONOMY

The calendrical numerology of the Maya inscriptions and codices is based on ritualized synodic periods of celestial bodies and the perceived geometry of space and the human body. Ultimately, man was the measure of all things.

The biaxial symmetry and sexual dimorphism of *Homo sapiens* form the basis of the duality which pervaded Mesoamerican cosmology. This duality was reinforced by the astronomical oppositions of Sun and Moon, day and night, "Morning Star" and "Evening Star," rise and set, and rainy and dry seasons. Life and death form the conceptual foundation for all of these dualities, human and cosmological. The sun, for example, was born in the morning and journeyed to its death at sunset; its skeletal, underworld form was reincarnated at dawn.

The *four-fold* symmetry of the human form yields the divisions of space and time into four quarters. The vigesimal counting system with its base-five subsystem undoubtedly reflects the five digits on each of the four human extrema. The 20-day "month" or *uinal*, the primary period of vigesimal completion, finds its formal counterpart in the Maya word for man, *uinic*. The four-part body along with the head, recapitulates the system of the hand and yields the five color-world-directions: the four "orthogonal" prime directions and the center.

Seven, nine, and thirteen were also vital numerological prime factors for the Maya and for Mesoamericans in general. Speculative geometrical constructions for all three numbers have been offered (see Hunt 1977:Ch. 6), but only the seven-fold symmetry of four directions, center, zenith, and nadir has real confirmation. The number seven probably relates directly to the seven classical planets or "wanderers": Sun, Moon, Mars, Mercury, Jupiter, Venus, and Saturn. The "magic numbers" nine and thirteen are associated with the

TABLE 19.1

NUMERICAL FACTORS AND PERIODS IMPORTANT IN MAYA CALENDRICS[a]

Note	Name of Period	# Days/ Other Units	Maya Equivalent	Factor Decompositions	Prime Factors
	(Prime-2)-Duality	2	.2	2	2
	(Prime-3)	3	.3	3	3
	Four World-directions	4	.4	2·2	2^2
	Five World-directions (Prime-5)	5	.5	5	5
	Seven Classical Planets? (Prime-7)	7	.7	7	7
*	Venus invisibility at Inferior Conjunction	8	.8	2·4	2^3
	Glyph G cycle Lords of the Night (Bolon-ti-ku)	9	.9	3·3	3^2
	Trecena (Oxlahun-ti-ku)	13	.13	13	13
	Veintena Uinal	20	1.0	2·10=4·5	$2^2 \cdot 5$
?	Two Trecenas (Half Calendar Round)	26	1.6	2·13	2·13
*	29-day Lunation	29	1.9	[29.530588] 29	29
*	30-day Lunation	30	1.10	[29.530588] 3·10	2·3·5
	Calendar Round	(52 yr.)	2.12	4·13=2·26	$2^2 \cdot 13$
	Five Trecena Cycle	65	3.5	5·13	5·13
	(Prime-73)	73	3.13	73	73
*	Palenque Moon Ratio [81 lun.=2392 days]	(81 lun)	4.1	9·9	3^4
*	Venus invisibility at Superior Conjunction	90	4.10		$2 \cdot 3^2 \cdot 5$
*?	[Mercury Synodic Period]	116	5.16	[115.88] 4·29	$2^2 \cdot 29$
	Cycle associated with Rain God [Mercury Synodic Period?]	117	5.17	[115.88] 9·13	$3^2 \cdot 13$
*	Venus visibility as Morning Star	236	11.16		$2^2 \cdot 59$
*	Venus visibility as Evening Star	250	12.10		$2 \cdot 5^3$
	Sacred Almanac "Tzolkin"	260	13.0	9·29-1=4·65=13·20	$2^2 \cdot 5 \cdot 13$
*	Tun Chronological Year	360	1.0.0	18·20	$2^3 \cdot 3^2 \cdot 5$
*	Computing Year	364	1.0.4	4·91=13·28=7·52	$2^2 \cdot 7 \cdot 13$
*	Haab, "Vague Year" Calendar Year	365	1.0.5	[365.2422] 4·91+1=18·20+5=5·73	5·73
*?	[Saturn Synodic Period]	378	1.0.18	[378.09]	$2 \cdot 3^3 \cdot 7$
*?	[Jupiter Synodic Period]	399	1.1.19	[398.88]	3·7·19
	Twenty Squared	400	1.2.0	20·20	$2^4 \cdot 5^2$
*	Maya Eclipse Cycle	(405 lun)	1.2.5	5·81=3·135	$3^4 \cdot 5$
*	Double Sacred Almanac Triple Lunar Half-Node Cycle	520	1.8.0	[173.30906] 3·(173.3) = 2·260	$2^3 \cdot 5 \cdot 13$
*	[Venus Synodic Period]	584	1.11.4	[583.92] 8·73	$2^3 \cdot 73$
	585-day Cycle	585	1.11.5	584+1=5·9·13=5·117	$3^2 \cdot 5 \cdot 13$
*	Half Year Drift Cycle	(754 yr)	2.1.14	½(1508 cal. yr.) = (14.5)·52 yr.	2·13·29
*	Triple Sacred Almanac [Mars Synodic Period]	780	2.3.0	[779.94] 3·260	$2^2 \cdot 3 \cdot 5 \cdot 13$
	819-day Cycle	819	2.4.19	3·273=13·63=9·91= 7·117	$3^2 \cdot 7 \cdot 13$

TABLE 19.1--cont'd.

Note	Name of Period	# Days/ Other Units	Maya Equivalent	Factor Decompositions	Prime Factors
*?	Year Drift Cycle	(1507 trip. yr)	4.3.7	11·137	11·137
*?	[1507·(365.2422) ≃ 1508·365] Year Drift Cycle	= 550420 (1508 cal yr.)	4.3.8	2·754=29·52	2 ·13·29
	Seven Sacred Almanac/ Five Computing Year Cycle	1820	5.1.0	7·260=5·364	$2^2 \cdot 5 \cdot 7 \cdot 13$
	Nine Sacred Almanac/ Twenty 117-day Cycle	2,340	6.9.0	=4·585 20·117=20·9·13 =9·260	$2^2 \cdot 3^2 \cdot 5 \cdot 13$
*	Palenque Moon Ratio [81 lunations]	2,392	6.11.12	81·(29.53086)	$2^3 \cdot 13 \cdot 23$
*	Five Venus Years/ Eight Calendar Year Cycle	2,920	8.2.0	5·584=8·365	$2^3 \cdot 5 \cdot 73$
	Four 819-day Cycle	3,276	9.1.16	9·364=4·819	$2^2 \cdot 3^2 \cdot 7 \cdot 13$
	Katun	7,200	1.0.0.0	20·20·18=20·360	$2^5 \cdot 3^2 \cdot 5^2$
*	Maya Eclipse Cycle [405 Lunations]	11,960	1.13.4.0	46·260=5·2392 5·81·(29.53086)	$2^3 \cdot 5 \cdot 13 \cdot 23$
	Twenty 819-day Cycle	16,380	2.5.9.0	=45·364 5·3276=20·819	$2^2 \cdot 3^2 \cdot 5 \cdot 7 \cdot 13$
	52 Tun/72 Almanac Cycle	18,720	2.12.0.0	52·360=72·260	$2^4 \cdot 3^2 \cdot 5 \cdot 13$
	Calendar Round 52 Haab/73 Almanac Cycle	18,980	2.12.13.0	52·365=73·260	$2^2 \cdot 5 \cdot 13 \cdot 73$
*	Great Venus Cycle 104-year Cycle	37,960	7.5.8.0	2·18980 = 65·584 2·52·365=104·365 2·73·260=146·260	$2^3 \cdot 5 \cdot 13 \cdot 73$
	Baktun	144,000	1.0.0.0.0	20·20·20·18=400·360	$2^7 \cdot 3^2 \cdot 5^3$
*	Half Calendar Year Drift Cycle	275,210 (754 yr)	1.18.4.8.10	365·754=365·26·29	2·5·13·29·73
*?	Calendar Year Drift Cycle: (1508 Calendar Years) (1507 Tropical Years) (14.5 Great Venus Rounds) (29 Calendar Rounds)	550,420	3.16.8.17.0	[(1507)·(365.2422)]≃ 1508·365 2117·260 4745·116 (942.5)·584 18980·29 29·18980 (14.5)·37960	$2^2 \cdot 5 \cdot 13 \cdot 29 \cdot 73$
*?	Double Calendar Year Drift Cycle: (29 Great Venus Rounds) (58 Calendar Rounds)	1,110,840	7.12.17.16.0	2·365·1508 = 2·550,420 1885·584 3016·365 4234·260 9490·116 18980·58 37960·29	$2^3 \cdot 5 \cdot 13 \cdot 29 \cdot 73$
	Thirteen Baktun	1,872,000	13.0.0.0.0	13·20·20·360	$2^7 \cdot 3^2 \cdot 5^3 \cdot 13$

[a]This table contains a listing of many of the numerical factors and periods of time that were important in Maya calendrical numerology. An asterisk in the "Note" column indicates that the number was probably inspired at least indirectly by some astronomical considerations. A question mark indicates those periods or numbers for which there is only tentative support regarding their use by the Maya. The numbers in brackets are the modern values for the astronomical periods to at least two decimal places. The "Maya Equivalent" column contains the number of days for each period of time in Maya Long Count notation.

nine Lords of the Night or Underworld (the Bolon-ti-ku that presumably form the G9 cycle) and the thirteen Lords of the Day or Heavens (the Oxlahun-ti-ku that form the "Trecena"). These temporal cycles are also related to a spatial system of vertical layering. Some accounts describe an underworld with nine layers, an earth with seven layers, and an upperworld of thirteen layers. An alternative description portrays this structure in pyramidal form: a five-layer step pyramid with nine steps up and down for the underworld and a seven-layer step pyramid with thirteen steps up and down for the underworld. The celestial bodies were supposed to follow this stepped course on their diurnal cycle of birth, growth, death, decomposition, and rebirth. All three of these prime structural factors--seven, nine, and thirteen--were combined to form the grand Maya dynastic cycle of 819 days.

Most of the greater Maya numbers and periodicities are composed of these low primary factors. The 260-day "Sacred Almanac" is generated by the 13 coefficients ("Trecena") cycling through the 20 named days ("Veintena"). It is the most pervasive and fundamental calendric cycle in this next echelon of complexity. The 260-day Almanac is not obviously astronomically inspired, but it does play an essential role in Maya astronomy.

It has been suggested independently by several students of the Maya that the 260-day Almanac originally derived from the 260-day interval between solar zenith passages as observed from a latitude of 15° North--the latitude, for example, of Copan and Izapa. This and other speculations connected with the zero date of the Maya Long Count seems unlikely to the present author. It seems more reasonable that the 260-day Almanac gained its ultimate prominence for two other reasons.

First, this interval is the approximate length of human gestation. In a ritual sense, the day of conception would have had the same name and associated auguries as the day of birth. Moreover, there was a virtually pan-Mesoamerican tradition of naming an infant after the birth date in the 260-day Almanac. This was of great consequence because of the cabbalistic and astrological divinations associated with that specification of the date. In this sense, the Almanac may also have functioned as a canonical conception horoscope, as it seems to function in some Maya villages today. Further support for this hypothesis of the Divinatory Almanac as gestation cycle comes from an understanding of the moon in the Maya world. The waxing and waning of the lunar goddess (Ix Chel) has always had a continuing association with conception and fertility. Neuenswander (1978) describes many "vestiges of early Maya time concepts in a contemporary Maya (Cubulco Achi) community." These include the chronology of periods of menstruation and pregnancy according to a lunar calendar. The 260-day Almanac is, after all, very close to nine lunations or nine female menstrual cycles (9 x 29 = 261). The contemporary Maya practice of counting the period of pregnancy in moons may shed light on the use of lunar data as a usual and important portion of any monumental calendric record. The events described in these texts, often births and other major rites of passage, are thus placed in the "lunar" context of the cycle of conception and birth.

In glyphs A and E of the Lunar Series, the symbol of the moon serves as a value of 20 or vigesimal completion--not the completion of a 29- or 30-day lunation (see Closs 1978). The moon, the symbol of the 20-day "month" or *uinal*, has become an intimate part of the 260-day cycle not as nine lunations but as thirteen *uinals*. It is interesting to speculate that this symbolic shift took place because of the constellation of associations of *9 moons* and *13 uinals*--the 260-day divinatory almanac--with the period of completion of conception through pregnancy to birth.

The second reason for the prominence of the 260-day period is astronomical rather than human. Quite fortuitously, it is neatly commensurate with several other important celestial periods. The Double Almanac (520 days) just happens to closely approximate three lunar or eclipse half-years (3 x 173.33 = 2 x 260 days). This division of the Double Almanac is an excellent approximation of the true value of the eclipse half-year: 173.30906 days. That the Maya noticed that "eclipse possible" periods came only at three evenly spaced times in this Double Almanac is evident in the structure of the Dresden Codex eclipse tables. And indeed, the correction strategies for these eclipse-warning tables show that the Maya imposed the Almanac structure on the natural eclipse data to such a degree that they apparently missed discovering the important "Saros" and "Metonic" cycles.

Furthermore, the Triple Almanac of 780 days corresponds closely and again most fortuitously to the synodic period of Mars: 779.94 days. There are Triple Almanac tables in the Dresden Codex that have been speculatively associated with Mars. The Maya can hardly have failed to notice and appreciate this numerological coincidence, but the connection between the Dresden tables and Mars is at best problematical--see Thompson (1974:91) for a discussion of this question.

Probably for these reasons, reinforced by yet further correspondences with astronomical periodicities (to be outlined later), the Sacred Almanac gained its position as the supreme calendrical cycle of social regulation in the Mesoamerican world.

CYCLES OF THE SUN, VENUS, AND MERCURY

The length of the solar (tropical) year was quite accurately known to the Maya. It existed in the approximate form of the 360-day *tun* (18 20-day *uinals*), the 364-day "computing year" (4 x 91 = 7 x 52 = 13 x 28 = 364 days), and the 365-day "Vague year" or *haab* (18 x 20 + 5 = 5 x 73 = 365). The 365-day *haab* is the closest whole number of days to the tropical year (365.2422 days). The Maya were aware that the 365-day year fell short by about a day in four years relative to the equinoxes, solstices, or solar zenith passage days. The "year drift" correction formula that they used most probably equated 1508 calendar years of 365 with 1507 tropical years; an important piece of evidence in support of this was first noted by Bowditch (1906). It involved the birth dates of the "Palenque Triad" of ancestral personages which are recorded in the three temples of the Cross Group. Each of these three dates falls within the 754th year from the start of the current Maya era--exactly half of the "year drift" cycle. This would place the contrived birth dates of these "ancestors" in the first calendar year since the start of the cycle where the positions of the equinoxes or solstices would be reversed with respect to the 365-day year (Lounsbury 1978:807).

Although the so-called "determinant" theory involving corrections to the recorded 365-day year (see Thompson 1971:Appendix V) is now highly suspect, the Maya were almost certainly aware of a formula equivalent to [(1507)(365.2422)] ≃ (1508)(365) = 550,420 days, and were aware of its fortuitous numerological significance. The number 1508 factors into 29 52-year calendar rounds or $2^2 \times 13 \times 29$. The full cycle in days is 550,420 = $2^2 \times 5 \times 13 \times 29 \times 73$--see Table 19.1. These prime factors are interesting because of the way this period corresponds to the synodic periods of Venus and Mercury.

The average synodic period of Venus is 583.92 days. The Venus Tables in the Dresden Codex plainly show that the Maya employed the canonical approximation of 584 days, which factors into $2^2 \times 73$. The structure of these tables also shows that the Maya recognized the celestial coincidence that five of these Venus years were equal to eight 365-day solar years; i.e., 5 x 584 = 8 x 365 = 2920 days. Furthermore, they created a Great Venus cycle of 37,960 days with a further commensuration of the 260-day cycle: 65 x 584 = 104 x 365 = 146 x 260 = 37,960 days. The knowledge of the natural commensuration between the Venus and solar years was demonstrably known to the Maya astronomers, and this must have given them great satisfaction and confidence in their system.

The average synodic period of the planet Mercury is 115.88 days. There is ample evidence that the Maya employed both cycles of 116 days ($2^2 \times 29$) and 117 days (9 x 13), though there is only speculation that either of these was associated with Mercury (Thompson 1971:215; 1972; 1974:91). Mercury's period is close to one-fifth that of Venus, and it is clear from the Codices that both of the periods 5 x 116 = 580 days and 5 x 117 = 585 days were of interest to the Maya numerologists.

Based on these observations, it seems possible that the Maya astronomers would have noticed the commensuration between the ritual periods of the two sun-following planets Venus and Mercury, the 365-day year, the 260-day Almanac, the Moon, and the 52-year Calendar Rounds, with the full Calendar Year Drift Cycle of 550,420 days ($2^2 \times 5 \times 13 \times 29 \times 73$). The numerology rests on the fortunate presence of the factor 73 in the 365-day Solar year and the 584-day Venus year, and the factor 29 in the 116-day Mercury Period and the 1508-year cycle. The factor 29 is also one of two Maya approximations to the lunation. These speculations are presented in a more readable form near the end of Table 19.1.

The final numerical game to play involves the Double Calendar Year Drift Cycle of 1,100,840 days ($2^3 \times 5 \times 13 \times 29 \times 73$). The Maya mathematicians routinely dealt with numbers of this magnitude and much greater. As shown in Table 19.1, this grand composite number contains a large number of important subcycles, now including both the Venus Year and its Great Cycle as integral multiples. As an elegant composition in numerology and astronomy, it would have been intensely interesting to the Maya had they conceived of it. I suggest that the Year Drift Cycle and its relatives are well worth searching for in the monumental inscriptions.

FURTHER SPECULATIONS AND CONCLUSIONS

Table 19.1 contains many more examples of Maya numerology. The more speculative examples are noted with a question mark. The periods which have astronomical origins are indicated with an asterisk. Table 19.1 shows periodicities for all of the naked-eye planets including Jupiter and Saturn. These two planets have not been previously mentioned because, as in the case of Mercury, there is no compelling evidence that the Maya observed or recorded them. However, as often noted, the absence of evidence is not evidence of absence.

The synodic periods of both Jupiter and Saturn possess low prime factors that would certainly have been pleasing to the Maya diviners. In fact, if we put ourselves "in the skin of the Maya priest-astronomer," we could construct Mercury, Mars, Jupiter, and Saturn tables similar in pattern and "natural" design to the Dresden Codex Venus and Eclipse tables. Such planetary calendars would certainly have delighted the Maya had they discovered them for themselves. Unfortunately, only three (possibly four) fragments of Post-Classic Codices that we know of have survived, and in these there are no such planetary calendars. Therefore, an analysis of the growing corpus of data from the monumental records offers the greatest chance of further discoveries in Maya calendrical astronomy.

The Maya system of reckoning the cyclical passage of time rested on the count of days. As they employed no fractions in their arithmetic, they developed a numerology of whole-day approximations to astronomical periods. In factoring these canonical ephemerides, they computed even grander calendrical cycles with a commensuration with the 260-day Sacred Almanac as a prime consideration. When these ritual calendars fell out of synchronization with astronomical reality, the Maya astronomers were forced to look at greater "drift" cycles and devise short-term algorithms to correct for these effects. Their "intercalations" were done in a characteristically unique fashion by shifting the base dates for the starting points of the cycles, while maintaining the appropriate nominal relationship with the Almanac.

The continuing elaboration of this system over the centuries was certainly encouraged by several fortuitous astronomical commensurations and by the fact that these periods happen to have low prime factors. For the masses, the divinatory calendar would have been one of the ultimate tools for social control. In the hands of the ruling elite, it was a cosmic mandate for their position in a dynastic scheme that stretched back into previous epochs of time. Yet they too were subject to the reign of Numbers. Presumably all of the human rites of passage were regulated by the auguries of the System, and this was subject to the manipulations of Calendar priests and diviners. The study of astronomy served this numerological system, not vice versa. The mathematical power to predict the motions of the planetary bodies must have been closely associated with the perceived power to prophesy, to manipulate fate, and control human destiny. From the small portion of this record that has survived, it would seem that the Maya astronomer was building for an eternity of cyclical order. This order was seemingly one of number more than of geometry and positional astronomy--numbers that were a procession of deities in human form. If we as Westerners are ever to appreciate the world-view of the Maya, we must learn to calculate the harmony of the universe with their living numbers.

20 The Celestial Bands in Maya Hieroglyphic Writing[1]

Beth A. Collea

The corpus of Maya hieroglyphic writing conveys astronomical information in a variety of ways. To date, scholars have relied primarily upon tables found in the codices listing astronomical cycles and the Supplementary Series found in the inscriptions recording the position of the moon in its synodic period at specific dates. The celestial bands also provide information about the heavens, but their exact meaning remains unclear. Förstemann (1895) was the first to identify the celestial bands as a distinct group of hieroglyphic passages in the Maya codices, but he discussed only the nine bands on pages 51-58 of the Dresden Codex. However, there are 60 other examples found in the Paris, Madrid, and Dresden codices.

The identification of a celestial band in this study rests on three criteria: (1) at least one of ten basic glyphs is used (shown as a-j in Fig. 20.2); (2) the glyphs comprising the bands are shown fused together rather than in separate cartouches; and (3) no affixes appear with the glyphs in the bands as they do when the same glyphs are used in the body of a text. (The final two criteria were first suggested to me by Michael P. Closs and John Justeson, respectively, in April and June of 1979.) The ten basic glyphs include all glyphs used in the celestial bands in the three codices which have already been identified as being either astronomical or celestial in nature in the broadest sense. They account for 95% of all the glyphs used in the 69 bands from the codices. The criteria were derived from the celestial bands in the Dresden, Paris, and Madrid codices; they were then used in recognizing other bands on pottery and in inscriptions.

Despite the frequent occurrence of celestial bands, they have yet to be thoroughly and systematically analyzed. The history of their interpretation is closely tied to the search for planetary tables in the codices. Indeed, their study seems to have been pursued only tangentially to this search. Only Förstemann (1895) dealt with the bands directly, although the work of Willson (1924), Kelley (1976), and Kelley and Kerr (1973) implicitly suggests alternative readings for some glyphs. I will review some of these interpretations of the celestial bands below.

My analysis of the bands employs three approaches, each designed to shed as much light as possible on these enigmatic passages. First, a taxonomic

[1]I would like to thank Anthony F. Aveni, Lyle Campbell, Michael P. Closs, Brian L. Foster, William H. Isbell, John Justeson, and Weldon Lamb for their helpful suggestions and encouragement; the American Philosophical Society for permission to use the illustrations of the Dresden Codex (Thompson 1972); and Akademische Druck- und Verlagsanstalt for illustrations of the Madrid and Paris Codices (Codices Selecti 1967, 1968).

approach examines the glyphs used, their frequency of occurrence, and the patterns of usage created with the nineteen component glyphs. Next, a contextual approach considers the symbolic use of the celestial band in the scene it accompanies. Finally, a historical approach finds Olmec antecedents and Mixtec analogues for the celestial bands. The present discussion is offered not as a statement of final results, but as a fuller articulation of the problem and a consideration of the kinds of insights further study may provide.

A brief review of the iconography of the bands must preface the discussion. The longer passages, containing four or more glyphs, are often displayed on the body of a serpent; Thompson (1970) identified these iguana and crocodile figures as the god *Itzam Na*, whose celestial manifestation resided in the heavens and sent rain to Earth. Shorter bands imply an association with *Itzam Na* through the presence of falling rain.

Förstemann (1895) approached pages 51-58 of the Dresden Codex in much the same way that he did the preceding pages (in which he discovered the Venus Table); he contended that pages 51-58 comprised a table relating Mercury's synodic period of approximately 115 days and the *tonalamatl* of 260 days. He garnered support for this interpretation from the fact that the total number of days in the table (11,960) equaled both 115 x 104 and 260 x 46. As Lounsbury (1978) reminds us, such a notion is consistent with the Maya penchant for integrating several celestial cycles into a single framework. The intervals of 177, 178, and 148 days were grouped into ten sections, each of which was separated from the next by a picture. These periods, which approximated five and six lunations, later led Meinshausen (1913) and Willson (1924) to interpret these pages as an eclipse table. This work has largely supplanted Förstemann's approach.

Three of Förstemann's identifications, however, relied on earlier work and are still accepted. First, the glyph for sun (*Kin*) had been previously identified by Brasseur de Bourbourg (Kelley 1962:33). Second, Förstemann (1906) deciphered the Venus glyph on the basis of its recurrence on pages 24 and 46-50 of the Dresden Codex. Kelley (1976:38) has expressed reservations about the specific identification of this glyph with Venus; this will be discussed later. Third, Förstemann's reading of the Crossbands as "conjunction" is now supported by linguistic evidence. Kelley (1976:152) read the Crossbands as marriage; Marcus (1976a:4) and Berlin (1968:15, 19) translated them as marriage or union. The Crossbands, then, may have "represented marriage literally and implied celestial conjunction symbolically" (Lyle Campbell, personal communication 1979).

Förstemann (1895) inferred the meanings of the glyphs he called Saturn and Jupiter from evidence for planetary cycles he found in the accompanying lists of intervals. These two readings have never been widely accepted, perhaps because of his alterations in the number sequence listed and the later interpretation of these pages as an eclipse table. However, it should be noted that the identification of an eclipse table on pages 51-58 did not necessarily negate Förstemann's reading of planetary glyphs. Recent studies (Smiley 1973) of the Venus Table demonstrate a similar situation: numbers important in lunar eclipse cycles (9360 and 11,960) are found separating the Venus Table correction dates. Indeed, the inclusion of other celestial cycles is especially consistent with the Maya desire to interrelate several astronomical phenomena.

Förstemann (1895) began his alterations by counting the first interval as 502 days. Since the zero date is not explicitly stated, he shifted the beginning day to the start of the first pictured interval. For this reason, he subtracted 502 from all subsequent cumulative totals. After making another twenty less-significant adjustments in the 138 numbers listed, he presented evidence for a synodic period of Jupiter of approximately 398 days. Six of the nine remaining sections of the table contained cumulative totals relatively close to a multiple of 398. Förstemann acknowledged, however, that there was a margin of error of between four and eighteen days in each case. Evidence for Saturn was found in his new final number for the table, 10,748 days, which he proposed as an estimate for 10,753, the planet's sidereal period.

Förstemann's (1895) identification of specific glyphs with Saturn and Jupiter was based on the total number of days contained in the table at the end of each pictured interval. Possible multiples of planetary periods were used as a justification for interpreting glyphs in the accompanying band as representations of that planet. For example, on page 53b, after making the alterations and subtracting 502, the total is 6762 or (17 x 398)-4 (Fig. 20.1). On this basis, Förstemann labeled the first glyph in the band Jupiter. He provided less rationale for the second glyph, which he identified as Saturn. Although his decipherment of the Jupiter and Saturn glyphs has never been widely accepted, Förstemann made an important contribution by recognizing the celestial bands as a distinct kind of hieroglyphic passage representing the heavens.

Fig. 20.1. Dresden Codex page 53b (from Thompson 1972)

Willson (1924) did not deal directly with celestial bands, since his primary interest lay in attempting to demonstrate the existence of a Mars table in Dresden Codex pages 43-45. Figure 20.2t shows the glyph Willson called Mars, or Mars at conjunction with the sun. There is only one celestial band on these pages and it does not include this glyph; it should be noted that no other band in the codices contains Willson's Mars glyph.

Although Gates (1931) did not add to the interpretation of the bands, he did make a contribution to their study. He collected all of the glyphs used in the bands in the codices for the first time and grouped them together in his dictionary.

Kelley and Kerr (1973) critically assessed the validity of all identifications of astronomical glyphs, and maintained that the only planetary glyph known with any certainty was Venus. Furthermore, they contended that the glyph frequently identified as Venus may only denote "star." Glyphs meaning "red star" or "great star" may have referred specifically to Venus. Kelley (1976:63) discussed the glyph for God C, which appears three times in the bands; he noted that it is often recognized, although without good reason, as the North Star. The similarity of God C's head and that of the glyph for north, combined with the representation of the god hanging from the sky by a rope, led to this unwarranted interpretation. Kelley's (1975) work on planetary cycles may eventually provide additional information with which to identify the glyphs used in celestial bands. However, his search for dates separated by significant

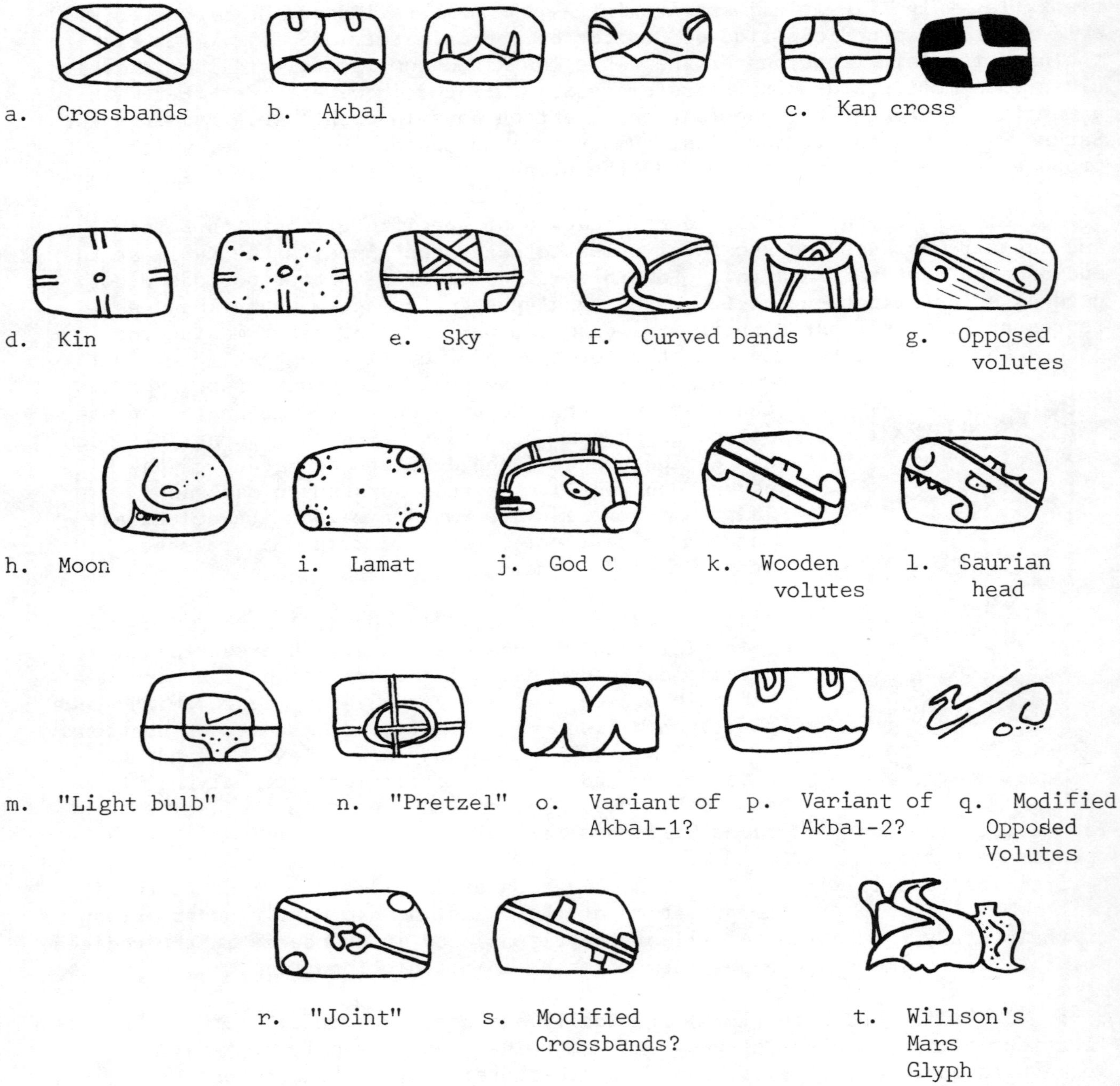

Fig. 20.2 (a-s) Glyphs used in the celestial bands
(t) Willson's Mars Glyph

astronomical periods has provided little aid thus far.

In the present paper, the taxonomic analysis began with a simple inventory of the glyphs found in the bands and their frequency relative to the total number of glyphs used in the bands. Table 20.1 summarizes the results. Figure 20.2 lists the component glyphs, with secondary examples demonstrating the degree of variation allowed.

TABLE 20.1

OCCURRENCE AND FREQUENCY OF GLYPHS USED IN CELESTIAL BANDS

Glyph	Number of Occurrences	Percentage
Crossbands	41	21.1%
Akbal	34	17.5%
Kan cross	30	15.5%
Kin	29	14.9%
Sky, T561	16	8.2%
Curved bands	9	4.6%
Opposed volutes	8	4.1%
Moon, T682b	8	4.1%
Lamat	6	3.1%
God C	3	1.5%
Notched bands	2	1.0%
Saurian head	1	0.5%
Light bulb	1	0.5%
Pretzel	1	0.5%
Variant of Akbal-1	1	0.5%
Variant of Akbal-2	1	0.5%
Modified Opposed volutes	1	0.5%
Joint	1	0.5%
Modified Crossbands	1	0.5%

We can see from Table 20.1 that the so-called celestial bands are made up of relatively few signs. Five glyphs constitute more than two-thirds of the total number of occurrences of all glyphs in the celestial bands; they are the Crossbands, Akbal, Kan cross, Kin, and Sky signs. An intermediate category of frequencies ranges between 4.6% and 1.5%. These glyphs may have more specialized meanings and their distribution suggests possible distinctions between the three Maya texts. The final nine glyphs appear only once or twice and may not even be strictly astronomical.

Table 20.2 presents the patterns of occurrence of these glyphs in matrix form. The vertical listing indicates which glyph occurred first in a left-to-right sequence, while the horizontal listing gives the second glyph. The second glyph is then listed in the vertical column against the third glyph in the horizontal row. Thus, the sequence  reads as follows in matrix form:

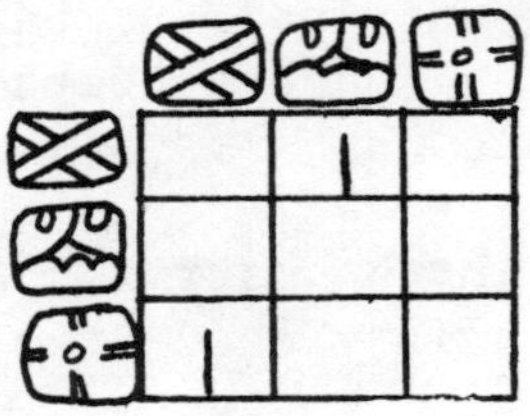

TABLE 20.2

MATRIX OF PATTERNS OF OCCURRENCE OF
GLYPHS IN CELESTIAL BANDS

	1	6	3	8	2				1								2		
	3		2	4	3	2				2									
	6	4	1	1	1	2	1	1							1				
	5	5	1	2	1		3	1											
		4	6	1															
		2							1										
			1	2		1				1						1			1
			1	1	1	1												1	
		2			1			2											
	2			1															
			1																
							1												
						1													
	1																		
							1												
	1																		
				2															
				1															

Astronomical and linguistic considerations motivated the taxonomic analysis. Other descriptions of the bands include constellation, planetary, or zodiacal bands. Implicit in these names is the notion of a fixed series of planet or constellation signs that may be traced through the entire corpus of bands. The present examination provides no evidence for such claims. The linguistic evaluation, which is still in progress, involves examining the sequential glyphic patterns following Kubler's (1967) analysis of Teotihuacán iconography. Glyphs are roughly divided into nominal, verbal, and adjectival categories in order to facilitate the study of usage patterns. The results of this investigation are as yet incomplete.

Fifty-eight bands are represented in the matrix. Eleven were excluded because of damaged texts or because of uncertainty about the reading order (as in the case of stepped bands). Madrid Codex pages 2b, 3b, 9, 10c, 32, 33a, 37, and 40, as well as Dresden Codex pages 37c and 66b and Paris Codex page 9 were excluded from the matrix. The method was initially undertaken in order to discover any repeated sequences in the glyphs used. No such patterns were found; however, several pairs of glyphs frequently occur together. For example, the Crossbands/Kan cross combination appears seven times. Following Thompson's (1960:276) suggestion that the Kan cross may be read as "precious," the pair may be translated roughly as "precious conjunction." Kan cross/Akbal, "precious darkness," occurs six times in the bands of the three Maya codices. The Sky glyph T561, and Kin, indicating "sunny sky," also occur six times. By examining solely the frequently-occurring pairs of glyphs, however, one can only partially decipher the content of the bands.

Separate matrices were tabulated for each codex. If one summarizes these results, one finds the glyph representing God C's head only in the Madrid Codex, where it occurs three times. Second, the Sky glyph, comprising 8.3% of the total number of glyphs used in the bands, never occurs in the Madrid Codex bands. Third, the Lamat and Moon signs are not used in Paris Codex celestial bands. Finally, the Opposed volutes appear seven times in the Dresden Codex bands, while only one uncertain example is found in the Paris Codex and none in the Madrid Codex.

In order to complement the taxonomic analysis, a contextual approach was implemented. This examined the placement of the celestial band relative to the figure or scene which it accompanied. Briefly, celestial bands imply four different contexts for these pictures: (1) heaven; (2) the edge of the sky; (3) earth; and (4) general or uncertain contexts. Figures 20.3-20.6 present sketches of all the bands which fall into each category.

Fifteen of the sixty-nine bands adorn a throne upon which a deity sits. Since the band iconographically represents the sky, the scene by inference takes place in heaven, above the sky (Fig. 20.7). If we examine the glyphic content, we find that ten of the fifteen bands used as thrones contain Crossbands. In the five remaining examples (Dresden Codex page 50, Paris Codex page 19, Paris Codex page 5, Paris Codex page 9, and Madrid Codex page 37), the first two contain the Curved bands, the third and fourth are partially effaced, and the last is a questionable identification. Robicsek (1975:38, 39) noted that the Curved bands may appear interchangeably with the Crossbands in the Pop glyph. Then, in every case with the exception of Paris Codex page 5, Paris Codex page 9, and Madrid Codex page 37, the Cross- or Curved bands appear. In light of the association between the Crossbands glyph and the use of celestial bands as thrones, the interpretation of the Crossbands as "conjunction" should perhaps be broadened to include "meeting." The scenes showing deities on thrones are reminiscent of passages from the *Chilam Balam of Chumayel* (Roys 1967) that record the auguries for the times ruled by certain deities; thus the Crossbands might indicate that the god had "met" his appointed time to rule. The celestial bands adorning thrones are also distinguished by the glyphs not included. The five most common glyphs are found,

Fig. 20.7 Celestial band used as a throne, Madrid Codex page 11c (from Codices Selecti, 1967)

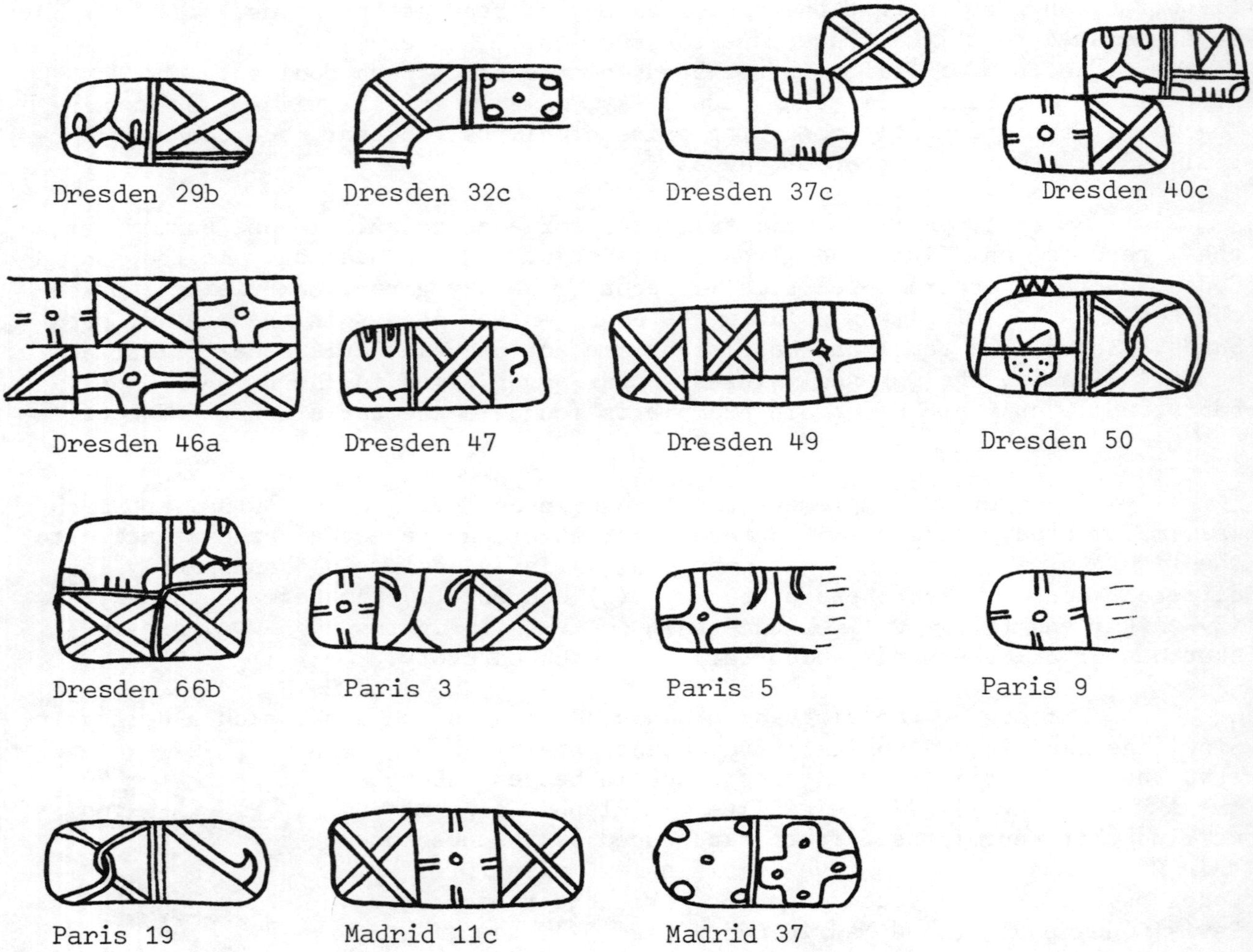

Fig. 20.3 Celestial bands used as thrones

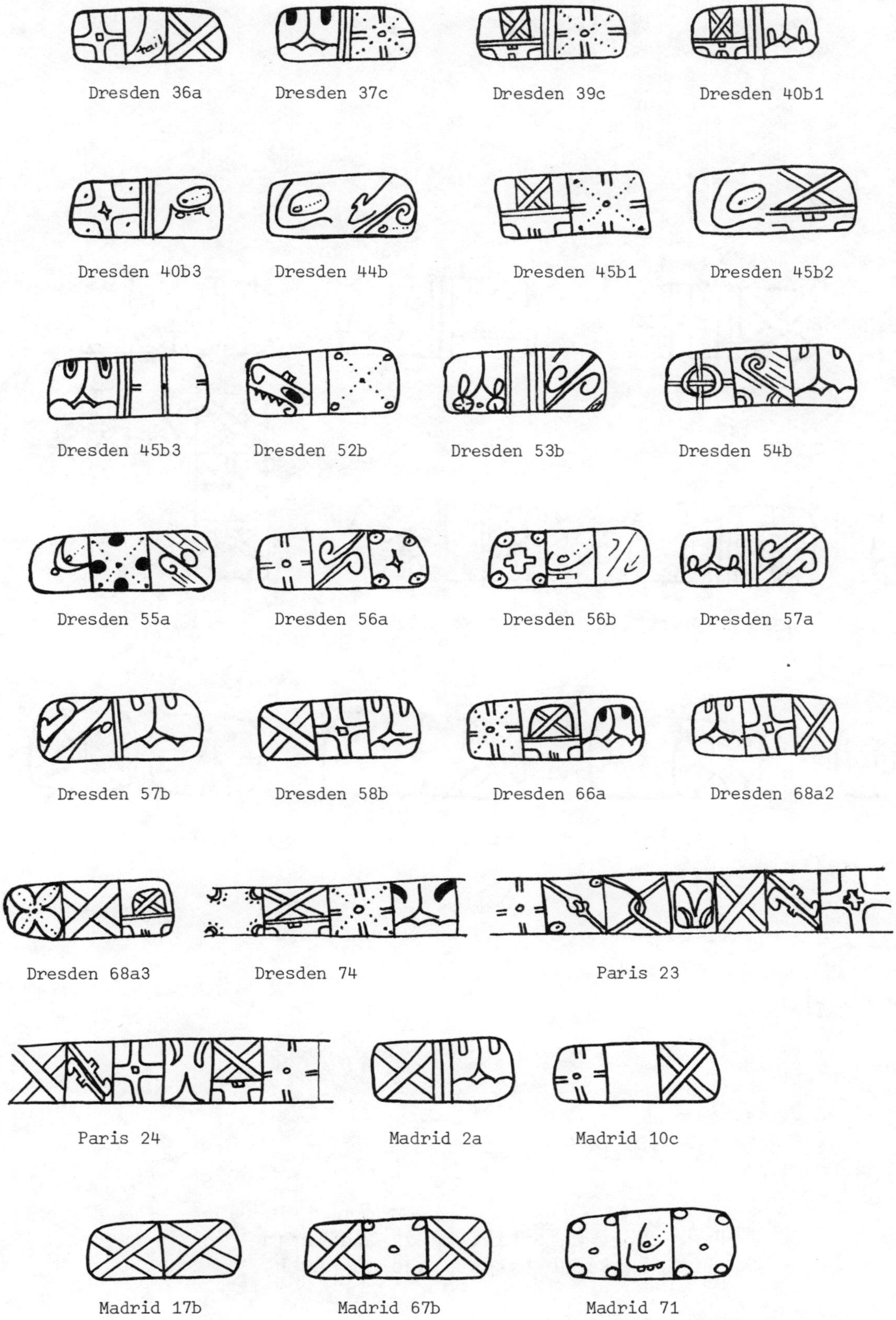

Fig. 20.4 Celestial bands with pendent figures and signs

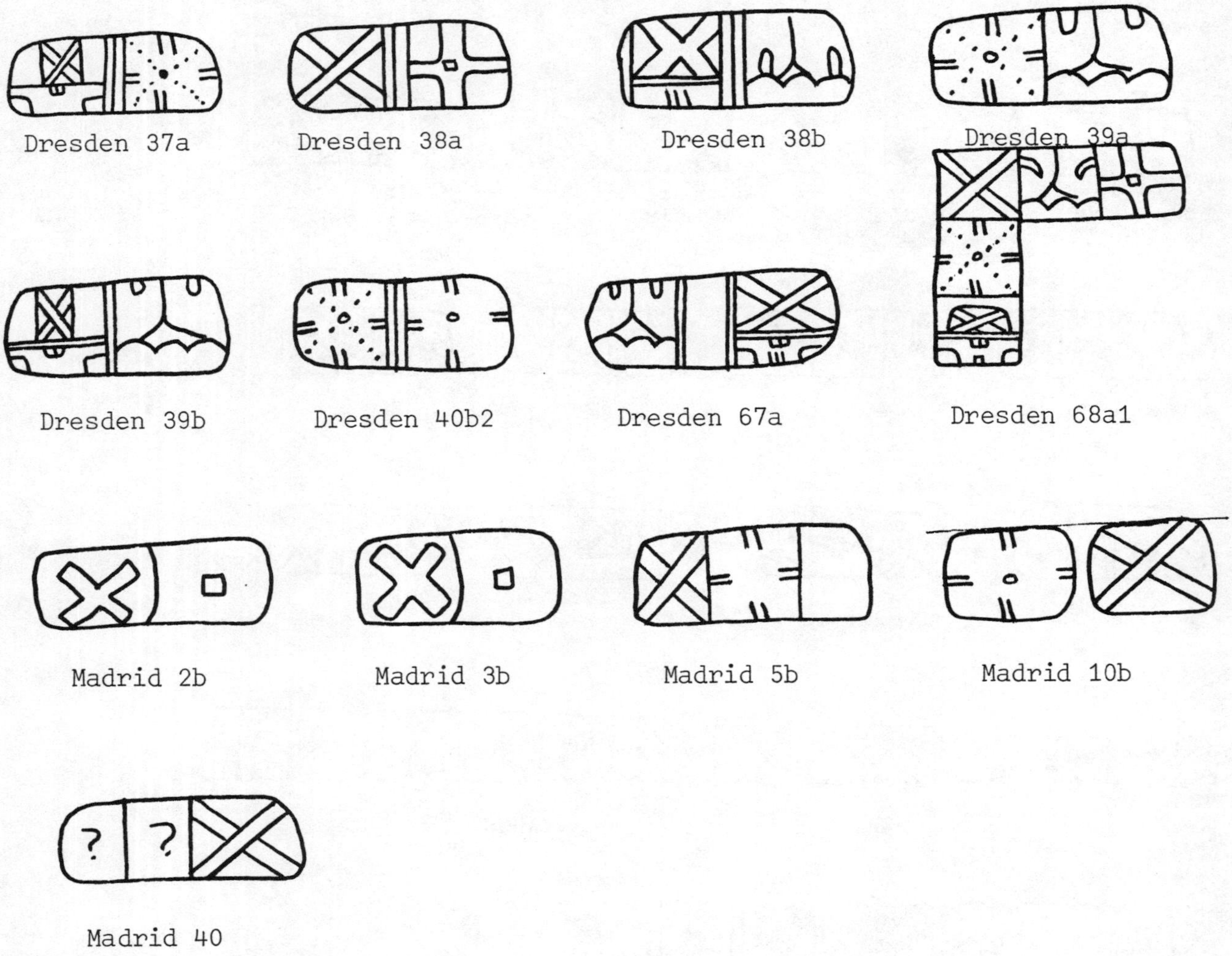

Fig. 20.5 Celestial bands indicating that the scene takes place on earth

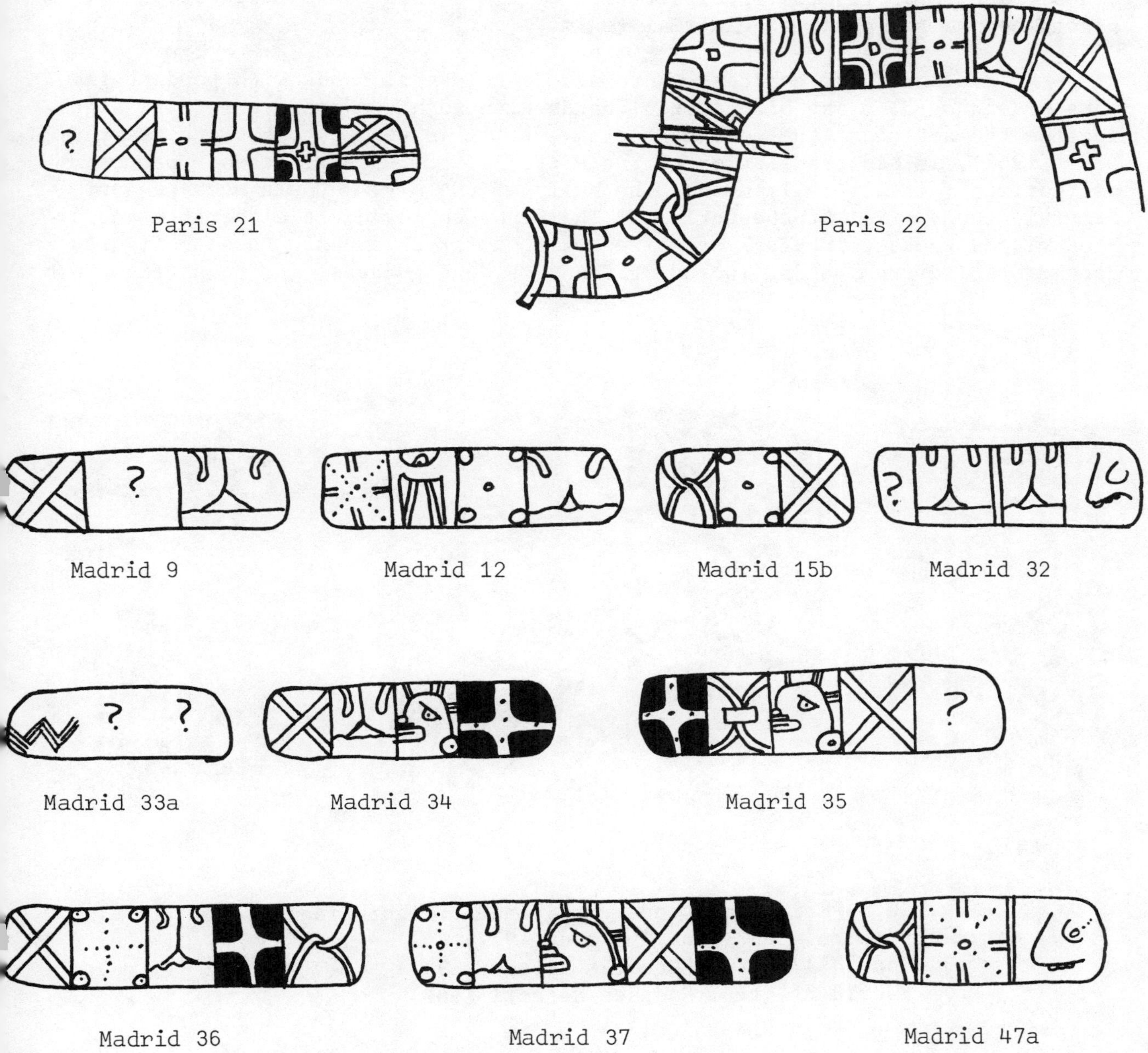

Fig. 20.6 Celestial bands with general or uncertain context for the accompanying scene

but the Moon and Venus glyphs are never used and the Opposed volutes appear in only one dubious example on Paris Codex page 19. The absence of Moon and Venus glyphs supports the interpretation of the scenes as representations of heaven beyond the visible celestial realm.

In the three codices, there are 29 celestial bands with pendent figures or signs. The use of celestial bands with such pendent figures and signs locates the action at the edge of the sky, still in view but above the earth. Roys (1965), in his translation of the *Ritual of the Bacabs*, reads *u nak caan* as "horizon"; it literally means "the belly of the sky." Could such pendent figures, then, exist at the horizon? The signs employed are either Kin and/or Moon glyphs encased in black or white ovals or larger sacks. Various figures dangle or dive from bands, including Chacs, a dead goddess, and dogs (Fig. 20.8).

(a) (b) (c)

Fig. 20.8 Celestial bands with pendent figures and signs
(a) Dresden 40B3 (from Thompson 1972)
(b) Madrid 17b (from Codices Selecti 1967)
(c) Madrid 71 (from Codices Selecti 1967)

An examination of the glyphic content shows that bands with pendent figures and those with signs contain nearly the same glyphs. If we compare these bands with the thrones, we see a greater variety in the meanings intended, given the number of different glyphs used. The pictures support this conclusion; some deal with eclipses, others with rain, and still others with drought.

The Opposed volutes are especially common in this group of bands. Their meaning remains undeciphered. Because of their extensive use in the Lunar Table, we might be justified in supposing that they refer to an action or phenomenon in the sky. Possibly they denote "a passing by or in front of." The Crossbands, meaning "conjunction," actually convey this through their form. For this reason, the notion of "passing by or in front of" is a sensible guess for the Opposed volutes, but it is no more solid than that.

There are thirteen bands which indicate that the accompanying scenes take place on earth (Fig. 20.9). The mundane activities portrayed, such as planting or walking in the rain, support this interpretation. As a reflection of the content in each codex, eight occur in the Dresden Codex, six in the Madrid Codex, and none in the Paris Codex. This group of bands contains the five most common glyphs. Akbal is used especially frequently because these bands often present auguries for rain. Since the scenes refer to the earth, it is not surprising that the moon glyph is used only once and Venus not at all.

Fig. 20.9 Celestial bands indicating that the scenes take place on earth
(left) Dresden 37a (from Thompson 1972)
(right) Dresden 40b2 (from Thompson 1972)

Eleven passages comprise the final category containing bands with uncertain or general contexts (Fig. 20.10). Nine occur in the Madrid Codex and two in the Paris Codex. Madrid Codex pages 34-37 fall into this category. The scenes below the bands contain several figures, and it appears that the two primary figures are offering corn to God C. Three of the bands are badly effaced: Madrid Codex pages 9, 32, and 33a. Their pictures are also partially destroyed, so any interpretation is most difficult. The two Paris Codex examples (pages 21 and 22) show figures both above and below the band; the figures below may be on earth or perhaps even in the underworld.

Two other observations on patterns of glyphic usage and the possible implications of these regularities for glyphic meaning were drawn from the taxonomic and contextual analyses. First, the Sky glyph, appearing sixteen times, is shown with rain in nine cases, and four other uses of T561 convey rain through diving Chacs and figures with axes, the symbol for thunderbolts. One of the remaining cases is an uncertain identification (Dresden Codex page 49) and two occur in very general contexts (Paris Codex pages 21 and 24). Perhaps T561 has a more specific meaning (referring to rain) than simply "sky." Second, the Crossbands and Curved bands, although interchangeable in some cases (Robicsek 1975), appear to possess a slightly different range of usage. The Curved bands do not occur either directly before or after Akbal and T561, both of which are found adjacent to the Crossbands. Furthermore, Akbal and T561 are frequently used to indicate rain. If we review the contexts in which Curved bands appear, we find that only two passages show falling rain. In each case, the Curved bands are followed by a Kan cross. These suggestions of possible patterns of glyphic usage must rely on a study of the books of Chilam Balam and of usage patterns in the body of the hieroglyphic texts for confirmation.

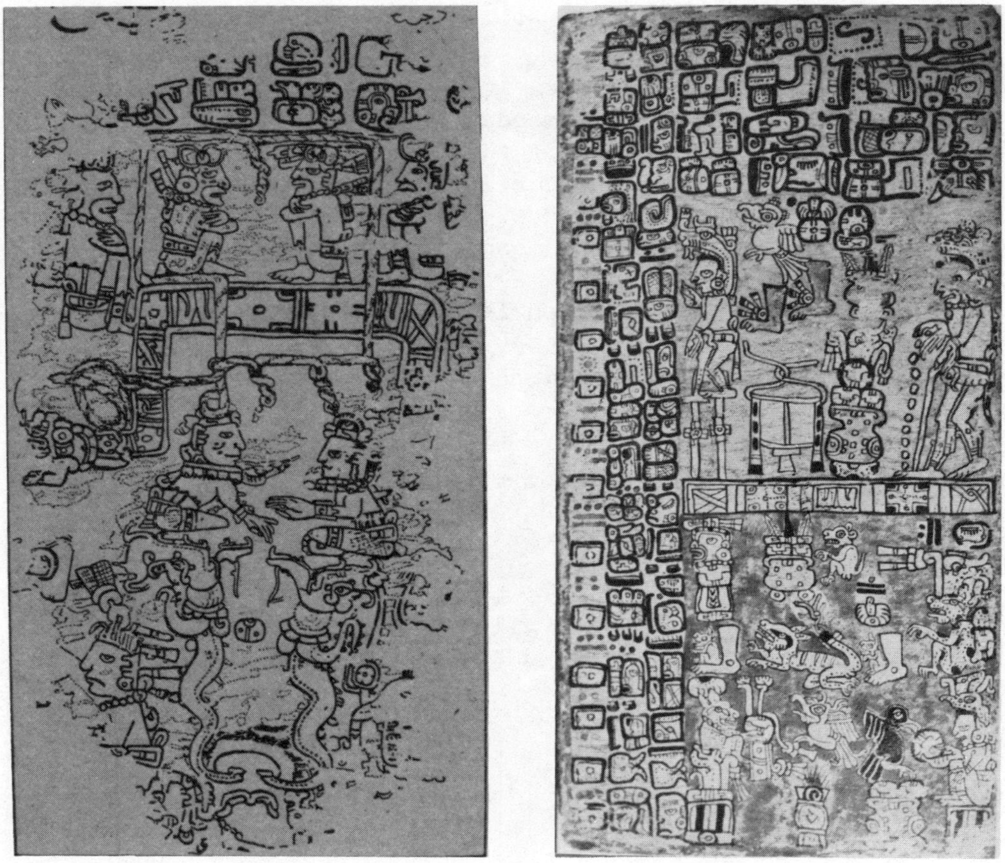

Fig. 20.10 Celestial bands indicating a general or uncertain context for the accompanying scene
(left) Paris 22 (from Codices Selecti 1968)
(right) Madrid 36 (from Codices Selecti 1967)

The historical approach considered the possible cultural continuity of the celestial band form in Olmec, Maya, and Mixtec cultures. The Olmec culture reached its florescence during the Middle Formative (800-400 B.C.) in the Gulf Coast region (Coe 1965:686). The hegemony of the Maya of the Yucatán, Chiapas, Guatemala, and Honduras occurred between A.D. 300 and 900. The Mixtecs occupied portions of the Central Mexican Highlands, and their celestial bands probably date from A.D. 1200 to 1400. As Coe (1976:111) notes, the only Olmec signs transmitted to Maya writing were of an astronomical nature. Although these signs were symbols in a complex iconographical system and not elements of the writing system directly antecedent to that of the Maya, they are of significance in this discussion since they represent the first appearance of these symbols (Marcus 1976b:43). The fact that Joralemon (1976:61) found evidence for the identification of the Olmec Dragon as the iconographic and mythological ancestor of the Maya deity *Itzam Na* strengthens the argument for an Olmec origin of the celestial bands.

The four Olmec signs that were transmitted were the Kin sign, Lamat sign, U-sign for Moon, and the Crossbands (Coe 1976). On further examination, one finds the Lazy-S (∞), which occurs on Maya pottery and in the inscriptions in celestial bands, also in Olmec carvings and pottery. Relief 1 from

Chalcatzingo even shows a figure holding the Lazy-S on a bar identified by Joralemon (1971:49) as a ceremonial bar. Here, then, is an example embodying both the same "glyphic" form and the same contextual use. It will be recalled that celestial bands often appear on Maya stelae adorning ceremonial bars; an example is Seibal Stela 10, shown in Figure 20.11.

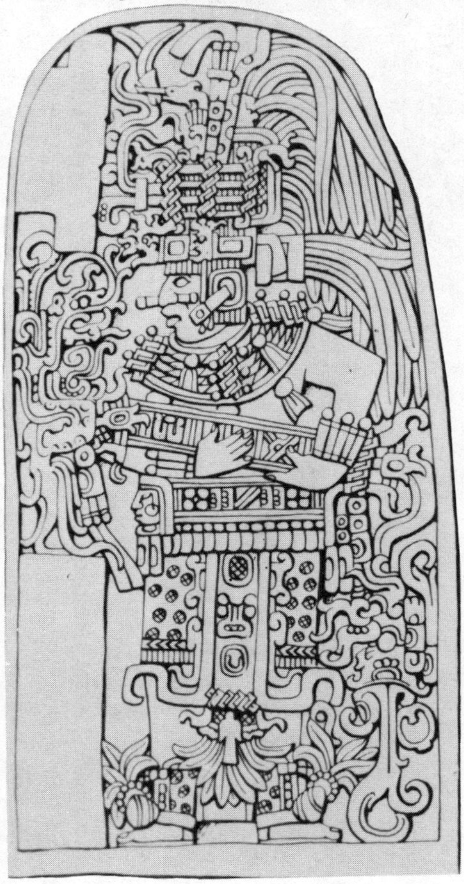

Fig. 20.11 Seibal Stela 10 (Courtesy A. F. Aveni)

Mixtec writings also provide a wealth of celestial bands. Although the same glyphs are not employed, the bands show a row of stars from which rain falls or figures hang, clearly reminiscent of celestial bands in Maya hieroglyphic writing. Figure 20.12a is taken from the Codex Nuttall; it shows three figures dangling from a band with a sun sign. Several other examples (Fig. 20.12) appear in the Codex Bodley, the Codex Selden, the Codex Vindobonensis, and in the Lienzos of Zacatepec (Smith 1973).

The books of Chilam Balam record similar kinds of divinatory prognostications, and provide another avenue of investigation concerning the celestial bands. Since they are written in transliterated rather than hieroglyphic form, they offer a different perspective on the patterns of usage and the meanings of individual glyphs.

In conclusion, this study tends to support the following hypotheses. First, the celestial bands are not merely captions for the pictures they accompany but have an iconographic role in the picture; indicating where the scene takes place, their position becomes as significant as their content. Second, the Lazy-S sign may be a fifth astronomical glyph found in Maya writing that was derived from Olmec iconography. Third, the following preliminary suggestions may be of assistance in the future decipherment of glyphs: (a) the Opposed volutes may represent a phenomenon in the sky, and perhaps even refer specifically to eclipses; (b) the Crossbands usually read as "conjunction" may also connote "meeting" in general; (c) the Sky glyph most commonly appears in passages referring to rain, and may therefore have a more specific meaning than simply "sky"; and (d) the Crossbands and the Curved bands are not completely interchangeable--a complementarity in meaning is suggested by their use in the Dresden Codex Lunar Table. Fourth, the presence or absence of certain glyphs in the bands of the three codices suggests differences in the subject matter of each text, although the significance of these patterns of occurrence remains unclear. However, general distinctions concerning the subject matter in each codex can be offered. The Dresden Codex bands are almost always fairly literal, and deal with divinatory subjects or very specific astronomical phenomena. The Paris Codex contains bands with very general contexts; the subject matter is perhaps broader and possibly mythological in nature. The

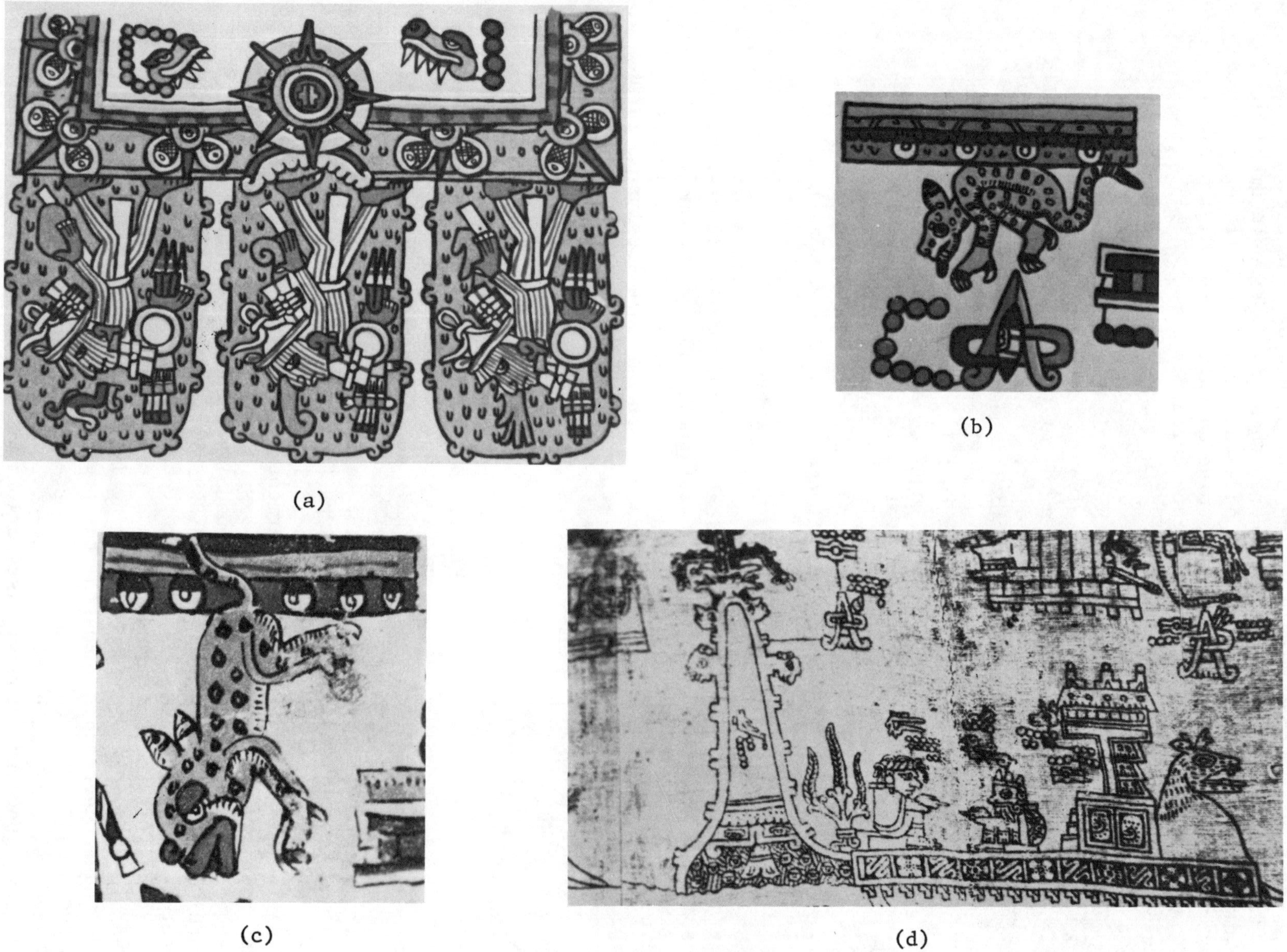

Fig. 20.12 Mixtec celestial bands: (a) Codex Nuttall 20 (from Nuttall 1975); (b) Codex Nuttall 27 (from Nuttall 1975); (c) Reverse of Vindobonensis 8 (from Codices Selecti 1963); (d) Lienzo of Zacatepec 1 (from *Picture Writing from Ancient Southern Mexico: Mixtec Place Signs and Maps*, by Mary Elizabeth Smith. Copyright 1973, University of Oklahoma Press).

Madrid Codex combines both the divinatory bands and the more general references.

The celestial bands provide an unusual vantage point from which to view Maya culture because their study has implications for such a wide variety of other topics. First, because they occur frequently and in a wide range of contexts and media, the celestial bands may bring new information to bear on astronomical studies (by clarifying the meanings of specific glyphs) and on art historical concerns such as iconography. Second, the celestial bands appear in both public and private art; they thereby permit comparisons of differences in presentation to each audience. For example, the bands generally appear on pottery as borders or as thrones, two uses very similar to those in the textual examples (Fig. 20.13). However, rulers used celestial bands on stelae to increase their personal prestige by having themselves portrayed holding a bar with a band on it; this symbolized their divine right to rule (Thompson 1970: 232). Third, the celestial bands may provide a very specific link between Olmec, Maya, and Mixtec cultures. Their long history, great degree of variability, and implications for other areas of research will surely continue to make the celestial bands a topic of study and debate for years to come.

Fig. 20.13 Celestial bands on Maya pottery (from Coe 1978, © by the Trustees of Princeton University. Reprinted by permission of the Art Museum, Princeton University. Photographs © Justin Kerr).
(left) Portion of scene on a vessel showing God N seated on a throne.
(below) Vessel from Naranjo.

21 Star Lore in the Yucatec Maya Dictionaries

Weldon W. Lamb

INTRODUCTION

Yucatec Maya dictionaries, especially those of the colonial era, are rich in starlore, although most of these astronomical terms apparently do not appear in any of the recent archaeoastronomical or anthropological literature. There are expressions for many different phenomena. Only the numerous entries about the sun and the times of day have been omitted here, since they are the subject of a future paper. The present work emphasizes the dictionary terms and supplements some of them briefly with names from other Mayan sources. The table at the end of the article lists every term as well as its gloss, its literal translation, and its sources.

MAYAN ASTRONOMICAL TERMS

Stargazers

The few surviving terms suggest different kinds or ranks of astronomers. *Ek* means "star" and "secular priest"; "secular" suggests that such a priest was not directly involved in rituals. *Ah kin* is literally "he of the sun," but it glosses as "astrologer." Perhaps the *ek* focused mainly on the stars, while the *ah kin* intently followed the sun. Each of the most accomplished watchers of the heavens might have borne the title *ah miatz*, "the sage" and "the scientist." This might be because (1) Edmonson (n.d.:19.1, n. 483), while making no mention of astronomers, describes an *ah miatz* as a person with the highest degree of training and expertise in religious and calendrical matters; and (2) because *ah miatz* is an integral part of the expression glossed with "star man (*estrellero*) or astrologer": *ah miatz tu xinbal ekob, tu zut ekob*, literally "he who is wise to the walking of the stars, to the revolving of the stars." From among the *ah miatz* came the two highest priests. The lesser was the *chilam balam*, "mouth or spokesman of the jaguar," and he also bore the title of *ah noh kin*, "the great sun priest" (Edmonson n.d.:19.1, n. 483; 25-21.1, n. 777) or "the great astrologer." The greater of the two highest holy-men was the *balam*, "jaguar"; he also enjoyed the office of the *ahau ca(a)n*. This name translates as "lord snake" or "rattlesnake" (Edmonson n.d.:25-25.1, n. 777) or "basilisk"; it glosses as "bishop," but literally means "lord of the sky," a fit title for the highest astronomer-priest.

Two Views of the Universe?

A passage from the Codex Perez mentions "the astrologers, acentric and concentric" (Barrera Vásquez and Silvia Rendón 1948:153; Craine and Reindorp 1979:89; Solís Alcalá 1949:182-183). Perhaps these terms designate two late colonial schools of thought on the structure of the cosmos. "Concentric" suggests the Ptolemaic system, in which the sun, moon, and planets each revolve around the central earth in a perfect circle; the term could apply as well to the early (1512-1609) Copernican model, according to which the earth and the other five known planets orbit in perfect, nesting circles around the central sun. "Acentric," if plausible as a corruption for "eccentric," might describe the later Copernican system, as proposed by Kepler in 1609: the orbits of the earth and the other planets are not concentric circles but ellipses. "Eccentric" is a technical term used to describe the degree of flatness of an ellipse. The word, however, might just as well be used in its more general sense of "uncommon, not normally accepted," because in the colonial literature which I have so far examined only the Ptolemaic system is illustrated or described.

Astrology

The same text from the Codex Perez which mentions the "acentric" and "concentric" astrologers also describes these men as great sages (*ah miatz*) who know all about the march of the sun, the moon, and the stars, and all about the whole of creation (Barrera Vásquez and Silvia Rendón 1948:153). The science of these savants was "astrology" or *(y)idzatil ah ahel (=ohel) tu xinbal caanob, tu xinbal ekob*, "the art of him who knows where the heavens walk, where the stars walk." No separate word exists for "astronomy" because astronomy was astrology, and men were the subjects of the stars (Barrera Vásquez and Silvia Rendón 1948:62).

Stars in General

Ek is "star" and "spots like those of the tigers and deer when small" and "spots that are on the body of the deer" and "carbuncle." The Maya equated stars with the jaguar's spots and the night sky with its skin (Thompson 1972: 164, n. 1). The Chilam Balam of Chumayel (Barrera Vásquez and Silvia Rendón 1948:137; Roys 1967:27, 94) describes a star as *u lol akab*, "the (big) flower of the night." Venus is *u lol zaz caan*, "the luminous (big) flower of the sky" (Barrera Vásquez 1965:52, n. 18; 58; 66). "The starry sky" is *chac nicen caan*, "great or red flowering sky." "The appearing of the stars" is *hokol ek*, "the stars sprout, rise, or emerge"; *tippil ek*, "the stars rise, emerge, appear above or pulsate"; and *kaatal eekob*, "the stars are put on high." Stars crammed the sky. *Tul ekil caan*, "starry sky," is "full of stars the sky"; and *hun tzotz ceh ek yan ti caan* glosses as "innumerable are the stars that are in the sky," but it is verbatim "(as many as) all the hairs of the deer are the stars in the sky." The lowest, visible layer of the heavens is *u nak caan*, "the belly of the sky."

The Planets in General and Saturn

Perhaps three terms denote the planets. One is *planetob*, simply the Spanish *planeta* pluralized à lá Maya (Coe 1975:19). The second is *chachac ek*, "red, red stars" (Roys 1967:150, n. 4) or "great, great stars." The third is *lacam ek*, "great star" or "banner star" or "somersault star"; the word "somersault" suggests the retrograde loops of a planet. Mejía (1969:43) writes that *x'ain ek*, "female crocodile star," names Saturn, but he gives no source. This could be rather important, for the earth monster, a cross between crocodile, iguana, and serpent, often displays the glyph for "star" (T510) and figures prominently in the iconography.

Morning Star, Evening Star, and Venus

Several names designate the luminaries of dawn and dusk. *Noh ek*, "big star," glosses as "luminary" and "luminary of the morning." *Chac ek*, "great or red star," is "luminary of the day." "Luminary" or "the morning star" is *chac noh ek*, "red, big star." Only *noh ek* and *xux ek*, "wasp star," explicitly refer to Venus as both morning and evening star; *chac ek* probably does, too. *Xux ek* recalls Roys' (1965:132) note that *chac ec*, "a reddish wasp that does not sting," sometimes replaces *chac ek*, and it recalls, too, Kelley's (1976:38) alternate reading of T510, the glyph for "star," as *kanal*; E. May defines *kanal* as a "large red wasp" (Roys 1965:132). *Hŏzan ek* glosses as "star of the evening" but verbatim means "tired-out star." In Landa *Hozanek* is the name of the black Bacab of the west, one of four brothers who hold up the corners of the sky; for Thompson (1972:67) Hozanek the Bacab is a form of Venus, brightest of all the evening stars; the title, however, could really apply to any evening luminary. The evening star is also *ah oczah kin*, "luminary of the night," literally "he who makes the sun enter (into the Underworld?)" and "he who sows or plants the sun." Another title is *ah ppiz akab*, glossed as "luminary that rises at first (hour of the ?) night and runs through all of it, such that it appears to be measuring it"; it literally reads "he who measures the night." This definition excludes only Mercury and Venus. "The luminary of the morning" is *ah ahzahcab (ek)*, "he (the star) that awakens the earth."

Bright Stars in General

Bright stars are important. Expressions list as *coch u pacat ek*, "broad-his-face star"; *zazil u pacat ek*, "luminous-his-face star"; and *zaztal ek*, "luminous star" (Genet in Tozzer 1941:133, n. 621).

The North Star and the South Star

Xaman ek, "north star," is very well documented, but only the Motul dictionary glosses it as "north star, guide of merchants." *Xaman* does not seem to translate in any sense other than "north." A rarely mentioned name is *chimal ek*, "north star"; literally the name reads "shield star." If the serpent on pages 12-18 of the Codex Madrid is Draco (Hatch 1975), the shield on page 16 might well be *chimal ek*, and the Chac(s) other parts of Ursa Minor and nearby stars. The southern Polaris, speculatively Canopus, is *nohŏl*, "south; greatest of all" and "the star of the south, counterpart of the north."

Ursa Minor

Chimal ek glosses with both "star of the north with its guards" and "guards of the north or trumpet." *Xaman ek* reads only as "guards of the north or trumpet." The "guards" and "trumpet" are names for Ursa Minor (Allen 1963:459). Specifically, the guards are beta, gamma 1 and gamma 2 of Ursa Minor (Allen 1963:459). The Vienna dictionary lists the following names for Ursa Minor, both glossed with "guards of the north or trumpet": *yah bakul (pach) xaman*, "those who circle in (the back of) the north" and *yah zutil (pach) xaman*, "those who revolve in (the back of) the north."

Xaman ek is not only the north star and Ursa Minor but also the guide of merchants, according to the Motul dictionary. The main Maya merchant deity is God M; he is called *Ek Chuah*, "Black Scorcher," the name for a large black scorpion (Barrera Vásquez and Silvia Rendón 1948:159, n. 10). God M is the jaguar lord of the 20-day month Uo and of the number seven (Kelley 1965; 102; 1976:72). The jaguar of seven might be *xaman ek*, for on a Classic monument at Pomona the pack of the merchant god replaces the feline head of the patron of Uo (Thompson 1970:307). A jaguar's skin can symbolize the starry sky (Thompson 1970:293; 1972:27, 68). The jaguar of seven is surely a star himself, for the glyph meaning "star" (T510) adjoins his feline face on Yaxchilan's Stelae 1, 4, and 10 and on the "Venus" mask at Copan (also Kelley 1976:39, Fig. 9). Furthermore, the feline seems to be GIII-"13 Death," the main god of the Temple of the Sun at Palenque (Schele 1976:11-12) and a major numen of the Late Classic (A.D. 650-850). The frontal face of the jaguar of seven is frequent on Classic war shields, such as that in the Temple of the Sun. This suggests some tie between the stellar feline and *chimal ek*, "shield star." Briefly, then, the logic is that *xaman ek* is God M, *Ek Chuah*, the starry jaguar of seven, GIII, and *chimal ek*.

Gemini-Orion and Scorpio

Literally "turtle star," *ac, ac ek* glosses as "the three stars together that are in the sign of Gemini, which with others make the form of a turtle"; for some, *ac* today is or includes Orion (Thompson 1971:111,116). The *ac* constellation might be referred to in glyph B4 of Lintel 3 in Tikal's Temple 4 (Kelley 1976:50); the inscription dates to about A.D. 740. *Ac* also means "dwarf" and "lead boar of a pack" and "grass for thatching." Well-known, too, is *zinaan ek*; verbatim it means "scorpion star," but it reads as "Scorpio, sign of the sky."

The Pleiades

The most famous of all Mayan asterisms is *tzab ek*, "snake rattle star" and "the kids, a constellation of seven stars"; the Kids are the Pleiades. *Tzab* is in the fifth layer of the sky (Roys 1965:7, 8) and seems to be as well the unidentified birds or insects *ko* and *oo* (Roys 1965:7, 9, 19-21). Other names for the Pleiades or perhaps for some of its stars are *Ix Hochan-ek*, "lady-scraped-clean-star" (Roys 1965:52); *Ix Hom-ti-tzab*, "she-who-sinks-into-the-rattles-constellation" (Roys 1965:37); *Ix Hun-ti-tzab*, "lady-unique-pulsating-rattles-constellation" (Roys 1965:52, 151); *Ix Ko-ti-tzab*, "lady-*ko* (-bird?)-in-the-rattles-constellation" (Roys 1965:155); *Ix Ti-ho-tzab*, "lady-at-the-five-rattles" (Roys 1965:8); and *Ix Ti-tzab*, "lady-in-the-rattles-constellation" (Roys 1965:9).

Gemini or the Hyades or Orion: The Elusive Astillejos

Near the Pleiades and in the sign of Taurus is the famous Aztec "Fire Drill" constellation *mamalhuaztli*, called by Sahagun the astillejos and mastelexos, "little sticks" (Coe 1975:25). The Nebrija dictionary (1951:n.p.) of about 1495 has this entry: "*astillejos--constelación del cielo. Orion.*" Of all the Yucatec wordbooks, *astillejos* appears in only the Motul: "*mehen ek: los astillejos, constelación del cielo*"; the Motul does not locate the *astillejos*. The Nebrija and the Motul both use *constelación* for this asterism and not *signo*, which was earlier reserved for the twelve signs of the zodiac, such as Gemini (*ac ek*) and Scorpio (*zinaan ek*). The present official *Diccionario de la Lengua Española* of the Royal Academy has this entry: "*Astillejos Cástor y Pólux, estrellas principales de la constelación de Géminis*" (1970:134). This last definition might have led Anderson and Dibble (1953:60) to see the *astillejos* in Castor and Pollux. Seler and Vigil place the asterism in the Hyades, while others (Coe 1975:25, 26; Robelo 1951:473, 474; Thompson 1972:68) identify it with the Belt, Sword, and Square of Orion.

The "Starry Ballcourt," Perhaps Ursa Major?

Tezozomoc describes the Aztec constellation *citlaltlachtli*, "starry ballcourt," as "the north and its wheel"; Seler places it among the circumpolar stars, but Coe (1975:23, 27) regards this as rather tenuous. Leal (1973:26) describes it as a circumpolar constellation including some of the stars of Ursa Major. It is very surprising to find *ekel ek* among the Yucatec words for "ball game." The name has no obvious relation to the game and could translate as "darkness stars." If the Maya had an equivalent of *citlaltlachtli*, *ekel ek* seems to be the best candidate. Just possibly *ekel ek* means "jaguar star" and refers to Ursa Major, a jaguar for the Aztecs (Icazbalceta 1891:213), because *ekel*, "spotty one," denotes "jaguar" (Kelley 1976:55); again, *citlaltlachtli* might include some stars of Ursa Major (Leal 1973:26). In the Yucatec village of Chan Kom, Ursa Major is today at least known as *noria ek*, "water wheel star" (Redfield and Villa Rojas 1934:206). This name recalls the description of *citlaltlachtli* as "the north and its wheel." To draw water from a *noria*, a horse or mule is hitched to a beam and goes round and round; the motion suggests that of a wheel. In short, then, *ekel ek* suggests a tie to the "starry ballcourt" and to the jaguar-Ursa Major, while *noria ek* links Ursa Major to the water wheel and thus perhaps to the "starry ballcourt" or *citlaltlachtli*, also called "the north and its wheel."

The Milky Way

One name for the Milky Way is *tamcaz*, "madness" and "fer-de-lance." The other is *ah poo (= ppoo) u*, which literally translates with "he, bulgy crystal-bead necklace" or "he who washes the moon." *Xibal bey* is the term which the Quiché Maya give the Milky Way in winter, when it is bifurcated; the bifurcation is identified with the Underworld, and most likely the Maya see the Milky Way as the road of souls journeying to the realm of the dead (Coe 1975:27, 28). *Xibal bey* means "frightful road" (Coe 1973:21) or "hell" and "similar roads" (León 1954:65, 66). In summer the Quiché Maya call the galaxy *sac be*, "white road" (Coe 1975:27, 28).

Comets

Several terms for "comet" are recorded. *Budz ek*, *budzal ek*, and *budzil ek*, literally "smoke(y) star," gloss as "maned comet like the one that appeared in the year 1577." "Big comet" is *kak noh ek*, "fire big star." "Small comets" is *chamal dzutan*, "cigar of the warlock." The expression for "tailed comet" is *ikomne*, "windy tail" or "wind foam tail." This is shortened to *ikom*, "windy" or "wind foam." Glyph B10 of Tikal's Stela 5 (A.D. 744) seems to name a comet (Kelley 1976:39, Fig. 9). *Kaktamay ek*, "fire ill-omened star," glosses as "carbuncle" and "fabulous animal"; at least its literal sense suggests that this term denotes comets.

Comets or Meteors?

Halal ek, "arrow star," is glossed as "comet that runs." *U halal dzutan*, "arrow of the warlock," is given as "ignited comet." The glosses use "comet," but "arrow" and "runs" suggest the speed of meteors. Anders (1963: 337) reports that the Lacandon Maya describe meteors as arrowheads.

The Moon and Its Glyphs

"Moon" itself is *u*, which also glosses as "collar or necklace or choker," which suggests the lunar crescent. The glyphs for the moon, T181 and T683, are essentially crescents. *U* is also "glass beads," which recalls the moon when it is not a crescent.

New Moon

The moon's appearing is *miz caan u* and *miz cuch u*, literally and respectively "the moon sweeps or cleans the sky" and "the moon sweeps or cleans its place." "At the new moon" is *hokzah u*, "to make appear the moon." "The new moon of one day" is given with various entries. One is *yax u*, "green, first-born or new moon." Another reads *dzay canu*, "serpent-fang moon" or "soapwort moon." A third is *chichaan u*, "flacid, sufficient, or showy-edge moon" or "very small moon" or "rinsed-out moon." *Pal u* literally reads as "child moon," while *paal u* is "child moon" or "crack moon" or "sherds moon." *Ne u* reads as "tail moon."

Waxing Moon

The waxing crescent is called *u nuchal u*, "the crescent of the moon"; it literally reads "its making fat the moon." The waxed moon is *nuc u*, "fat or old moon."

Full Moon

"Full moon, opposition" is *yiih u*, literally "old moon" and "ear-of-maize moon"; and *yih u*, "mature moon." *Num caan u* is given as "full moon that lasts all the night" but verbatim is "full sky moon." The name is probably due to the fact that the moon crosses the whole night sky for the first time when it becomes full. *Oxlahun caan u* is another expression for the full moon, meaning word-for-word "thirteen heaven moon"; the thirteenth heaven is the highest, presumably because the new crescent appears in the east and the old one vanishes in the west, with the full moon in between, both in time and in space.

Waning Moon

"Waning moon" reads as *yih u*, "mature moon," and as *yiih u*, "old moon" and "ear-of-maize moon." Another term is *mucul u*, "hidden or buried moon." "The very waning moon that almost does not appear" is *benel u, bene u tu ch'en*, "the moon goes, the moon goes to its cenote."

Conjunction

"Moon at conjunction" is given as *(u) (y)ich luum yan u*, "within the earth is the moon." Another way of saying this is *u nup(p)tanba*, "the moon (and earth?) collide with or enclose each other." *Up(p)ahtanba u* glosses as "conjunction of the moon" and literally translates as "the moon (and earth?) dent or break or enclose each other." *U paktamba* glosses the same and reads verbatim as "the moon (and earth?) join with each other and mix." *Zipit caan* literally reads "the sky allows (the moon) to withdraw," or maybe "(the moon) withdraws from the sky." *Helep u* is "change moon." *Hun balan u* translates as "roped moon" (Brinton, in Kelley 1976:45) or "wholly hidden moon." The last expression is *binaan u, binan u ti ch'en*, literally "gone the moon, gone the moon to its cenote"; the gloss reads "moon, moon (the days?) it is not seen."

CONCLUSION

This extensive survey shows how rich the dictionaries are in starlore. The word books, literature, folklore, and ethnographies of the many other Mayan languages are probably just as rich. The inscriptions and the codices should also be scoured for all instances of the star glyph in order to discover not only new names, but also Classic and Postclassic antecedents for the colonial ones. No source can be neglected. Maya star lore, so fascinating in itself, might well give us many insights into the complex workings of the Mayan calendar, into the correlation question, and into obscure but intriguing beliefs about the afterlife.

TABLE 21.1

LIST OF TERMS IN THE TEXT ALONG WITH THEIR GLOSSES, TRANSLATIONS, AND SOURCES

Abbreviations used in the table:

A	Pío Pérez n.d.	S	Swadesh et al. 1970
BM	Vienna Dictionary 1972	SF	San Francisco Dictionary 1870
C	Charcency 1884	V	Vienna Dictionary 1978
CA	Pío Pérez 1898	n	note
MH	Motul Dictionary 1929	r	*recto*, front of page
MS	Motul Dictionary n.d.	s	Spanish-Maya section
P	Pío Pérez 1866-77	v	*verso*, back of page

After each term comes first the underlined gloss, then my literal translation of the term. Normally a term in BM is in V, and if it is in MS, it is in MH. In some cases I did not note parallel page numbers for entries in these works. In a very few instances, there are no parallels.

A. STARGAZERS

1. *(y)ahau ca(a)n*
   rattlesnake — A3 MS5r P3 V34
   basilisk — C9
   bishop — MS212v P4
   lord of the sky

2. *ah kin*
   astrologer — C7
   priest — MS20r SFs96 V45
   he of the sun

3. *ah miatz*
   sage — A8 BM103v C74 MS23r P6 S65,105 V47,402
   scientist — S65,105

4. *ah miatz tu xinbal ekob, tu zut ekob*
   star man or astrologer — BM103v V47,551
   he who is wise to the walking of the stars, to the revolving of the stars

5. *ah noh kin*
   the great sun priest — Edmonson n.d.:19.1, n. 483; 25-25.1, n. 777
   the great astrologer

6. *balam*
   jaguar — A20 MS44v P18 SF15 V70
   highest priest — Edmonson n.d.:25-25.1, n. 777

7. *chilam (balam)*
   spokesman (of the jaguar) — A56,117 MS144r P73 V41, 197 Edmonson n.d.:19.1, n. 483; 25-25.1, n. 777

8. *ek*
   star — A69 BM103r C34 MS162v P99 S113 SF88,s187 V224
   secular priest — V224

B. ASTROLOGY

1. *(y)idzatil ah ahel (= ohel) tu xinbal caanob, tu xinbal ekob*
   astrology BM22r V303
   the art of him who knows where the heavens walk,
   where the stars walk

C. STARS IN GENERAL

1. *chac nicen(hal) caan*
   the starry sky BM103r,v MSs113v SFs187 V186
   (make) red or great flowering sky

2. *ek*
   star A69 BM103r C34 MS162v,s113v P99 S113 SF88,s187 V224
   carbuncle A69 BM103r C84 MSs162v P99 S50,117 SF88 V224
   spots like those of the tiger and deer when small BM163r V224
   spots (that are on the body) of the deer MS162v S50,124

3. *hokol ek*
   the appearing of the stars MSs113v SFs187 V256
   the stars sprout, rise, or emerge

4. *(hun) tzotz ceh ek yan ti caan*
   innumerable are the stars that are in the sky MH265 MS120r
   [as many as] (all) the hairs of the deer are the stars in the sky

5. *katal eekob*
   the appearing of the stars BM183v V329
   the stars are put on high

6. *tippil ek*
   the appearing of the stars BM183v MS424,s113v SFs187 V540
   the stars rise, emerge, appear above, or pulsate

7. *tul ekil caan*
   starry sky BM103r V553
   full of stars the sky

8. *u lol akab*
   flower of the night Roys 1967:27,94 Barrera Vásquez and Silvia Rendón 1948:137

9. *u lol zaz caan*
   luminous flower of the sky Barrera Vásquez 1965:52, n. 18; 58,66

10. *u nak caan*
    the sky or layer that we say of the sky, that which appears to be seen of the sky BM135r MH660 MS321v V421
    belly of the sky

D. PLANETS

1. *chachac ek*
   planets Roys 1967:150
   red stars Roys 1967:150, n. 4
   great or red stars Barrera Vásquez and Silvia Rendón 1948:71
   red, red stars

2. *planetob*
   planeta pluralized à lá Maya Coe 1975:19

3. *lacam ek*
   great star   MH534 MS257r
   banner star
   somersault star

4. *X'ain ek*
   Saturn   Mejía 1969:43
   female crocodile star

E. MORNING STAR, EVENING STAR, AND VENUS

1. *ah ahzah cab (ek)*
   luminary of the morning   MH70 MS4r V64
   he (the star) that awakens the earth

2. *ah oczah kin*
   luminary of the night   MH100 MS24v
   he who makes enter (into the Underworld?) the sun
   he who sows or plants the sun

3. *ah ppizba hab*
   error in MH105 for *ah ppiz akab* in MS28v
   he who measures the year

4. *ah ppiz akab*
   luminary that rises at first (hour of the?) night and runs through all of it, such that it seems to be measuring it   MS28v
   he who measures the night

5. *chac ec*
   a reddish wasp that does not sting   Roys 1965:132; a pun, it replaces at times *chac ek*
   red wasp

6. *chac ek*
   luminary of the day   M134v S42,119 SFs242
   great or red star

7. *chac noh ek*
   luminary or morning star   BM103v V186
   red, big star

8. *hózan ek*
   star of the evening   C34
   tired-out star; name of the black Bacab of the west

9. *kanal*
   large red wasp   E. May in Roys 1965:132
   star   Kelley 1976:38

10. *noh ek*
    luminary and luminary of the morning   BM103v CA224 MS145r SFs187,242 V431
    luminary of the morning and star of the evening, Venus   BM138v MH679 MS330r V431
    Venus   S69,132
    luminary and big star   A146 P245
    star, luminary of the day   MSs113v
    big star

11. *xux ek*
luminary of the morning   MH935   MS466r   S92,119
luminary of the morning and star of the evening, Venus   BM138v   V607
wasp star

F. BRIGHT STARS

1. *coch u pacat ek*
star of great clarity   BM103v   V115
broad-his-face star

2. *lacam ek*
big star   MH534   MS257r
big tumor
banner star
somersault star; a planet?

3. *zazil u pacat ek*
star of great clarity   BM103v   V133
luminous-his-face star

4. *zaztal ek*
luminous star   Genet in Tozzer 1941:133, n. 621

G. NORTH STAR AND SOUTH STAR

1. *chimal ek*
star of the north   MSs113v   P74   SF72,s187
shield star

2. *no ek*
Lamb 1979:22:error, delete

3. *nohól (ek)*
south; greatest of all and star of the south, counterpart of the north   MH680   MS330v
star of the south   S69,113
south (star)

4. *xaman ek*
star of the north   A232   BM151r   CA237   MH916   MS455r,s113v   P389   S90,113   SFs187
star of the north, guide of merchants   MH916   MS455r

H. URSA MINOR

1. *chimal ek*
star of the north with its guards   BM151r   V198
guards of the north (or trumpet)   MB115r,151r   MH311   MS144r   V198
ursa minor   S44,122; S122 reads major but must be minor
shield star

2. *xaman ek*
guards of the north (or trumpet)   MH311   MS144r
north star

3. *yah bakul (pach) xaman*
guards of the north or trumpet   BM115r   V34,591
those who circle around in (the back of) the north

4. *yah cuchil*
Lamb 1979:22:error, delete

5. *yah zutil (pach) xaman*
   guards of the north or trumpet   BM115r   V40,591
   those who revolve in (the back of) the north

I. GEMINI-ORION AND SCORPIO

1. *ac (ek)*
   the three stars together that are in the sign of Gemini, which with others make the form of a turtle   MH66   MS1r
   star of Gemini   S33,113
   turtle-, dwarf-, leading boar- or thatching grass-star

2. *zinaan (ek)*
   Scorpio, sign of the sky   MH237   MS104r   S95,112
   scorpion (star)

J. THE PLEIADES

1. *Ix hochan-ek*
   Lady-scraped-clean-star   Roys 1965:52

2. *Ix-hom-ti-tzab*
   she-who-sinks-into-the rattles-constellation   Roys 1965:37

3. *Ix-hun-tip-tzab*
   lady-unique-pulsating-rattles-constallation   Roys 1965:52,151

4. *Ix-ko-tip-tzab*
   lady-ko (-bird?)-in-the-rattles-constellation   Roys 1965:155

5. *Ix Ti-ho-tzab*
   lady-at-the-five rattles   Roys 1965:8

6. *Ix Ti-tzab*
   lady-in-the-rattles-constellation   Roys 1965:9

7. *ko*
   unidentified bird or insect   Roys 1965:7,9,19-21,25

8. *oo*
   unidentified bird or insect   Roys 1965:7,9,19-21

9. *tzab (ek)*
   the kids, a constellation (of seven stars)   BM31v   CA157,161   MH254   MS113v,s42r   S85,104   SF71,s15?   V153
   the kids of the sky, constellation   A213
   the seven kids of the sky   P359

K. GEMINI OR THE HYADES OR ORION: THE ELUSIVE ASTILLEJOS

1. Castor and Pollux:*Diccionario de la Lengua Española* 1970:134; Anderson and Dibble 1953:60

2. Hyades:Seler and Vigil, in Coe 1975:25-26

3. *mehen ek*
   the astillejos, constellation of the sky   MH62   MS303v
   son or semen star

4. Orion:Nebrija Dictionary; Coe 1975:25-26; Robelo 1951:473-474; Thompson 1972:68

L. <u>THE STARRY BALLCOURT, PERHAPS URSA MAJOR</u>

1. *ekel ek*
   <u>game of ball</u> A69 CA25 P101 S50,118
   darkness stars
   spotty one or jaguar stars

2. *noria ek*
   <u>water wheel star</u> Redfield and Villa Rojas 1934:206

M. <u>THE MILKY WAY</u>

1. *ah poo (= ppoo) u*
   <u>Milky Way</u> MH103 MS26v
   he the bulgy crystal bead necklace
   he who washes the moon

2. *tam(a)caz*
   <u>Milky Way</u> S80,132
   <u>Milky Way or Road of St. James</u> MH834 MS399v
   <u>fer-de-lance</u> Coe 1975:27
   <u>madness</u> A185 MH834 MS399v P317

3. *sac be*
   <u>white road</u> Coe 1975:27

4. *xibal bey* (Quiché Maya)
   <u>Milky Way</u> and <u>similar roads</u> and <u>hell</u> León 1954:65,66
   <u>frightful road</u> Coe 1973:21

N. <u>COMETS</u>

1. *budz (ek)*
   <u>maned comet like the one that appeared in the year 1577</u> MH160 MS58v
   <u>comet</u> SF125 S37,106
   smoke (star)

2. *budzal (ek), budzil (ek)*
   <u>maned comet like the one that appeared in the year 1577</u> MH160 MS58v
   <u>comet</u> SF125 V90
   smokey (star)

3. *chamal dzutan*
   <u>small comets</u> MH296 MS137r S42,106
   cigar of the warlock

4. *ikom*
   <u>comet</u> MSs52v
   windy or wind-foam

5. *ikomne*
   <u>(tailed) comet</u> MH468 MS225r SF125
   windy tail or wind-foam-tail

6. *kak noh ek*
   <u>big comet</u> BM42v V321
   fire big star

7. *kak tamay (ek)*
   <u>ulcer</u> Roys 1965:154
   <u>carbuncle</u> A102 P166
   <u>fabulous animal</u> A102 P166
   fire ill-omened (star)

O. COMETS OR METEORS?

1. *halal ek*
   comet that runs BM42v V239
   arrow star

2. *u halal dzutan*
   ignited comet BM42v V239
   arrow of the warlock

P. THE MOON AND ITS GLYPHS

1. *u*
   moon A220 MH882 MS436r,s144r P369 S87,119 SF277,s243 V567
   collar, necklace, choker A220 MH882 MS436r P369 S87 SF277,s243 V566
   glass beads A220 CA81 P369

Q. NEW MOON

1. *chichaan u*
   the new moon of one day V567
   flacid, sufficient or showy edge moon
   very small moon
   rinsed out moon

2. *dzay can u*
   the new moon of one day BM138r V167,567
   serpent-fang moon
   soapwort moon

3. *hokzah u*
   at the new moon V255
   to make appear the moon

4. *miz caan u*
   for the moon to come out when its brightness shows on the horizon MH629 MS306r
   moon sweeps or cleans the sky

5. *miz cuch u*
   for the moon to come out when its brightness shows on the horizon MH629 MS306r
   moon sweeps or cleans its place

6. *ne u*
   new moon C50
   tail moon

7. *paal u*
   the new moon of one day or new moon BM138v MH750 MS366r,s144v V473,567
   child moon
   crack moon
   sherds moon

8. *pal u*
   the new moon of one day P369 SF277
   child moon

9. *yax u*
   the new moon of one day   BM138v
   green, first-born or new moon

R. WAXING MOON
   1. *nuc u*
      to be grown the moon   BM50v   MH686   MS333v   V435
      fat moon

   2. *nuchal u*
      to grow the moon BM50v   V435
      to make fat the moon

   3. *u nuchal u*
      the crescent of the moon   BM50v   V435
      its making fat the moon

S. FULL MOON
   1. *num caan u*
      full moon that lasts all the night   MH691   MS335v
      full or whole sky moon

   2. *oxlahun caan u*
      full moon, opposition   MH724   MS353v
      thirteen heaven moon

   3. *yih u*
      full moon, opposition   SFs243
      mature moon

   4. *yiih (u)*
      full moon, opposition   MH454   MSs144v
      old moon
      ear-of-maize moon

T. WANING MOON
   1. *benel u, benel u tu ch'en*
      the very waning moon that almost does not appear   MH146   MS50v
      moon goes, moon goes to its cenote

   2. *mucul u*
      waning moon   BM138v   MH637   MS310v,s144v   P226   SF177   V408,567
      hidden, buried moon

   3. *yih u*
      waning moon   P369   SF277
      mature moon

   4. *yiih (u)*
      waning moon   SF309
      old moon
      ear-of-maize moon

U. CONJUNCTION
   1. *binaan u, binan u ti ch'en*
      moon, moon (the days?) it is not seen, between moon and moon   BM94r   V81,567
      gone the moon, gone the moon to its cenote

2. *helep u*
   conjunction of the moon   MH381 MS183r
   change moon

3. *hun balan u*
   conjunction of the moon   MH406 MS196v
   roped moon   Brinton in Kelley 1976:45
   wholly hidden moon

4. *(u) (y)ich luum yan u*
   moon in conjunction   BM45v,138v CA168 MH465 MS224v,s55v SF353 V305,567
   within the earth is the moon

5. *u nup(p)tanba*
   conjunction of the moon or moon in conjunction   SFs107,243
   the moon (and earth?) collide with or enclose each other
   moon enemy

6. *u paktamba*
   conjunction of the moon   A159 CA168 P268
   the moon (and earth?) join with each other and mix

7. *up(p)ahtanba u*
   conjunction of the moon   MSs55v
   the moon (and earth?) dent or break or enclose each other

8. *zipit caan*
   conjunction of the moon   BM45v V145
   the sky allows (the moon) to withdraw
   (the moon) withdraws from the sky

22 Architecture, Astronomy, and Calendrics in Pre-Columbian Mesoamerica

Vincent H. Malmstrom[1]

INTRODUCTION

It has long been recognized by archaeologists working in Mesoamerica that the great pre-Columbian ceremonial centers of this region were the consciously-planned products of societies with highly developed architectural sensitivities. The design and construction not only of individual buildings or complexes but also, on occasion, of entire cities are known to have been carried out with careful attention to such concepts as harmony, balance, and proportion. On the other hand, only relatively recently has it begun to be appreciated that many of the architectural alignments found in Mesoamerica appear to be related to the religious and/or cosmological beliefs of the early civilizations which built them. This tardy recognition of the correlation between architecture and religion is all the more surprising when we pause to consider that most of the monumental structures which these societies erected, be they pyramids, temples, or ball-courts, were meant to serve a religious function. And because an almost obsessive preoccupation with the calendar and astronomy was central to the spiritual lives of all Mesoamerican cultures (Miles 1965:284), it is only reasonable to expect that many structural orientations within this region should reflect such astronomical events as the equinoxes, the solstices, or the zenith passages of the sun. In spite of this, an alignment which has frequently been encountered on the Mexican plateau but not adequately explained to date is one that is oriented to an azimuth approximating 285°, or about 15° north of west.

The foremost example of this alignment (which is skewed some 15° clockwise from the cardinal points) is that provided by Teotihuacán, located approximately 50 km northeast of the present city of Mexico. In its heyday, which lasted from approximately the time of Christ until the mid-seventh century of the Christian era, Teotihuacán was the largest urban agglomeration anywhere in the New World. Indeed, synchronous with Rome in Europe and Peking in Asia, it may have ranked as one of the three most populous cities in the world at the time. It owed its greatness to a congenial site on the shore of a large lake which afforded both ease of transportation and irrigation water for the rich alluvial soils of its semi-arid mountain basin. Moreover, it was strategically located on a pass leading across the mountains which fringe the Mexican plateau so that it could effectively command the exchange of goods between the

---

[1]The author is especially indebted to Paul A. Dunn, Dartmouth Class of 1981, for his assistance in conducting measurements in the field, and to Mark S. Fagan, Dartmouth 1980, for preparing the final maps.

cooler, drier uplands to the west and the warmer, moister lowlands to the east. At its peak, Teotihuacán was a thriving metropolis of well over 50,000 inhabitants (some estimates place the city's population as high as 100,000 to 200,000), but a city which--as Millon has emphasized--was as much a religious center as it was an economic node (Millon 1973:54-59).

As the configuration of this great city was revealed during the Teotihuacán Mapping Project in the 1960s and 1970s, it was realized that its principal thoroughfare--a grandiose avenue called the "Street of the Dead" by the Aztecs--was oriented 15.5° east of north, and that a grid system having this same alignment had been superimposed on all of the surrounding countryside. Even the drainage ditches separating the productive, alluvial farm-plots or *chinampas*, that had been reclaimed from the south end of the lake near present-day Xochimilco, display the same well-ordered orientation (Coe 1964:96). Here was urban and regional planning on a massive scale, but what was the principle behind it? Why had the Teotihuacános chosen to orient everything 15.5° off of north, rather than to the cardinal points?

As early as 1945 Macgowan, in a brief note published in *American Antiquity*, had commented on the fact that most pre-Columbian ceremonial centers in Mesoamerica appeared to have been oriented east of north (Macgowan 1945:118). He observed that orientations seemed to cluster in three general groups: true north, about 7° east of north, and about 17° east of north. From field measurements begun in 1971, Aveni was able to confirm that 50 of 56 sites which he examined were indeed oriented east of north, and that a 17° "family" of alignments did in fact seem to exist (Aveni 1975:166). Aveni also concluded that this particular pattern of orientation seemed to be most prevalent in the Valley of Mexico, and he specifically mentioned Tenayuca, Tepozteco, and Tula as sites where this alignment was encountered. (It should be pointed out that Aveni's "17° family" of alignments includes orientations as far off as 15.5°, for he includes Teotihuacán in this "group".)

Ever since the 15.5° alignment of the Street of the Dead at Teotihuacán--and the consequent orientation of the front of the Pyramid of the Sun toward an azimuth of 285.5°--was first recognized, numerous explanations have been advanced to explain it. In 1967, Dow suggested the setting position of the Pleiades as a possible motive for such an orientation (Dow 1967:326-344), following the same line of thought as Marquina and Ruíz had presented for the orientation of the Tenayuca pyramid in 1934 (Marquina and Ruíz 1934:101-106). However, Dow also considered the rising positions of both Sirius and Dubhe as possible factors in the orientation of Teotihuacán as well. Heyden and Gendrop (1975:39) stated that the Pyramid of the Sun is oriented toward the sunset position on the day that the sun passes vertically overhead at Teotihuacán, though this can easily be shown to be 291.0° rather than 285.5°. In the same year, Heyden offered another possibility, suggesting that the Pyramid of the Sun might be aligned in the same direction as the mouth of a cave which had recently been discovered beneath the structure (Heyden 1975:131-147). Summarizing the various hypotheses on orientation which had been advanced up to 1975, Aveni also mentioned an unpublished idea of Drucker that the Pyramid of the Sun might be oriented to solar positions which utilize local topographic features (Aveni 1975:170), but he himself concluded (Aveni 1977b:5) that "they (the Pleiades) must remain the prime candidate for an astronomical motivation in the orientation of Teotihuacán." Aveni reached this conclusion after admitting in his earlier work that (1) the Pleiades would have been "invisible well before they reached the horizon of Teotihuacán" and (2) "the precessional motion of the Pleiades is so rapid that a mistake of 100 years in the dating of the baseline is equivalent to a shift of ½° in the azimuth of the setting point" (Aveni 1975:169). Recognizing the latter point as a particularly critical flaw in the "Pleiades hypotheses," Aveni has had to dismiss all later sites built on the Mexican plateau having similar orientations to Teotihuacán (such as Tenayuca, Tepozteco, and Tula) as being simply "non-functional imitations" of the great Mesoamerican metropolis (Aveni 1975:170).

More recently (1978), Aveni and his associates have investigated the distribution of "pecked cross symbols" in Mesoamerica, most of which are located on the Mexican plateau near Teotihuacán, and some of which seem "to have been modified to fit the overriding architectural plan" (Aveni et al. 1978: 279). On the basis of these observations, Aveni concludes that "Teotihuacán North" was more important than "astronomical north" and that, apart from the "surprising" discovery of similar symbols in the Petén region of Guatemala, the "data are consistent with an origin in and diffusion from the classic Teotihuacán empire" (Aveni et al. 1978:279). The fact that some of the symbols are composed of approximately 260 elements further suggests that they may have been related to the ritual count of 260 days which was used throughout pre-Columbian Mesoamerica (Aveni et al. 1978:267-269).

As a result of field work completed in early 1979, the present author has been able to confirm alignments to an azimuth of ≈285° not only at several sites on the Mexican plateau but in the Yucatán, Petén, and Oaxaca regions as well. These observations were refined and augmented by data kindly supplied by Professor A. F. Aveni (personal communication 1979) and are tabulated in Table 22.1. Although no claim can be made for the completeness of the survey, the principal conclusions which can be drawn from the distribution which emerges are that (1) virtually all regions of Mesoamerica are represented and (2) sites ranging in age from 1000 B.C. to well into the Classic Period (A.D. 300-900) are included. In the discussion to follow, a brief resumé is given of only the major sites which have been identified elsewhere than on the Mexican plateau.

Thanks to the work of Matheny (1976:639-640), it is now known that the ceremonial center of Edzná in the southwestern part of the Yucatán peninsula was a large and thriving urban agglomeration as early as 150 B.C. It was thus essentially synchronous with Teotihuacán. It is interesting to note that the major structure at Edzná--a five-story pyramid known in Spanish as "Cinco Pisos"--is squarely oriented to an azimuth of 285.5°, and as the map prepared by Andrews reveals (Aveni 1975:257), a similar orientation is shared by many other structures at the site.

At Chichén Itzá, Toltec reconstruction has altered most of the buildings on the northern side of the site, but several of those on the southern side, such as Chicchan Chob and the House of the Deer, show a 285° orientation. The observatory, or Caracol, is an especially interesting case, for the older Maya core of the building has the mid-line of its front window (Window 1) oriented squarely toward 285°, whereas the encircling platform and front stairway added by the Toltecs (Gendrop 1972:151) faces directly toward the summer solstice sunset.

Although individual structures at a number of smaller sites demonstrate orientations approximating 285°, it is perhaps significant that in many

TABLE 22.1

STRUCTURES ORIENTED TO ≈285°[a]

General Region	Site	Time-Frame	Structure	Orientation	Measurement
Mexican plateau	Teotihuacán	A.D. 0-650	Pyramid of the Sun	Front	*285.5° (M)
	Tula	A.D. 950-1150	Temple B	Side	287.1° (A)
	Tenango	A.D. 650	Principal Structure	Side	283.7° (A)
	Xochicalco	A.D. 650	Temple of Plumed Serpent	Front	286.8° (A)
Yucatán	Edzná	150 B.C.-A.D. 850	Cinco Pisos	Front	*285.5° (M-D)
	Sayil	A.D. 600 (?)	Main Palace	Side	282.4° (A)
	Labná	A.D. 600 (?)	Temple of Columns: Mirador	Side	282.1° (A)
	Kabáh	A.D. 600 (?)	Codz Pop	Front	283.6° (A)
	Mayapan	A.D. 900-1300	Observatory: (152)	Front	282.8° (A)
			Temple of Sacrifices (151)	Side	282.8° (A)
	Izamal	A.D. 600 (?)	Platform, Pyramid of Kinich-Kakmo	Side	287.2° (A)
	Chichén Itzá	A.D. 700-1300	Chicchan Chob	Front	285.0° (M-D)
			House of the Deer	Side	285.0° (M-D)
			Caracol (mid-line of Window)	Front	*284.9-285.7° (A)
	Cobá	A.D. 600	XV, XVI	Side	285.0° (M-D)
	Kohunlich	A.D. 600	Ball-court	(Main axis)	285.0° (M-D)
	Chicanná	A.D. 450	Groups C and E	Side	283.0° (M-D)
	Becán	A.D. 450	East Group (I, IV)	Side	283.0° (M-D)
			II, III	Front	283.0° (M-D)
Belize-Petén	Altun Ha	A.D. 600 (?)	Structure A-5	Front	285.0° (M-D)
	Tikal	A.D. 200-900	Alignment of Temple I to Temple IV	Front	*285.0° (M-D)
			Alignment of Temple I to Temple V	(Right Angle)	15.0° (M-D)
Oaxaca	Monte Albán	600 B.C.-1100 A.D.	Mound Y	Front	285.0° (M-D)
	Huamelulpan	300 B.C.- ?	Main Pyramid	Front	285.0° (M-D)
Gulf Coast	La Venta	1000-600 B.C.	Stirling "Acropolis" & "Plaza"	Side	*285.0° (B)

[a]The present author has chosen to re-define Aveni's "17° family" of alignments as the "15.5° family" inasmuch as the principal structures in the major sites of Mesoamerica appear to center on this value; see asterisked items above.

[b]Sources of measurements: A--Aveni; B--Bernal; M--Millon; M-D--Malmström and Dunn. Inasmuch as the author and his assistant carried out their measurements with a surveyor's compass, wherever data obtained by more precise instrumentation are available, they are cited instead--rounded to a tenth of a degree.

of the later and more peripheral ceremonial centers of the Maya cultural realm fewer and less important structures seem to share this alignment, as though suggesting that the "vogue" for such orientations gradually tended to die out both with time and distance. For example, at such recent centers as Palenque and Comalcalco on the western edge of the Maya area, no structures at all were found which were aligned to 285°.

THE ASTRONOMICAL MATRIX AT TIKAL

The largest and most impressive ceremonial center constructed by the Maya is undoubtedly Tikal, located in the rainforest of Petén, the northernmost department of Guatemala. It is dominated by five great pyramids, all of which reach or exceed 60 m in height. Although each of these structures is architecturally interesting in its own right, it is their spatial relationship to each other that is of paramount importance to any appreciation of the astronomical significance of the site. Radiocarbon dating of the lintels found in the pyramid superstructures reveals that Temples I, II, and V were all completed about A.D. 700, whereas Temple IV, the highest, was completed about 741 and Temple III was finished about 810. The range in completion dates, although spanning more than a century, should not obscure the possibility that the entire ensemble may have been planned as an integrated complex. The reason for thinking this is that Temples I and II face each other across the 80-m wide Great Plaza on an axis which runs 9° clockwise of a true east-west alignment (i.e., 99-279°), but Temple I is offset just enough to the north of Temple II so as to afford an unobstructed view of Temple IV some 730 m to the west. The roof-comb of Temple II is also the lowest of the five pyramids, suggesting that although the structure lends symmetry to the Great Plaza, it was not intended by either its height or its distance from Temple I to serve any astronomical function. (The fact that the edifices fronting on the Great Plaza are oriented 9° off of the cardinal points may reflect a magnetic alignment extant in the eighth century; today the magnetic variation in the Petén averages 5° E.)

Once the spatial arrangement of the four most widely separated pyramids is mapped, it becomes apparent that Temple I served as the anchor-point of a relatively sophisticated astronomical matrix. A sight-line between the doorways of Temples I and IV marks an azimuth of 285°, whereas one between the doorways of Temples V and I marks a right-angle to this, namely 15° (Fig. 22.1). Thus the placement of both Temples IV and V appears to commemorate the same recurrent alignment found in the ceremonial centers of the Mexican plateau and of the Yucatán peninsula. In addition, however, a sight-line between the doorways of Temples I and III demarcates a true east-west line and therefore the axis of equinoctial sunrise and sunset. Moreover Temple III is so located (ca. 280 m west of Temple I and 400 m southeast of Temple IV) that the middle of its roof-comb breaks the horizon precisely at an azimuth of 115° as seen from the doorway of Temple IV, permitting an accurate calibration of the sunrise position at the winter solstice. Thus, two of the great pyramids at Tikal are keyed to alignments related to 285°, and a third apparently is situated so as to serve both as an equinoctial and solsticial marker. When one is aware that Tikal constitutes not only the Mayas' crowning achievement in architecture, engineering, and artistry but also a brilliantly conceived and superbly executed astronomical and calendrical matrix, one cannot but stand in yet greater awe of their outstanding abilities and accomplishments.

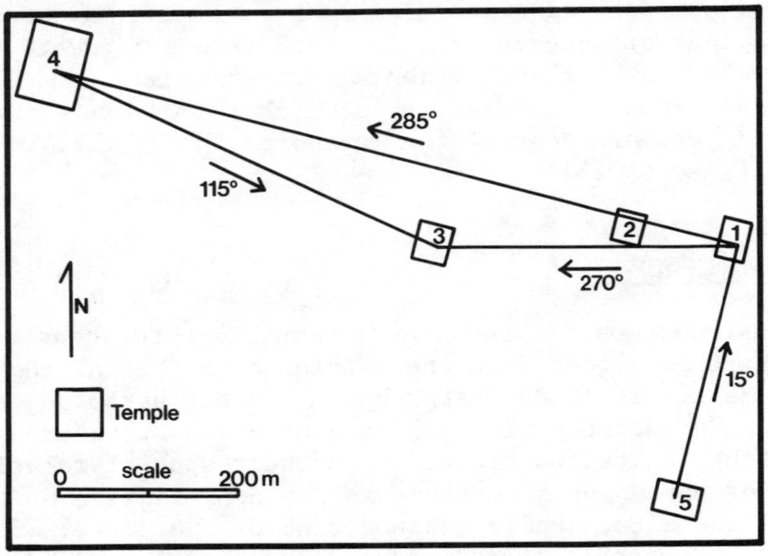

Fig. 22.1 The "Astronomical Matrix" at Tikal (Simplified from National Geographic Society *Archaeological Map of Middle America*, 1972)

SITES IN THE OAXACA AND GULF COAST REGIONS

Having confirmed the presence of architectural alignments toward an azimuth of 285° at both extremities of Mesoamerica (i.e., on the Mexican plateau in the west and in the Yucatán and Petén regions in the east), the author was encouraged to investigate the possibility of similar orientations in intervening geographic areas such as the highlands of Oaxaca and the coastal lowlands bordering on the Gulf of Mexico. The former region constitutes the core area of both the Zapotec and Mixtec peoples, and one key site in each cultural realm was selected for field study. For the Zapotecs, the site selected was Monte Albán, located on the top of a mountain overlooking the modern city of Oaxaca; for the Mixtecs, the ancient site of Huamelulpan in the district of Mixteca Alta, some 100 km northwest of Oaxaca city, was chosen for study. In the Gulf Coast region, the site of La Venta (the "capital" of the Olmecs), was also examined for structural orientations with the help of a map.

Monte Albán dates back to at least 600 B.C. and has been rebuilt in whole or in part several times during its long history. Today virtually all of the mountain-top site conforms to a rigid symmetry oriented to an axis running about 5° east of north. The only structure which deviates markedly from the overall layout is an arrowhead-shaped edifice known as Mound J. Because of its strange configuration and its lack of conformity with the remaining structures in and around the Great Plaza, Mound J has been interpreted as an "observatory," and Aveni has argued that its orientation toward an azimuth of 45° commemorates the heliacal rising of the star Capella, which heralds the passage of the vertical sun over Monte Albán. In support of his contention, Aveni observes that there is a zenith-tube located in Mound P, a structure situated directly to the northeast of Mound J. Without challenging this explanation, one can only wonder what Mound J originally looked like, for it is known to have been rebuilt about the time of Christ and was also probably involved in the total renovation of the site which occurred about A.D. 600. In any event, neither Mound J nor any other structure in and around the Great Plaza demonstrates an alignment toward 285°. Nevertheless, lying half a kilometer to the northeast of the Great Plaza on a lower ridge of the same mountain is a structure known as Mound Y; its orientation is visibly at variance from the general plan of the site even at a distance. Although some restoration of Mound Y has taken place, the largest stones (comparable to those used in the construction and reconstruction(s) of Mound J) which form the foundation platform for the temple have quite obviously not been moved since they were first set in place. Thus Mound Y appears to have been one of the earliest buildings to have been erected at Monte Albán but, unlike Mound J, it does not seem to have been rebuilt on the two occasions that the Great Plaza was reconstructed. It is in all likelihood the oldest structure on the site which still retains its original alignment. Significantly, the front of the platform of Mound Y was found to face squarely toward an azimuth of 285°.

The site of Huamelulpan in the Mixteca Alta district is a far more remote and modest site than Monte Albán. In fact, it consists today of no more than one partially excavated and restored temple. Although the upper walls of the temple have been largely destroyed through centuries of neglect and agricultural activity, the lower foundation is still solidly in place. The southeast corner of the temple is especially interesting, because it is composed of several massive, well-dressed stones measuring up to 2 m in length, almost 1 m in width, and about 0.5 m in thickness. There one finds a beautifully-preserved carving of a lizard, underlain by calendrical symbols which pre-date the initiation of the Long Count (i.e., at least going back before the third century B.C.). Once again the front of the temple foundation was found to be oriented precisely toward an azimuth of 285°.

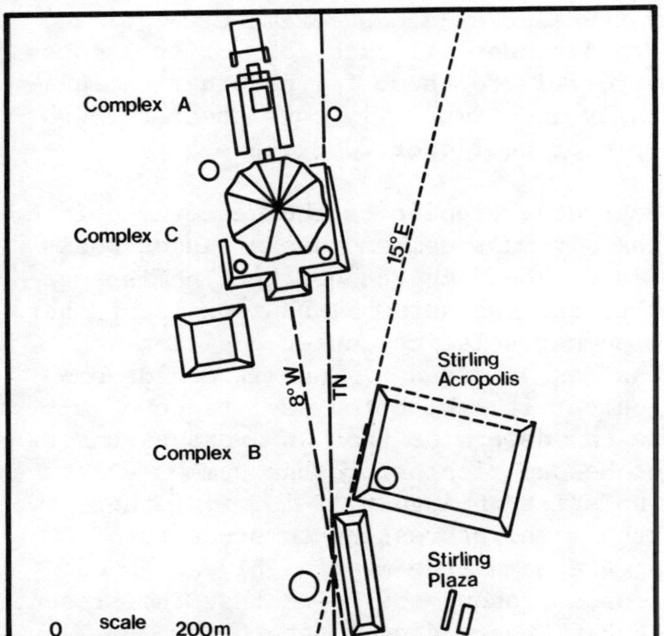

Fig. 22.2 Structural alignments at La Venta (from Bernal 1969:34)

Because accurate field studies of some of the most ancient ceremonial centers of Mesoamerica (namely those erected by the "Mother Culture" of the region, that of the Olmecs) were precluded in this survey, it was decided to utilize the site map of La Venta, an accessible reproduction of which is found in a publication by Bernal (1969:34). Although it has been known ever since the site was first excavated by Stirling that its principal orientation lay 8° *west* of true north, what has apparently been overlooked is the fact that the structures that comprise the so-called Stirling "Acropolis" and Stirling "Plaza" are aligned some 23° to the east of that axis. In other words, the Stirling "Acropolis" looks 15° east of north, while its northwestern face is oriented once again toward an azimuth of 285° (Fig. 22.2). Although the discovery of this alignment at Monte Albán pushed the origin of this "innovation" back at least as far as 600 B.C., its presence at La Venta means that even the Olmecs were commemorating this orientation as long ago as 1000 B.C.!

THE MEANING OF THE 285° ALIGNMENT

Because many of the sites listed in Table 22.1 were either synchronous with Teotihuacán or preceded it, the possibility that the pattern of orientation toward an azimuth of 285° originated in or diffused from that great pre-Columbian metropolis is thereby ruled out. Furthermore, the geographic distribution of 285° orientations occurs through such a broad zone of latitude (roughly 5-8°, depending on which sites are being included) that a stellar association of any kind must be totally discounted. Similarly, they are found through such a wide longitudinal range that a uniform magnetic alignment is likewise precluded. The variation within Mesoamerica currently exceeds 5° between the easternmost reaches of the Yucatán and the westernmost areas of the Mexican plateau. A lunar association with an azimuth of 285° is also quite unlikely, because only the most extreme moonset positions are easily definable, and in the northwest quadrant of the sky these occur at approximately 290° and 300°. Therefore, unless one is willing to discount all such 285° orientations as coincidences, one must ascribe a solar origin to them, for no other explanation can be advanced which is both spatially and temporally consistent through the whole of Mesoamerica during a time span of well over 2000 years.

Because the azimuth of 285° demarcates a point on the western horizon some 15° north of west, it is apparent that it must commemorate a sunset position at a time of the year when the sun is overhead in the northern hemisphere--i.e., between the vernal equinox (March 21) and the autumnal equinox (September 21). Obviously, because this is not the summer solstice sunset position--an azimuth which averages 295° throughout the Mesoamerican region--it could commemorate a date either before June 22 or after it. With the assistance of a solar ephemeris or a nautical almanac one can determine that an average sunset position of 285.5° is recorded throughout Mesoamerica *before* the summer solstice on April 30-May 1, and *after* the summer solstice on August 12-13.[2] Of these two possibilities, the latter is definitely the more interesting to students of Mesoamerican calendrics. This is because the Maya--the most highly advanced of all Mesoamerican peoples in the realm of astronomy--believed that the present era of the world began on August 13, 3114 B.C., according to the Goodman-

---

[2]At 0° horizon altitude, sunset positions on these dates vary from 285.4° at 13° N latitude to 286° at 0° N latitude.

Martínez-Thompson correlation between the Maya and Christian calendars. Ironically, Thompson himself dismissed the correspondence between the beginning date of his correlation and an astronomical origin for the most ancient of Mesoamerican calendars, the 260-day ritual almanac, as being coincidental (Thompson 1960:98). However, recurrent building alignments throughout Mesoamerica appear to commemorate a date which can only have had astronomical (i.e., solar) origins or, at the very least, could only have been established by astronomical means--i.e., by counting the number of days from some fixed point in time. (In the latter regard it should be pointed out that the August 13 sunset position could easily have been defined simply by counting 52 days from the summer solstice.) In the light of the evidence at hand, one can conclude that far from being just one more in a series of "coincidences," the correspondence of building alignment azimuths throughout Mesoamerica and the beginning date of the Goodman-Martínez-Thompson correlation should more realistically be considered as additional proof of the validity of the latter.[3]

If the 285° architectural alignments indeed lend support to the Goodman-Martínez-Thompson correlation, they also perforce support the argument that the "beginning of time" as defined by the Maya actually commemorates an astronomical event--not necessarily a one-time "happening" but some solar occurrence which is replicable every August 13. It is highly unlikely, however, that sunset on that day, in and of itself, would be sufficient cause to initiate a calendar such as the 260-day sacred almanac. In that case it would be far more likely that the explanation first put forward by Nuttall in 1928 is the correct one. She argued that the 260-day count probably could be traced to the interval between vertical sun positions at the latitude of Copán, the Mayas' principal astronomical center, located just south of the fifteenth parallel of latitude in the mountains of western Honduras (Nuttall 1928). This argument won further support from Merrill who remarked on the correspondence between the first zenith passage of the sun at Copán (i.e., August 13) and "the beginning of time" as calculated by the Goodman-Martínez-Thompson correlation (Merrill 1945:307-311). However, both arguments were dismissed by Thompson, who preferred to believe that the 260-day almanac was either (1) an approximation of the human gestation period, or (2) simply a permutation of the numbers 13 and 20, both of which had magical significance to the Maya (Thompson 1960:98-103). (Obviously, if one adopts either of the latter explanations, an August 13 beginning for the sacred almanac *has* to have been a coincidence.) Even so, one is left wondering *whose* gestation period was so important as to commemorate it, and specifically how did one decide *when* it actually began? Or similarly, *what day* was so remarkable that a series of permutations of 13 numerals with 20 day-names was initiated on that date and none other?

Although both Nuttall and Merrill were correct in pointing out that the sacred almanac could be accurately calibrated at Copán, they were wrong in hypothesizing that the 260-day count had actually begun there. In an earlier paper, I have demonstrated that both geographical and historical arguments can be raised against such a contention (Malmström 1973:939). On the first score, they appear to have overlooked a very cogent observation made by the German

---

[3]Although at least nine other correlations have been proposed, none of them comes within 40 days of the beginning date of the Goodman-Martínez-Thompson correlation. In terms of sunset azimuths, the closest approximation is some 8° away. The Spinden correlation, which is the best known of the other systems, employs a starting date of October 15, when the sun is vertically overhead at approximately 8° south latitude and its sunset azimuth is about 262°.

naturalist Gadow in 1908 and cited by Thompson (Gadow 1908:303). Gadow argued that the 260-day count must have begun in a tropical *lowland* region because several of its day-names commemorate animals such as the alligator, monkey, and iguana; none of these is native to the highland ecological niche of oak and pine forests where Copán is situated. On the second score, they both seem to have been unaware that Copán was a relatively recent ceremonial center, whereas the sacred calendar was a very ancient creation. Morley, for example, points out that the oldest dated monument at Copán goes back *only* to A.D. 465 (Morley 1946:59), while also contending that the ritual almanac must have been devised at least as early as the fourth century B.C. (Morley 1946:47). Even if one accepts Morley's relatively conservative estimate of the calendar's age, one is confronted with a hiatus of at least eight centuries between the time when the count supposedly started and the ceremonial center of Copán was founded. It was for these reasons, therefore, that I was prompted to look for a tropical *lowland* site on the *same parallel* of latitude as Copán but of *sufficient antiquity* to qualify as the birthplace of the 260-day sacred almanac. This I identified as the large ceremonial center known as Izapa, located on the Pacific coastal plain of Mexico adjacent to the Guatemalan border. Here a tropical rainforest environment surrounds a site whose earliest radiocarbon date goes back to 1500 B.C. (Ekholm 1969:1).

Therefore, if the alignments of buildings throughout Mesoamerica to an azimuth of 285° do indeed commemorate the sunset position on August 13, they not only help to confirm the validity of the Goodman-Martínez-Thompson calendar correlation, but they also testify to an astronomical origin for the calendar on which it is based. Moreover, they conclusively restrict the geographic origin of that calendar to a location where the date of August 13 was locally significant--in other words, just south of the parallel of 15° north latitude. Ironically, the 285° alignments provide precisely the kinds of clues that Thompson argued the 260-day almanac did not provide. For example, Thompson wrote that there were "serious drawbacks" to the thesis that the 260-day count represented the interval between zenith sun positions because this interval varies across Mesoamerica from 260 days on the south to 311 days on the north: "One must assume, then, that the cycle of 260 days originated on the periphery of the area in which it was current, and that, spreading northward and westward, it was eagerly adopted by peoples for whom it had no solar significance" (Thompson 1960:98). That, in fact, seems to be exactly what happened, for orienting one's buildings and ceremonial centers to an azimuth that had no meaning except at the latitude of Izapa must have been a learned pattern of behavior which diffused along with the ritual almanac itself. Thompson wrote further that "there is absolutely no evidence that the 260-day cycle originated in the vicinity of Copán or anywhere along latitude 14°30' which is on the periphery of the area it covered" (Thompson 1960:98). Yet that evidence has lain undiscovered and unappreciated all these years in the very alignments of the great religious edifices and ceremonial centers themselves. Indeed, there are probably few other regions of the world where the principles of architecture, astronomy, and calendrics found so intimate and dramatic a blending as they did in pre-Columbian Mesoamerica. Today, as we gaze at the grid pattern rigidly superimposed by Teotihuacán on the valley of Mexico or the majestic pyramids rising out of the rainforest at Tikal, we must marvel anew at the complementary roles which science and religion played in the lives of these now-dead civilizations.

Finally, we must recognize in this diffusion of almanac and architectural alignments a spatial-temporal scenario which is perfectly consistent with an Izapa-based point of origin. In an earlier publication (Malmström 1978:105-116), I presented a reconstruction of the chronology of the Mesoamerican calendrical systems which now gains further support from these architectural findings. For example, if La Venta was commemorating August 13 at 1000 B.C., it is obvious that the calendar must already have been well established by then. Indeed, my reconstruction postulated an origin in the fourteenth century B.C. (Malmström 1978:108). This suggests that a fairly rapid diffusion had taken place through the lowland corridor of the Tehuantepec Gap into what has been called the "metropolitan area" of the Olmecs (Bernal 1969: 16), and that thereafter it had spread into the highlands of Oaxaca, reaching the Mexican plateau sometime before the birth of Christ. About the same time it was showing up in Edzná, the first great urban center of the Maya; it then gradually diffused east and southwards to reach such peripheral areas as Chichén Itzá and Copán. (Fig. 22.3 presents a hypothetical reconstruction of the diffusion pattern of the 260-day sacred almanac.)

Because no calendrical inscriptions have been found at La Venta, we must turn instead to Monte Albán, which has the oldest recorded dates of any site in Mesoamerica (ca. 600 B.C.). It was for this reason that Caso was led to hypothesize that the calendar had, in fact, first been developed there (Caso 1971:333-348). But neither at Monte Albán nor at Huamelulpan (where the dating has been given as 300 B.C.) has any evidence of the Long Count been found (Broda de Casas 1969:80, 83). The latter represents the fusion of the 260-day almanac with the 265-day secular calendar in such a way that it records both a number and a day-name in each of the counts. (For example, in the Long Count a day would be known as 4 Ahau 8 Cumku, the first number and name being its designation in the sacred almanac and the second being its position in the secular calendar; prior to the initiation of the Long Count the same day would have been recorded simply as 4 Ahau.) The earliest Long Count dates yet discovered record events around the time of Christ, suggesting that it did not come into use until some time between 300 B.C. and A.D. 0; again, this conclusion is entirely consistent with my earlier reconstruction, which had pinpointed a date of 235 B.C. for its inception (Malmström 1978:108). Unlike the 260-day almanac, however, the Long Count was never adopted in the highlands of Oaxaca or on the Mexican plateau, and remained confined almost entirely to the Maya-dominated areas of Mesoamerica. Similarly, the notational system used to record calendrical dates shows a corresponding variation between the Oaxaca and Mexican plateau regions to the west--both of which gradually came to employ only dots for numbers--and the Maya culture realm to the east, where the original notational system employing both dots and bars (the latter signifying a value of 5) continued in use throughout pre-Columbian times.

CONCLUSION

It has long been known that the sacred 260-day calendar was utilized by all of the pre-Columbian civilizations of Mesoamerica, from the Maya in the south and east to the Aztecs and Tarascans in the north and west. Only recently has it been recognized that recurrent architectural alignments centered on an azimuth of 285.5° are also spatially distributed over a similarly extensive geographic area, ranging from the rainforests of the Yucatán and Petén on the one hand to the semi-arid basins of the Mexican plateau on the other. Because these alignments involve structures spread over nearly 10° of latitude and almost 20° of longitude and were constructed during a span of more than 2000 years, any astronomical significance they may have would appear to

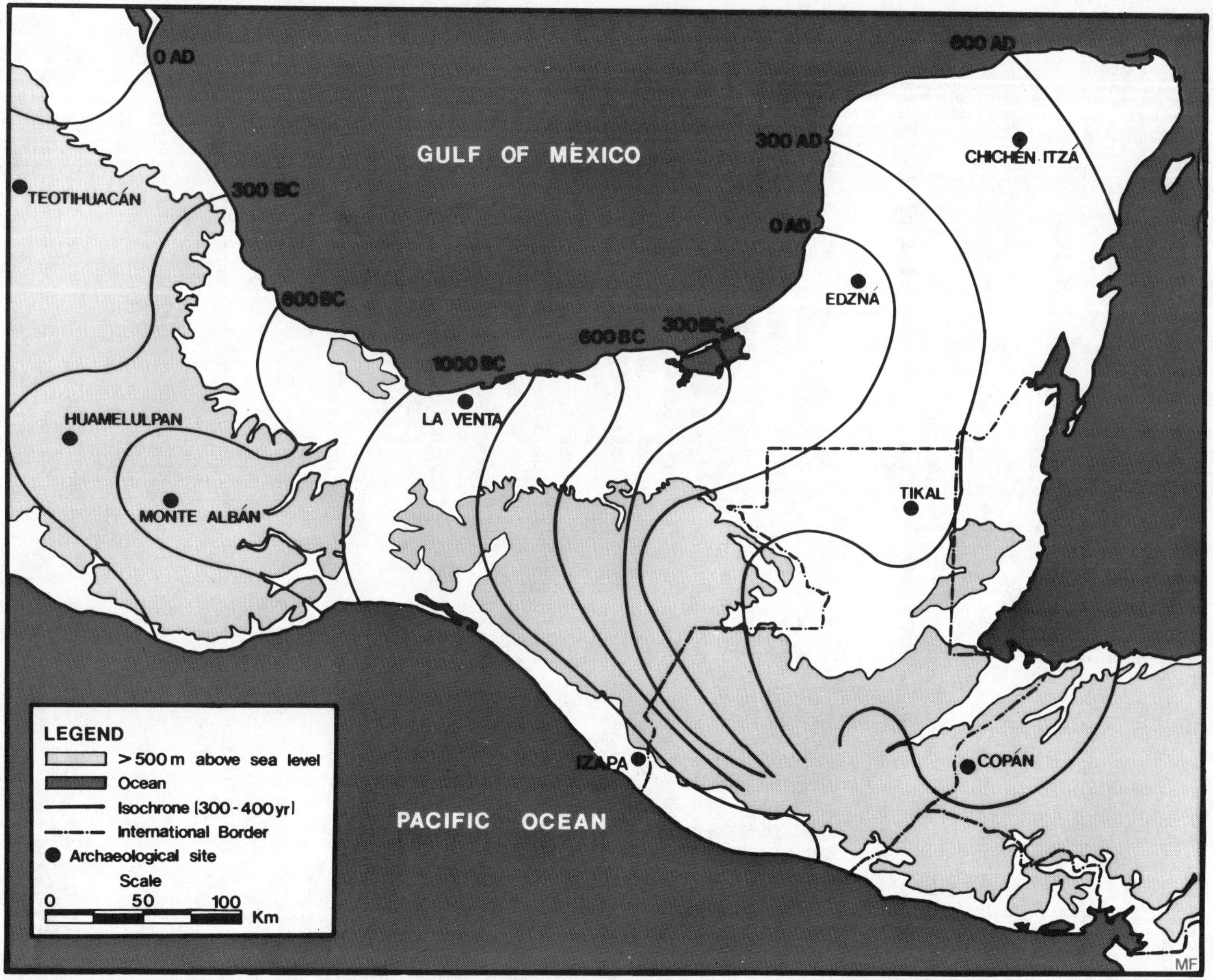

Fig. 22.3 Postulated diffusion of the 260-day ritual almanac

commemorate a recurring event which remained essentially unchanged throughout this expanse of space and time. This event seems to have been sunset on the thirteenth of August, a date which the Maya equated with "the beginning of time" in the Goodman-Martínez-Thompson correlation. Because August thirteenth also marks the southward passage of the zenith sun over the fifteenth parallel of north latitude, initiating a 260-day period which ends on the following May first when the northward passage of the zenith sun occurs at the same place, there is the strong likelihood that these alignments not only testify to the astronomical origins of the 260-day sacred calendar but also to its specific geographic birthplace at the Early Formative site of Izapa on the coastal plain of southeastern Mexico. A diffusion of the sacred calendar from Izapa beginning in the fourteenth century B.C. is perfectly consistent both spatially and temporally with what is known of the ages of alignments of key structures at a number of major archaeological sites distributed throughout the length and breadth of Mesoamerica.

23

Astronomical Imagery in the Serpent Sequence of the Madrid Codex[1]

Susan Milbrath

INTRODUCTION

A number of scholars agree that the fundamental nature of the ancient Mesoamerican pantheon is astronomical (Coe 1975:28; Brundage 1979:8; Barthel 1964), but records from Conquest-period sources provide evidence for associating only a few gods with planets and stars. The rich content of the Maya codices is only partially understood, but those passages that have been deciphered appear to focus on astronomy. And even though the Venus and eclipse tables have been deciphered in terms of the calendric counts, many of the symbols on these pages still remain mysterious, as do large sections of the manuscripts. Analysis of the astronomical content of the codices may provide an important key to reading the Maya glyphs. In this study, it will be suggested that the Pleiades, an important calendar star-group in modern and ethnohistorical Maya sources, are clearly represented in the Madrid Codex as the *Chicchan* serpent. In addition, it will be argued that God B may be equated with Castor and Pollux, twin stars with calendric importance in the ethnohistorical sources.

A recent analysis of an interesting sequence from the Madrid Codex suggests that pages 12 to 18 may represent a seasonal sidereal calendar (Hatch 1975). According to Hatch (1975:286, 289, 296), the repeated serpent images on these pages, and their changes in form and position, "exactly correlate with the path of Draco" through a period that begins on the winter solstice with the setting of Eta Draconis and terminates in August with its upper meridian transit.[2] Her analysis is intriguing, but flawed in certain respects. Although visual correspondence is important in her argument, her charts are difficult to use and do not provide good visual evidence. Hatch (1975:310) indicates that the direction the serpent faces (left or right) corresponds to the position of Draco to the left or right (west or east) of the meridian (a north-south line passing through the pole star, dividing the heavens into eastern and western

---

[1]This research was funded by a fellowship from the Tinker Foundation. I wish to thank Michael Coe, George Kubler, Weldon Lamb, Gary Urton, Flora Clancy, Horst Hartung, Marion Hatch, and Tony Aveni for reading and commenting on earlier versions of this paper. Madrid Codex page photographs are from Villacorta (1930).

[2]Hatch (1975:296) points out the serpent may not conform to the entire constellation Draco as we know it, but specifies that three stars are significant: Beta and Gamma Draconis marking the head, and Eta Draconis marking the hump. The curved arrow in my figure represents the direction of the remainder of the constellation, which may be the serpent tail.

segments). When the positions of Draco are mapped out and compared to the corresponding serpent positions as specified by Hatch, certain inconsistencies become apparent (Fig. 23.1). For example, pages 17 and 18 of the Codex show a left-facing serpent when Draco is still east of the meridian. Nevertheless, given the sequential format of the pages, and the rows of day signs overlying the serpents, it seems quite plausible that the serpent pages are a form of calendar used in astronomical calculations.

In her analysis, Hatch argues that because Eta Draconis events (lower meridian transit, eastern elongation, upper meridian transit) are spaced equally between major solar events (winter solstice, spring equinox, summer solstice), the movements of that star would most certainly have been recorded by the Maya. In addition, Hatch (1975:289) notes that Eta Draconis would have been astronomically significant because it had an unchanging right ascension for a millennium (and thus a fixed relationship to the solar seasons, unlike other stars) due to its unique relationship to the pole of the ecliptic. The fact that Eta Draconis has these specific astronomical characteristics does not, however, demonstrate that the Madrid Codex refers to this star. Furthermore, the dates for the unchanging right ascension (1800 B.C. to A.D. 500) predate the manuscript by about a millennium. And perhaps most significantly, there is no evidence from Mesoamerica to suggest Draco was visualized as a serpent, nor is there evidence that stars in Draco were observed, or are observed today, in the area.

Some methodological considerations should be discussed before proposing the Pleiades as an alternate identification for the possible serpent asterism. The date of the Madrid Codex is not clearly established, but it probably falls in the range of A.D. 1200-1400, and its provenience is believed to be northern Yucatán around 20° latitude (Hatch 1975:284). The star maps employed here (Figs. 23.1, 23.2) offer only the general appearance of the Yucatecan sky between 18° and 22° latitude north, circa A.D. 1400. Should the Codex prove to date prior to A.D. 1400, one can estimate that the rising and setting dates for an individual star would fall an average of two days earlier for every 100 years. A provenience difference of 5° latitude would not significantly affect the dates of rising and setting for specific stars (a difference of about one day only). The dates of solar events would remain the same regardless of the latitude, except for the solar zenith and nadir, which are determined by latitude. The date of the solar zenith would differ by about three weeks in the northern and southern extremes of the Maya area. The significance of this difference, and the problem of adjustments to the solar calendar to allow for leap days, are discussed later.

The serpent pages of the Madrid Codex depict a unique set of six related serpent figures, each varying in pose and iconographic detail. The changing appearances of the serpents (Fig. 23.2) form a sequence exhibiting two directional changes (facing left and then right, and then left again) and one posture change (head down, then head up). If the serpent represents a star group, it may depict the movement of that asterism through the course of the year. The apparent pattern (as Hatch recognized) does suggest a parallel with the movement of star groups around the pole star. Only star groups near the pole would be visible during a major part of their circuit around the pole star in Mesoamerican latitudes, changing their direction of movement from east-west to west-east while still visible above the horizon, and rotating position in the process. Since the Madrid serpents show posture and directional changes, the Codex may refer to a star group closer to the pole than to the celestial

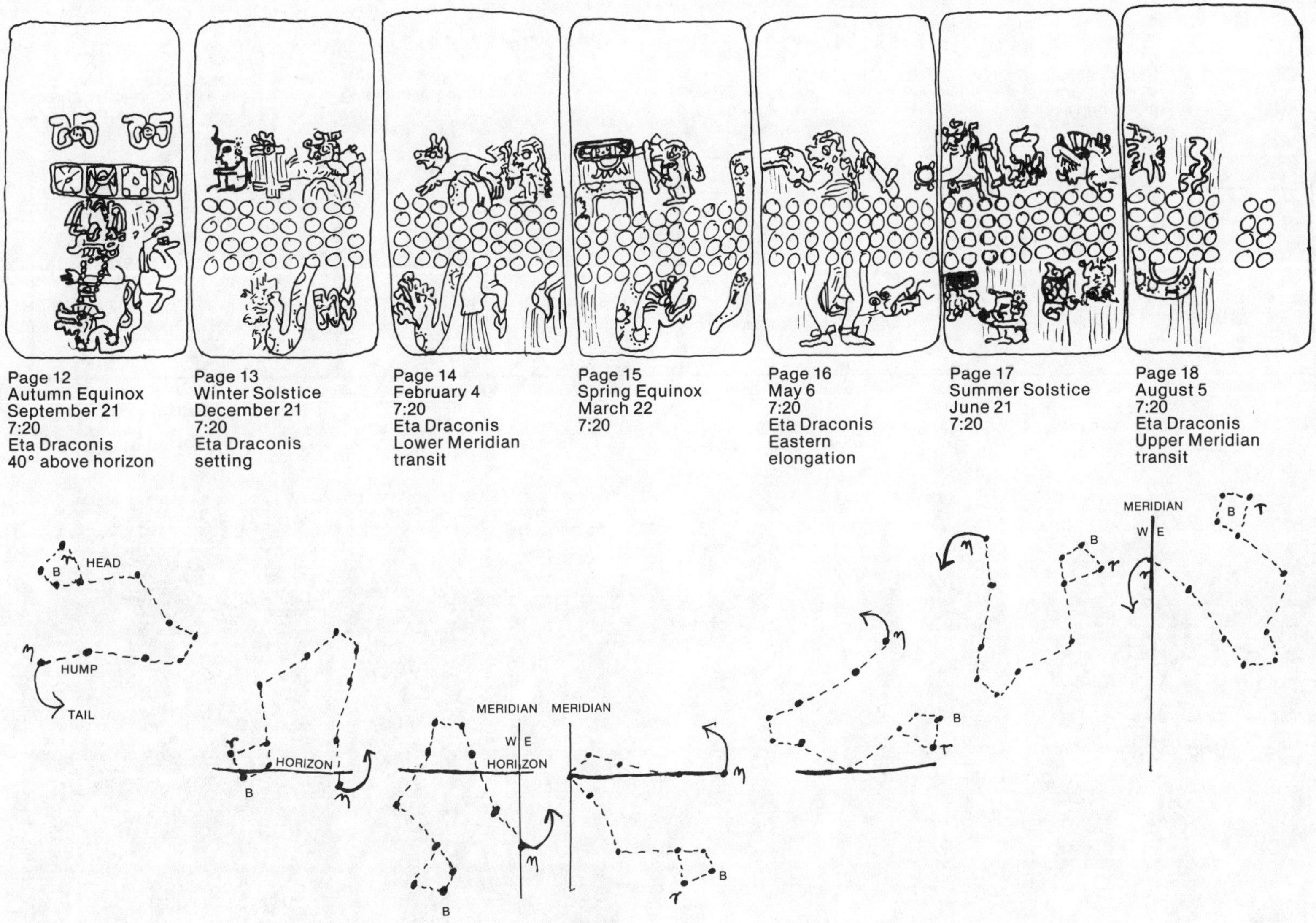

Fig. 23.1 Madrid Codex Serpent pages and movement of Draco

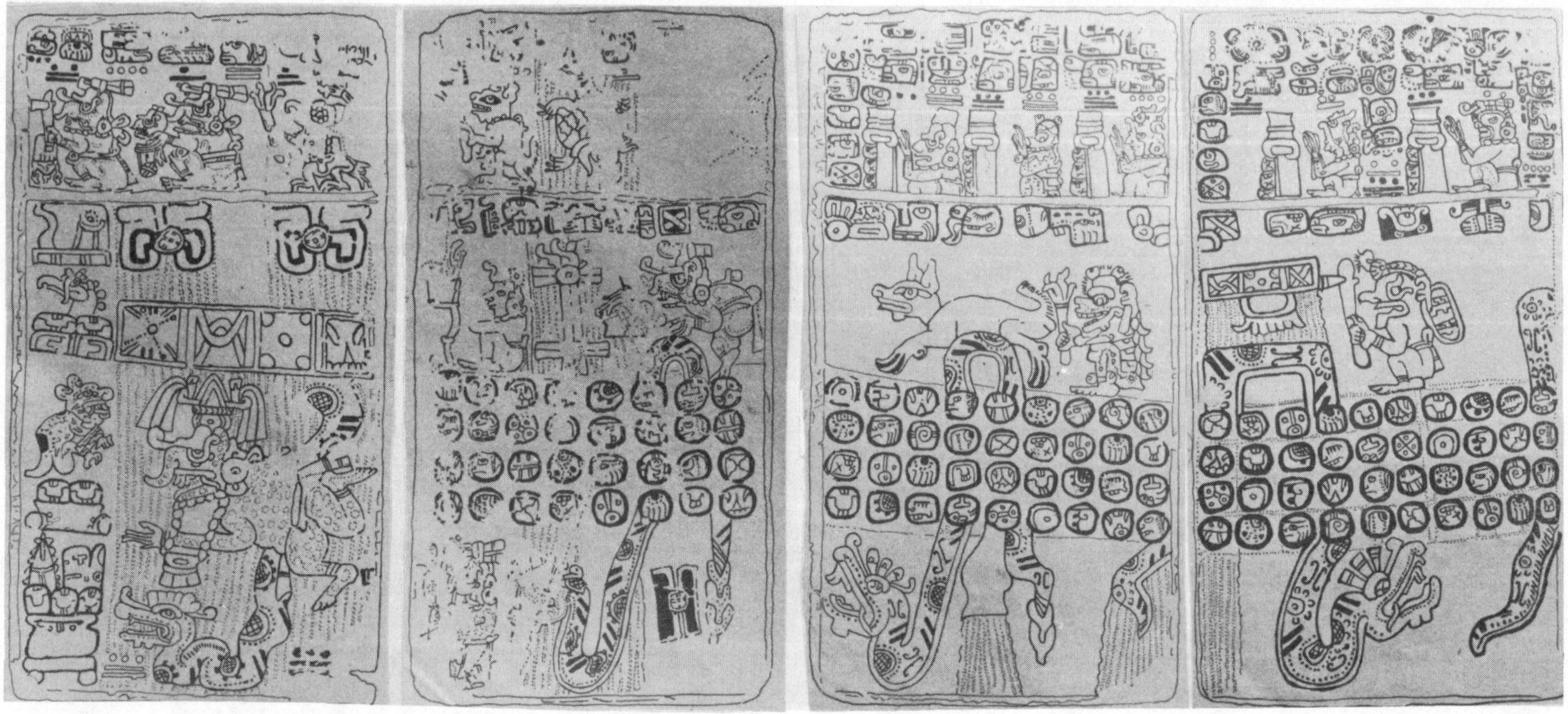

Page 12 November-January
29 Day Period
(December 23-January 21)
10a

Page 13 January 22-March 16
52 Day Period
10b

Page 14 March 17-May 7
53 Day Period
10c

Page 15 May 8-June 30
53 Day Period
10d

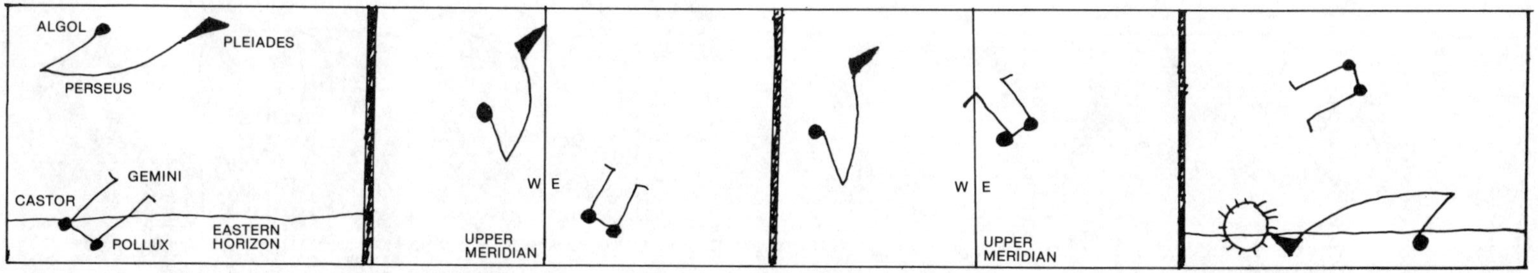

December 21 Dusk
Winter Solstice
Castor and Pollux Rise

Late January Dusk
260 Day Count Begins
Pleiades at Zenith

March 21 Dusk
Spring Equinox
Gemini at Zenith

May 21/22
First Solar Zenith
20° Latitude
Pleiades Set

Fig. 23.2a Madrid Codex, pages 12-15

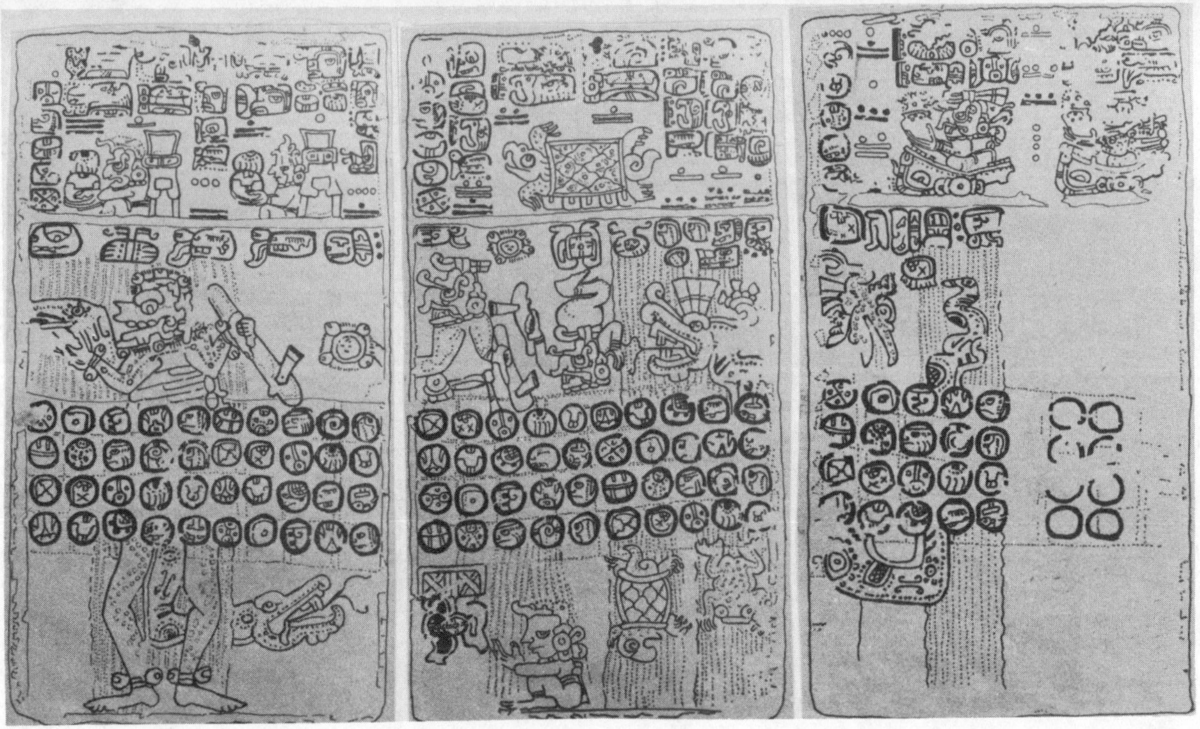

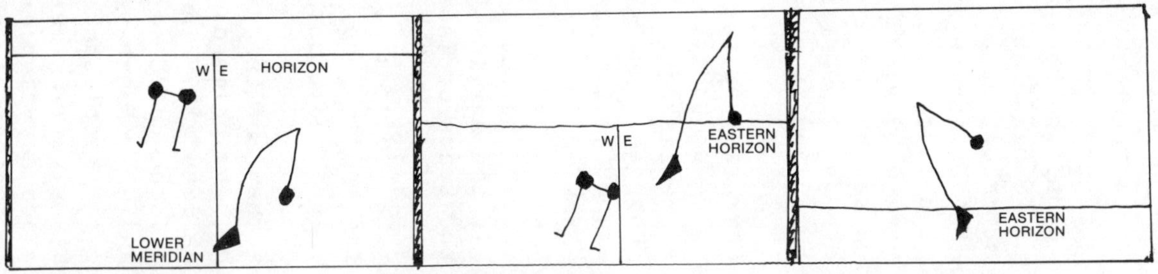

Fig. 23.2b Madrid Codex, pages 16-18

equator (the stars close to the celestial equator would appear to move only east to west, without any apparent rotation in position). This star group may have been recorded as a sequence of serpents representing its position throughout the year.

THE PLEIADES AND THE FEATHERED SERPENT

Among the Yucatec Maya at the time of the conquest the Pleiades were known as the rattlesnake's rattle (*tzab*), a name preserved today in Yucatán, and found also among the Lacandon Maya (Motul Dictionary 1929; Tozzer 1941: 133; Villa Rojas 1945:156; Bruce et al. 1971:15). The Pleiades are still considered the most important asterism in Maya agricultural cycles, and are said to go from behind the sun to in front of the sun when the rains begin in May (heliacal set to heliacal rise; Remington 1977:83; Girard 1962:78). Since the Pleiades continue to be visualized as a rattlesnake's rattle among the Yucatec Maya (as opposed to the western view of the group as part of Taurus, the bull), this suggests a possible correspondence to the Madrid Codex rattlesnake (Fig. 23.2; only pages 13 and 14 show the rattles).

The Maya view of the Pleiades as a rattlesnake's rattle would indicate that they were probably part of a larger group of stars forming a serpent. If the Pleiades are linked to Taurus, the resulting form in no way suggests a serpent form. However, the Pleiades are actually closer to the bright stars of Perseus, and when joined to that star group form an image in a hairpin form with the very bright binary star Algol at one end and the Pleiades at the other end. This form resembles the unusual hooked form of the Madrid Codex serpent (Figs. 23.3-5). Linking the Pleiades to Perseus is not totally without precedent, for in the earliest Babylonian astronomy the Pleiades were considered separate from the stars of Taurus and were conceptually linked to Perseus, since the moon was exalted in both star groups (Gleadow 1968:170). Although it will be argued that the Pleiades *were* linked to Perseus, the calendric correlations suggested for the Madrid Codex sequence would be equally valid if the serpent referred only to the position of the Pleiades.

The serpent of the serpent pages has long been recognized as the *Chicchan* serpent, the primordial water serpent associated with rain and fertility (Thompson 1960:75). The serpent embodies the fifth day-sign of the Maya divination count, the day *Chicchan*. In Yucatec Maya astrological texts, the day signs *Kan* and *Chicchan* have the attribute for prognostication of *ah tzab ti can*, meaning the rattlesnake's rattle in the sky (Lamb, this volume). This provides a substantive link between the Pleiades (*tzab*) and the day-sign *Chicchan*. The Maya day-sign *Chicchan* is equivalent to the fifth day *Coatl* (serpent) in the Mexican divination count, ruled by the water goddess *Chalchiuhtlicue* (Fig. 23.6). Both the day-signs *Chicchan* and *Coatl* are associated with the concept of water, and the figure for the day-sign *Coatl* resembles the *Chicchan* serpent in that it is plumed and has a distinctive hump in the midsection of the body--a hairpin pattern like that of Perseus. Kelley (1976:96, 109, 319) defines *Chicchan* as twisted snake, and links the serpent to the Yucatec feathered serpent *Kukulcan* and the Mexican feathered serpent *Quetzalcoatl*, both plumed water serpents connected with Venus. *Chan* means serpent in Mayan, but the meaning of *chic* is not clear, though Hatch (1975:337-338) suggests it involves the idea of rotation. *Chichaan u* has the meaning "serpent-fang moon" (Lamb, this volume), which suggests that *chic* means fang.

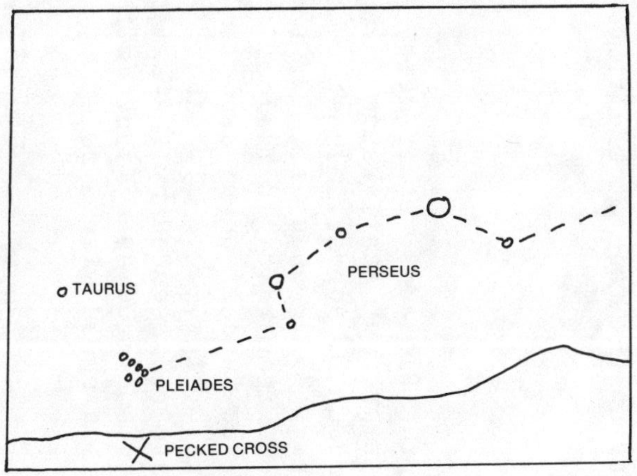

Fig. 23.3 Teotihuacan view of Pleiades setting over Cerro Colorado

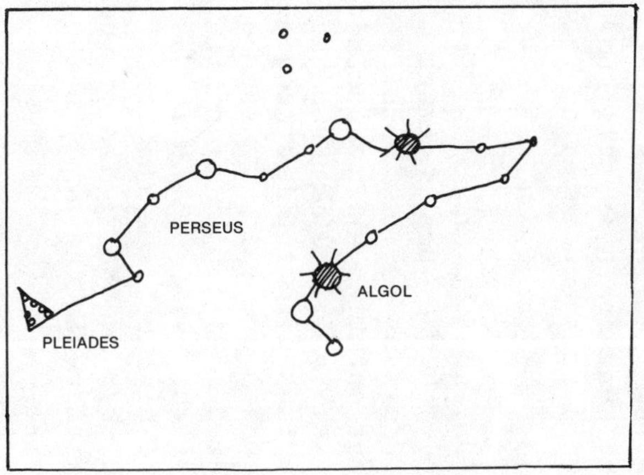

Fig. 23.4 Pleiades linked to Perseus

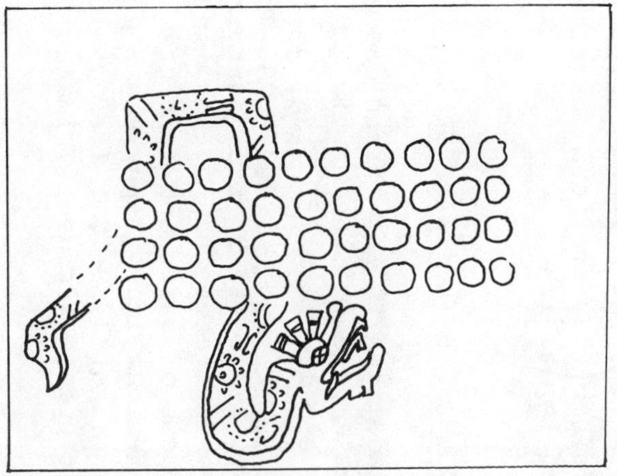

Fig. 23.5 Madrid Codex, page 15

Fig. 23.6 Chalchiuhtlicue, the Water Goddess and the Day-Sign Serpent

The fact that the feathered serpent complex is associated with Venus does not exclude the possibility that the *Chicchan* serpent may be identified specifically with the Pleiades. A connection between Venus and the Pleiades may have developed from a Venus count begun at the conjunction of the Pleiades and Venus as either the morning or evening star, an event recurring every eight years (the 104-year calendar round of the Dresden Codex Venus Tables would be equivalent to 13 successive conjunctions). The idea that a planet has a special connection to a specific star group along the sun's path is well known in Western astrology, and oddly enough Venus is specifically connected to Taurus, and the Pleiades.[3]

The clear identity of *Quetzalcoatl*, *Kukulcan*, and the *Chicchan* serpent as celestial rattlesnakes suggests a connection with the Yucatec image of the Pleiades. Indeed, Dow (1967) believes that the Pleiades may be identified with *Quetzalcoatl*. The fact that all of these serpent entities are connected with the concepts of fertility and celestial water (Kelley 1976:109; Brundage 1979:104-106) would have a natural basis if they are linked to the Pleiades, for when the sun moves into conjunction with the Pleiades the rainy season begins. Among the Cakchiquel and Chorti Maya, the Pleiades in conjunction with the sun are considered a sign announcing the onset of the rainy season in May (Girard 1962:78, 148; Remington 1977:83). The Chorti also believe that the *Chicchan* serpent of the north (i.e., in the northern sky, or the zenith of the sky) is chief among the four sky *Chicchans* (four positions in the sky?) responsible for sending the rains (Wisdom 1940:410-411). Since the Pleiades are one of the most northerly of the ecliptical constellations (only Castor and Pollux of Gemini are positioned closer to the pole), and the star group does achieve a true zenith in Mesoamerican latitudes in conjunction with the noonday sun at the beginning of the rainy season, the astronomical associations of the group would readily correspond to the northern rain-bearing *Chicchan*. Although the Chorti never say the *Chicchan* serpent is the Pleiades, they may be linked esoterically since they are both described as lifting up the rain at the time of the first solar zenith.

CASTOR AND POLLUX AND THE RAIN GOD

In addition to the serpent as an image of the Pleiades, other deity figures on these pages of the Madrid Codex may also represent star groups. The fanged God B, usually identified as the Maya rain god *Chac*, appears almost as frequently as the *Chicchan* serpent on these pages (Fig. 23.2; pages 12, 13, 17, 18). *Chac* is believed to be the counterpart of the Mexican rain god *Tlaloc*, who is most easily recognized by his fangs, goggle eyes, and frequent association with streams of water and serpents in iconography (Figs. 23.7-8). In Post-Classic Mexican iconography these goggle eyes are represented by the same form as stars (Figs. 23.9-10); *Tlaloc* would thus be visualized as a figure with a pair of star eyes. Frequently images of *Tlaloc* are crowned with a solar

---

[3]According to Sahagún, many Aztec deities were equated with planets. In his scheme *Cihuacoatl* is Venus, and *Quetzalcoatl* is Hercules (an atlantid, not a planet; Brundage 1979:80). It seems here that Sahagún is equating the quality of each god with a Roman counterpart, rather than indicating that informants told him *Cihuacoatl* was actually the planet Venus (which would contradict the identification of Venus with *Quetzalcoatl* he makes elsewhere in his work).

Fig. 23.7 Teotihuacan Tetitla, Lightning Tlaloc

Fig. 23.8 Tlaloc, the Water God

Fig. 23.9 A Mexican observatory outlined by stars

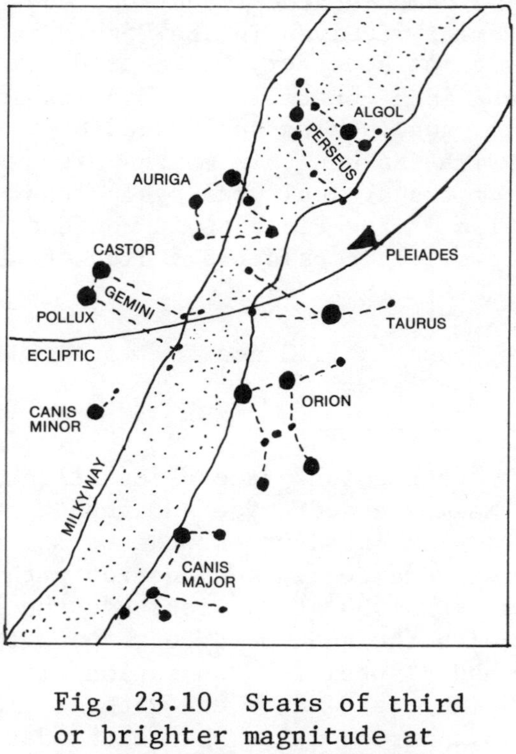

Fig. 23.10 Stars of third or brighter magnitude at intersection of the Milky Way and the Ecliptic

ray, suggesting a star group in the sun's path. *Tlaloc* has a directional association with the north (Codex Fejevary Mayer 1), indicating a star group in the northern sky or possibly one that is close to the zenith in Mesoamerican latitudes. There is evidence from the Maya area that north and south refer not to topographical space, but to above and below (Brotherston 1976; Gossen 1974b:33). Castor and Pollux, the very bright twin stars of Gemini, are close to the zenith in Mesoamerica, and mark the northern extreme of the sun's position at the time of the summer solstice and at the beginning of the heaviest rains in late June. Thus it seems that Castor and Pollux, in conjunction with the solstice sun, may have symbolized the rain god, just as the Pleiades in conjunction with the zenith sun symbolized the celestial rattlesnake (Milbrath 1980a; 1980b).[4]

CALENDAR STARS

Gemini, or more specifically Castor and Pollux, and the Pleiades are mentioned as the most significant star groups in both Maya and Mexican ethnohistorical sources (*ac ek* and *tzab* in Yucatec, Tozzer 1941:133; *mastelejos* and *cabrillas* in Aztec sources using Spanish names, Sahagún 1953:60). Both the Yucatec Motul Dictionary and Sahagún emphasize that the former constellation (usually translated as Gemini) is composed of three stars, suggesting the Belt of Orion, and in the Nerbrija dictionary the correlate term *astillejos* is Orion (Weldon Lamb, personal communication). Stars of Orion are also called turtle stars among some Yucatec (Thompson 1960:116), which is the same designation (*ac ek*) Gemini receives in the Motul Dictionary. Among modern Maya the Pleiades are the sign of November, Orion of December, and Gemini of January; for in the given month each constellation has its longest visibility, rising at dusk and setting at dawn (Remington 1977:83-84). The period of invisibility during conjunction with the sun (six months after opposition) may be equally significant in terms of the agricultural cycle. As noted above, among various Maya groups the position of the Pleiades in conjunction with the sun (just before the heliacal rise) is considered a sign of the onset of the rainy season (Remington 1977:83; Girard 1962:78).

There is evidence for worship of the rain god and feathered serpent

---

[4]Precession affects the alignment of the stars with the solar seasons, making a star rise an average of one day later every 69 years. Castor rose heliacally (at dawn) at the summer solstice, June 21, ca. 200 B.C. By 100 B.C. it remained in conjunction with the sun and was thus invisible on this date. By A.D. 1500 at 20° latitude the last visibility of Castor prior to conjunction with the sun would correspond to the summer solstice. Thus the star may have had a specific association with the summer solstice as far back as 200 B.C. The position of the Pleiades in relation to the first solar zenith sun is a bit more complex, since the date of the solar zenith is dependent on latitude. At 15° latitude (the area of Copan) the first solar zenith falls on May 1. The conjunction between the sun and the Pleiades at this date would have been valid from 1000 B.C. through the present. At 20° latitude (the area of northern Yucatán) the first solar zenith takes place on May 21 or 22. The sun would not have come into conjunction with the Pleiades on that date until around A.D. 500, but between A.D. 300 and A.D. 500 the heliacal rise of the Pleiades would have corresponded to the first solar zenith (Aveni 1972b; Colgate University Computer Center, n.d.).

at the time of the Conquest in the same months now associated with the Gemini and the Pleiades. Around A.D. 1550, Bishop Landa noted that the principal Yucatec festival to *Chac* (*Oc Na*) took place in the months of *Chen* or *Yax*, on a day set by the priests (Tozzer 1941:161). These months were dated from December 23 to January 11 and from January 12 to 31 in the Julian calendar (each month being 20 days; Pio Perez 1842), corresponding to January 2 through February 10 in our Gregorian calendar (N.S.). Since Landa specified that the priests selected the day of the ceremony, it may well have been astronomically timed. The fact that Gemini is now considered to be a sign of January suggests a correspondence with the timing of *Chac*'s festival ca. A.D. 1500.

Among the Aztec, *Chac*'s counterpart *Tlaloc* was honored in the festival of *Atemoztli*, dating from December 21 to January 9, N.S. (Nicholson 1971: Table 4; Julian dates corrected to Gregorian). *Atemoztli* overlaps with the Yucatec month *Chen* but is earlier, and would mark the first day after the winter solstice, the point at which Gemini would first move into opposition with the sun. December 24, A.D. 1500 (20° latitude, 0° horizon) is the day Castor rose in opposition to the setting sun (Aveni 1972b). Although Gemini would be visible in preceding months, it would rise later in the evening. During late December and January the twin stars would be seen on the eastern horizon just at dusk, and by February at dusk they would already be well above the horizon. Another major *Tlaloc* festival fell in the month of *Etzalcualiztli*, which celebrated the death of *Tlaloc*, from June 3 to June 23, N.S. (Nicholson 1971:Table 4). This death ceremony correlates with the last visibility of Castor and Pollux, prior to their disappearance from the night sky due to conjunction with the sun. The births and deaths of gods may be astronomical metaphors. I have argued elsewhere (Milbrath 1980a) that the birth of *Huitzilopochtli* corresponds to the nadir sun and the first appearance of Scorpio after conjunction with the sun. The time of birth follows the invisibility of a star in the womb of the earth goddess, and the point of death is the last visibility prior to conjunction.

Just as the timing of ceremonies to the rain god in both the Aztec and Maya festival calendars suggests a correspondence to the significant positions of Castor and Pollux, there are festivals celebrating the feathered serpent that appear to correspond to the movements of the Pleiades. The Maya festival of *Chic Kaban* celebrated the descent of the feathered serpent *Kukulcan*. The festival fell on the sixteenth to twentieth day of *Xul*, corresponding to November 8 to 12 in the Julian calendar (Pio Perez 1842), or November 18 to 22, N.S. Landa mentions that a new fire was lit and banners were placed on the temple during this ceremony (Tozzer 1941:158). This suggests a relationship to the banners and new fire of the Aztec New Fire Ceremony, celebrated in November in the month of *Panquetzaliztli* ("banner raising"). Although the principal gods of the New Fire Ceremony and the month *Panquetzaliztli* are those associated with the fire serpent (the seasonal complement of the feathered serpent, and a probable representation of Scorpio located opposite to the Pleiades; Milbrath 1980a), *Quetzalcoatl* is also present at the ceremony (Codex Borbonicus 34). The New Fire Ceremony was timed by the midnight zenith of the Pleiades (and their longest visibility in opposition to the sun), an event that took place on November 16 at the time of the Conquest (Broda 1980). Thus the *Chic Kaban* ceremony, falling just after November 16, would mark the time when the Pleiades were past the zenith and descending toward the horizon when viewed at midnight; if a dawn observation were employed, it would mark the time when the Pleiades set at dawn.[5]

There are essentially three important observation positions for a star group: at the two horizons, and at the upper meridian. There are also three important observation times corresponding to those sky positions: dawn, dusk, and midnight. It has been shown above that the ceremonies discussed do relate to important positions of Castor and Pollux and the Pleiades at the key observation times. Therefore, not only are there iconographic reasons for associating the rain god and the feathered serpent with specific asterisms, but the timing of festivals associated with these gods also supports such an identification. Further, modern Maya practice suggests that the star groups in question are now considered a sign of the month corresponding to those festivals at the time of the conquest. Thus November is today the month of the Pleiades, just as November was the month celebrating the feathered serpent at the time of the Conquest, and January is now the month of Gemini, just as late December through January was the time of festivals to the rain god.

THE MADRID CODEX SEQUENCE

If the Yucatec Maya Madrid Codex represents a seasonal or star calendar, it would seem most likely Castor and Pollux and the Pleiades would be represented, given their locus at seasonally significant points along the ecliptic, framing the celestial crossroads of the Milky Way and ecliptic as it achieves zenith. Indeed, Gemini and the Pleiades seem to have been especially important in the context of telling time among the Yucatec Indians at the time of the conquest. Circa A.D. 1550 Landa said that the Maya used these two star groups as a sky clock, and in A.D. 1601 Herrera noted that "they regulated themselves by night, in order to know the hour, by the morning star and the Pleiades and Gemini" (Tozzer 1941:133,220).

The Madrid Codex appears to depict a series of events marked by the disappearance and reappearance of figures, as well as changes in positions among the figures, through a set of seven pages. Most striking are changes in the position of the serpents, and changes in their iconographic detail. In two pages the serpents have rattles; in the remaining pages the rattles are lacking (Fig. 23.2). On one page (16) the serpent lacks feathers, and on another (15) the serpent has a volute in its eye. Among the Chorti, four sky *Chicchan* serpents are connected with bodies of water and rain making, with the *Chicchan* of the north acting as chief (Wisdom 1940:392-396). The question remains as to whether these are really four different *Chicchan*s, or the same *Chicchan* in four different positions in the sky. Similarly, in the Madrid Codex there are variant forms of the *Chicchan* serpent, but there are six serpents, which possibly represent the four sky *Chicchan*s and two earth *Chicchan*s. I would argue that these variations may represent the different sky and underworld positions of a single astronomical entity.

In the Venus pages of the Dresden Codex, Venus has four different directional aspects--the north and south directions are associated with the celestial positions above and below (Brotherston 1976). In addition, the

---

[5]The feathered serpent may have been visualized as visiting the earth in the descent of the Pleiades below the horizon in the dawn sky.

images are divided into five different groups representing the five Venus cycles (equivalent to eight solar years), with the five sidereal positions of Venus perhaps beginning the five Venus counts. A number of Mesoamerican deities in fact have different directional aspects (the *Tezcatlipocas*, the *Bacabs*, the *Chacs*) which may reflect the various sky positions of a single astronomical body, rather than different deities.

In the analysis of the Madrid Codex sequence it will be demonstrated that there is a strong correlation between the periodicity of major agricultural dates (usually coinciding with major solar positions at dusk), the positions of Castor and Pollux and the Pleiades, and the positioning of the *Chacs* and *Chicchan* serpents. The actual form of the *Chicchan* serpent may well reflect the differing appearance of the Pleiades-Perseus configuration when viewed through the course of the year (Fig. 23.2). The changes in the direction and the elevation of the head to body may be literal designs indicating the position of bright stars in this group. Thus the head of the serpent (which I equate with Algol of Perseus) faces to the left when Algol is to the west of the Pleiades (representing the serpent tail) and to the right when it is east of the Pleiades.[6] When the head of the serpent is lowered it appears to depict the position of Algol below the Pleiades, relative to the horizon; when the head is raised it appears to designate the position of Algol above the Pleiades. Except for the opening page, where the serpent image is cut by a sky band, the hairpin shape of the serpent remains invariable, suggesting the hooked pattern of stars in Perseus.

MADRID CODEX PAGE TWELVE AND THE OPENING OF THE SEQUENCE

The Chamula Maya initiate the agricultural calendar at the time of the winter solstice (Gossen 1974a:247). According to Landa, at the time of the Conquest the Yucatec Maya began cultivation in January (Tozzer 1941:97), which suggests the winter solstice may have been the astronomical event by which they calculated the time to begin preparing the fields. The modern Yucatec practice is to begin clearing the fields in December or January, followed by burning in March or April and planting in late May or June (Roys 1972:38). Thus the opening page may refer to the winter solstice in the Yucatec month of *Mol*. Cohodas (1974) has argued that a fixed *haab* or 365-day agricultural calendar with a fixed relationship to the seasons may go back to Classic Maya times. I support this position, though I would argue that the beginning date of the calendar falls either on the winter solstice in December or on the second solar nadir in late January, based on analysis of the Madrid Codex calendar.

The opening page appears to depict a 29-day period in the upper panel (12, 6, and 11 in black numerals placed horizontally), probably corresponding to one lunar month. Thirteen days are also shown in the lower panel, exactly the number of days needed for intercalation to bring the 365-

---

[6]The drawing of the constellation for page 18 inaccurately renders Algol to the right of the Pleiades.

day calendar into renewed alignment with solar events after 52 years, or one full calendar round. The Aztecs celebrated the close of their calendar round with a major astronomical ceremony timed by the midnight zenith of the Pleiades. Since the subsequent 52-year cycle was the time when the divination calendar and the festival calendar of 365 days were realigned (both beginning a new count at that point), any corrections to the calendar could have been inserted prior to the new cycle. Edmonson (1976) notes the Post-Classic Maya and Mexican calendars lend themselves to an intercalation of 13 days every 52 years. The Aztec New Fire Ceremony was astronomically timed by the Pleiades zenith and the first solar nadir in November (Milbrath 1980a); consequently, stellar and solar times were also coordinated at this time. Thus the Madrid introductory page appears to refer to an intercalation period bracketed by the first solar nadir (November 22 or 23 at 20° latitude) and the winter solstice, which would correspond to the Yucatec months *Yaxkin* and *Mol* (November 23 to December 12, N.S. or Gregorian, and December 13 to January 1, N.S.).

The Madrid Codex calendar apparently records a full 365-day calendar that begins in *Yaxkin* and ends in *Xul*. *Xul* means "end" and *Yaxkin* means "new sun" (Thompson 1960:110). "New sun" relates to the concept of the rebirth of the solar god *Huitzilopochtli* in the Aztec month *Panquetzalitzli*, corresponding in date to the Maya month *Yaxkin* (Milbrath n.d.). In both *Panquetzaliztli* and *Yaxkin* there is a general renewal of household artifacts, and thus a reaffirmation of continuity appropriate to the calendric renewal.[7]

On page 12 the *Chicchan* serpent is suspended from a band of symbols, and a large *Chac* appears to ride on the serpent's back (Fig. 23.2). Only the symbol on the far left of the band (the *kin* sign standing for sun) has been definitively deciphered. The other glyphs have been thought to represent planets, but my analysis indicates that they may refer to a celestial locus on the path of the sun, and may consequently represent time periods. The second symbol is equivalent to the Mexican day sign *ollin*, and signifies movement or earthquake in the Mexican day-sign series. The *ollin* glyph is associated with Aztec images of the nadir sun, such as the Aztec Calendar Stone (Milbrath 1980a), and I would suggest that it may refer to the zenith-nadir axis, the position of the sun at the time of the Aztec New Fire Ceremony in November. The next glyph, the *kan* cross, is variously referred to as the symbol for green or turquoise and is associated with the southern direction. Based on data from the Chorti Maya (Girard 1960:45, Pl. 8), and on the appearance of the glyph on two pages of the Madrid sequence corresponding to the winter and summer solstices (pages 12 and 15), I would suggest that it refers to the solstice suns. The glyph following the *kan* cross is *akbal*, meaning night or darkness (Thompson 1960:89). *Akbal* is frequently paired with the sun glyph, and is interpreted by Thompson (1960:260) as darkened sun. I would suggest that here, following the *kan* cross, it signifies that nights predominate--qualifying this as the winter solstice when nights are longer than days. Thus the complete symbol-band may denote the period of the first solar nadir and winter solstice.

The *Chicchan* serpent emerges from the glyph for darkness, which is appropriate to a dusk rising of the Pleiades and their longest period of visibility, when they would appear to dominate the night sky. One month later the dusk rising of Castor and Pollux would have been celebrated concurrently

---

[7]In *Yaxkin* all household artifacts were anointed with blue bitumen (Tozzer 1941:159).

with the winter solstice.[8] The prominent *Chac* probably refers to the rise of Castor and Pollux, although, based on the confusion of Orion and Gemini in the ethnohistorical sources and the modern Maya practice of assigning December to Orion, there is a small possibility that the *Chac*s should be equated instead with Orion.

MADRID CODEX PAGE THIRTEEN AND THE BEGINNING OF THE 260-DAY COUNT

The remaining pages of the serpent sequence depict the glyphs of the 260-day divination count overlying the sequence of serpents. In the Maya area, from Chiapas to Guatemala, there is definite evidence of a 260-day agricultural count (Girard 1962:327-328; Laughlin 1969:179; Cohodas 1974:97). The Tzotzil begin their count on February 21, and the midpoint of the count is the summer solstice, which for reasons of calendric symmetry is celebrated ten days after the solstice. A fixed 260-day ritual count also extends north into the Chatino area of Oaxaca, where the 260-day count begins on February 25 and ends on November 12 (Greenberg 1978:170). The count is divided into four groups of 65 days, with the first group ending on May 1 with a rain ceremony. The 260-day agricultural count is clearly distinct from the 260-day divination count since it is fixed in relation to the seasons, rather than shifting in a continuous repeating cycle like the divination count. However, days of the divination count are used in prognosticating agricultural activity today, just as they were at the time of the Conquest (Tozzer 1941:97). Since a specific day and number combination (20 day signs and 13 numerals) of the divination count could not recur for 260 days, at the beginning of the 260-day fixed agricultural count, it would be possible for the priests to forecast the luck of the days throughout the agricultural season.

The day signs lack numeral coefficients in the layout of days in the Madrid Codex, but a numerical value is implied by their positioning in four rows of 65 days (the last part of the count is incomplete, however). Beginning on page 13 (Fig. 23.2), the days of the vertical column in the first position would all be numbered one, the second two, and so on until the numeral one repeats after the thirteenth vertical column. Lacombe and Block (1975) have noted that this arrangement is especially useful in calculating heliacal rising dates of Venus. Since the heliacal rising of Venus was considered an unlucky event, this aspect was probably calculated along with the fortune attributed to each day, to arrive at a propitious day for agricultural activity.

Among the Chorti Maya, the fixed 260-day agricultural count begins on February 8, rather than on February 21 as among the Tzotzil, or on February 25 as among the Chatino. At this time the Chorti say that the sun has begun to move from its resting place at the winter solstice (Girard 1962:7). Although at the time of the Conquest the Yucatec Maya calendar began in July (concurrent

---

[8] The dusk rise and longest visibility of Pollux at 20° latitude would exactly correspond to the winter solstice (December 21) ca. A.D. 1000, whereas around A.D. 1500, the star would not rise until December 28. On the other hand, Castor would rise exactly on the winter solstice between A.D. 1300 and 1400, the presumed date of the manuscript (Colgate University Computer Center, n.d.).

with the second solar zenith), Landa recorded a 260-day count beginning in the month of *Yax* (Tozzer 1941:151-152). *Yax* is dated from January 22 to February 10 in our Gregorian calendar, and thus would specifically overlap with the beginning of the 260-day agricultural count preserved today among the Chorti.

Before embarking on a page by page analysis of the iconography and related astronomical events of the Madrid Codex sequence, the calendar periods for each page must be defined. The upper panels of pages 14 to 17 provide evidence that each page defines a 52-day period. The black numerals placed horizontally across the upper panels refer to 53, 53, 47, and 55 days in this sequence of pages, averaging 52 days; the horizontal row of page 13 is effaced. The row on page 12 indicates a 29-day period, and that on page 18 a 27-day period. Presuming that page 13 originally displayed a 52-day period, the total would be 29 + (5 x 52 = 260) + 27, which would seem to bracket a 260-day period between a synodic lunar month and a sidereal lunar month.[9]

Whereas the horizontal rows appear to refer to *trecena* counts of 13 days (4 x 13 = 52), the vertical columns of five day-signs might refer to 20-day *ventienas* or so-called "months." When read top to bottom, each sign is spaced 12 days from the one below it in the columns on pages 14 to 17, yielding five sets of 12, or 60 days per page (or three Maya months of 20 days). However, if every day sign were intended to be read with the same numeral coefficient as the top sign (often effaced), there would be 52 days between each sign, and each column would show a total of 260 days. The total would then be 6 x 260, or 1560 days, which is four 365-day years plus 100 days. Actually the count may be more complex, for the first day in the column on page 15 is 4 *Ahau*, the 160th day. Thus the total count would then be 160 plus the four times 52 days remaining in the column, or 368 days.

Since there is considerable variation in the beginning date of the 260-day fixed agricultural count--as early as January 22 (Landa's count in *Yax*) and as late as February 25 (the Chatino)--it may be that lunar phases are studied to time the beginning of the count. Many aspects of the modern Yucatec agricultural calendar are timed by lunar phases (Perez Toro 1942) and among the Quiché the 260-day count is considered a lunar count (Lothrop 1930:653). The apparent correspondence between the Madrid Codex page 12 and the months *Yaxkin* and *Mol*, in addition to the reference to a synodic month in the upper panel, makes it appear likely that the beginning of the 260-day count on page 13 was timed by a specific lunar phase following the winter solstice in *Mol*. *Mol* is followed by the month *Chen*, spanning January 2 to January 21, N.S.[10]

---

[9]Hatch (1975:308) also suggests that 52-day periods are shown in the horizontal rows, but indicates they should be read as 5 x 52, presuming the column of five glyphs on the left indicates the 52 days should be multiplied by 5, yielding a total of 260 days for each page.

[10]Landa's Yucatec festival calendar was composed around A.D. 1550. The Madrid Codex is from the same area, but probably dates at least 150 years earlier. The dates of Landa's calendar would be almost 40 days later than those of the corresponding months at the time of the Madrid Codex, if no intercalation were employed. If an intercalation were made, the difference between the months would be not more than 13 days. The astronomical events important to each page all fall in the first half of the 52-day period defined for each

January 21 is the day of the second solar nadir at 20° latitude, when the sun passes directly "below" the earth as reflected by the zenith of the ecliptic at dusk overhead. At this time the ecliptic would be like a giant hoop standing precisely at a vertical axis, with the Pleiades at dusk zenith, followed by Gemini at zenith and then Leo at zenith at midnight, when the sun passed to its lowest position in the underworld. The beginning of the count was probably timed by a variety of astronomical events: the second solar nadir, the dusk zenith of the Pleiades (the most important calendar stars), and also a specific lunar phase. Landa recorded a count beginning in the month *Yax*, and January 22 (the first day of *Yax*) would be the day after the solar nadir (or the day of the solar nadir, for zenith and nadir dates can vary by one day from year to year). This date would also correspond to the dusk zenith of the Pleiades ca. A.D. 1400. This was probably the stellar and solar date the Maya calculated by; they then chose the closest propitious lunar phase after that date as a beginning of the count. The reference to a 29-day period or synodic month prior to the 260-day count that begins on page 13 of the Madrid Codex indicates that lunar phases were coordinated with the 260-day count.

The Codex pages would be laid out according to 52-day periods which would correspond to agricultural work-periods as well as meteorological events. Beginning the count on January 22, at the beginning of *Yax*, the layout is as follows:

Page 12: November 10 to January 21 (29, 30, 31 days): complete harvest

Page 13: January 22 to March 16 (52 days): clear fields before rains

Page 14: March 17 to May 7 (52 days): burn, then plant with rains

Page 15: May 8 to June 30 (52 days): complete planting, then weed

Page 16: June 31 to August 20 (52 days): weed, second planting

Page 17: August 21 to October 12 (52 days): double maize stalks

Page 18: October 13 to November 9 (27 days): harvest begins

Similarly, the Yucatec Maya months and ceremonies appear to show a parallel with individual pages, as do specific astronomical events of the solar year.

Page 12: first solar nadir. Last day of *Xul* or first of *Yaxkin*, winter solstice in *Mol*; *Yaxkin* means reborn sun; in *Yaxkin* prepare for ceremony to all gods in *Mol*.

Page 13: second solar nadir. Last day of *Chen* or first of *Yax*; principal festival to *Chac* in *Chen* or *Yax*; incense burned and idols renewed.

Page 14: spring equinox. Last day of *Ceh* or first of *Mac*; fire ceremony and hunting ritual in *Mac*.

Page 15: first solar zenith. Last day of *Muan* or first of *Pax*, summer solstice in *Kayeb*; Black god *Ek Chuah* and *Chac* worshipped in

---

page except for the summer solstice, which might have been celebrated at the beginning of the period defined by page 16, rather than the end of the period on page 15. There is evidence that the modern Maya practice is to celebrate the solstice on July 1 (among the Tzotzil), which would correspond to the first day of the 52-day period initiated on page 16.

Muan, Red Puma in *Pax*.

Page 16: second solar zenith. Last of *Uayeb* or first of *Pop*; ceremonial year ends in the five days of *Uayeb*; fire and cord stretching ritual in *Pop*.

Page 17: autumn equinox. Last day of *Zip* or first of *Zotz*; fire ceremony to *Kinich Ahau Itzamna* or *Uo*; in *Zip* a ceremony to *Ix Chel* (moon); ritual hunting and pole erected.

Page 18: close of 260-day agricultural cycle begun in *Yax* (on page 13) in month of *Xul*, which means "end"; celebration of descent of *Kukulcan* in last five days of *Xul*.

The imagery of each page is in accordance with the ceremonies described by Landa, the solar events corresponding to those months, and the positions of the Pleiades and Castor and Pollux, as will be shown below. Furthermore, each page except the first and last (12 and 18) covers a 52-day period totaling 260 days, corresponding to the fixed count that would begin at a specific lunar phase coordinated with the second solar nadir and the first day of *Yax*. The count would begin no later than 30 days after *Yax* begins, and would end no later than *Xul* (meaning end), which dates from November 3 to November 22, N.S. Thus the end of the count would be linked to the first solar nadir (last day of *Xul*), just as the beginning of the count was linked to the second solar nadir and the month *Yax*.

During the period covered by page 13, the major ceremony to *Chac* was held on a day specified by the priests in either *Chen* or *Yax*--January 2 to January 21 (N.S.) and January 22 to February 10 (N.S.). This suggests the ceremony was astronomically timed. Three events would occur during this period: the dusk zenith of the Pleiades, the second solar nadir, and the dusk rise of Castor and Pollux of Gemini, celebrated as a sign of January by the modern Maya. In the drawing on page 13, the prominent serpent rattles might refer to the dusk zenith of the Pleiades, and the twin *Chac*s might refer to the dusk rise and longest visibility of Castor and Pollux. Because their declinations were close to the latitude of the observer, Castor and Pollux and the Pleiades are zenith stars in the latitude of Mesoamerica (ca. A.D. 1400, Castor was at 32°, Pollux at 29°, and the Pleiades at 22°). They also mark the position of the sun at the winter solstice and at the first solar zenith, and are located at the intersection of the Milky Way with the ecliptic; these are more than adequate reasons why these stars were especially important in Mesoamerica.

In addition to the correspondence between the *Chac*s on page 13 and the festival to *Chac* in *Chen* or *Yax*, Landa mentions that incense was burned to the gods in *Yax* (Tozzer 1941:161); appropriately, an incense brazier is burning in front of a seated god on this page.

MADRID CODEX PAGE FOURTEEN AND THE SPRING EQUINOX

The month of *Ceh* (March 3 to 33, N.S.) corresponds to the spring equinox. *Ceh* literally means deer, and appropriately a deer without horns is pictured, suggesting a direct link to the month of March when deer in Yucatán shed their horns (Leopold 1959:510-512). Activities performed in the month *Mac* (March 23 to April 11, N.S.) are also shown on page 14. In *Mac* there was a fire ceremony in connection with a hunting ritual (Tozzer 1941:163), and the representation of a death god holding a torch behind the deer seems apt.

In a number of Maya communities from Guatemala to Yucatán, a new fire ceremony coincides with the spring equinox or Easter week, and falls on the first full moon after the spring equinox (Redfield 1941:104; Tozzer 1907:106; Girard 1962:65). In Yucatán they begin burning the fields in preparation for planting around the spring equinox on a full moon (Perez Toro 1942:14, 43). Thus the timing of both the modern fire ceremony and the burning of the fields appears directly linked to the same lunar phase, and no doubt the ceremony is a formal celebration of the role of fire in preparing the fields.

MADRID CODEX PAGE FIFTEEN: THE FIRST SOLAR ZENITH AND SUMMER SOLSTICE

The black god M is a dominant image on page 15. He has been firmly identified with the merchant god *Ek Chuah*, meaning black or star scorpion (Kelley 1976:72, 85). According to Landa, *Ek Chuah* is one of the main gods worshipped in the month *Muan* (Tozzer 1941:164). God M sometimes wears a scorpion tail, and sometimes is shown drilling for fire (Kelley 1976:Figs. 23, 64). These features suggest a parallel to the Mexican fire god *Xiuhtecutli*, who is similarly associated with the scorpion and fire drilling. I have argued elsewhere (Milbrath 1980a) that Scorpio is the fire serpent, which is especially important as a marker of the position of the nadir sun during the November New Fire Ceremony at the beginning of the dry season. The fact that a scorpion is shown among the 13 animals suspended from sun symbols (or eclipse signs) in the Maya Paris Codex suggests that the Maya also recognized a star scorpion in the path of the sun. Kelley (1976:Table 6) identifies the scorpion with Scorpio, and Coe (1975:26) believes Scorpio is the star scorpion on the basis of ethnographic and ethnohistorical evidence from the Maya and Huichol. Thus the black god may be associated with Scorpio, and his celebration in *Muan* (page 15) corresponds to the dusk rise of Scorpio in May.

It will be noted that the serpent on page 15 has reversed direction, lost its tail rattles, and bears a volute in its eye. All these iconographic changes may refer to the conjunction of the Pleiades with the first solar zenith sun. Because the Pleiades-Perseus configuration sets in a rotated position at this time, it provides a good correspondence to the position of the serpent (Fig. 23.2d). The lack of rattles probably refers to the invisibility of the Pleiades as it travels in conjunction with the sun at the time of the first solar zenith. The cross in the eye may reflect an association of the zenith sun with the concept of the cross. Even today a number of Indian communities celebrate the Catholic Day of the Cross (May 3) as a planting ceremony linked to the arrival of the rains at the time of the first solar zenith (Brinton 1896:95; Garibay 1957; Girard 1949:422-423). The Pleiades setting is clearly linked to a cross at Teotihuacán, since they set over the pecked cross on Cerro Colorado when viewed from another cross near the center of the city (Fig. 23.3; Aveni 1977c:179, 182). Since the Pleiades and the zenith sun are located at the great crossroads in the sky formed by the intersection of the Milky Way and ecliptic (Fig. 23.10), the cross concept may have a broader meaning that refers to this celestial locus (see Girard 1949:459, 573-574, 603 for further discussion).

The band of symbols above the serpent parallels that on the opening page, just as the astronomical events on this page (first solar zenith and summer solstice) complement those on page 12 (first solar nadir and winter solstice). Although the sun glyph is missing, the sequence *ollin* and *kan* cross is the same, but *akbal* (night) is replaced by the St. Andrews cross, usually read as sky. I would suggest the *ollin* glyph refers to the zenith-nadir axis and the *kan* cross to the solstice, with the St. Andrews cross qualifying it as the summer solstice, when the days are longer than the nights.

MADRID CODEX PAGE SIXTEEN AND THE SECOND SOLAR ZENITH[11]

The serpent on this page is similarly placed with respect to preceding pages, which is appropriate to the reversed position of Algol relative to the Pleiades. Algol is now to the east rather than to the west of the Pleiades, as on pages 12 through 14. The serpent lacks rattles, plumes, and a nose volute, iconographic changes that may indicate an altered state in the underworld. The underworld is signified by a huge death god, behind which the serpent passes. The god may have the significance of the lower meridian, for at the time of the second solar zenith (July 26) the Pleiades-Perseus serpent is halfway across the underworld at the dusk lower meridian. Although there is no evidence of fire or cord-stretching ceremonies such as those described by Landa for *Pop* (the month initiating the ceremonial year), the figures in the upper panels represent *kan-imix* glyphs. Since *Imix* is the first day of the 260-day count, a new divination count may begin on page 16 at the time of the second solar zenith. This count differs from the fixed 260-day agricultural count, which could begin on any day of the 260-day divination count.

At this point I should raise a sticky problem. All of the offerings on opening page 12 (ascribed to November and December) look exactly like the types of offerings made at the New Year ceremonies in the Dresden (25-28) and Madrid (34-37) codices, which presumably fall in the month of July as specified by Landa, concurrent with the second solar zenith. But all of the rest of the imagery on pages 12-18 correlates with the sequence of months *Yaxkin* to *Xul*. *Xul* does mean end, and *Yaxkin* does mean new sun, which suggests the year began and ended in late November. Since the Aztec New Fire Ceremony was also in November (an astronomically-timed ceremony), as was the Maya festival celebrating the descent of *Kukulcan*, it may be that the astronomical year began (and ended) in November. Whereas the festival calendar began in late January or early February (concurrent with the second solar nadir?) among the Aztec, it began at the time of the second solar zenith among the Yucatec Maya. Whether the New Year ceremonies of the Dresden and Madrid codices are more appropriate to November or to July is a topic that cannot be explored at this point.

MADRID CODEX PAGE SEVENTEEN AND THE AUTUMN EQUINOX

The imagery of page 17 is complementary to that on page 13, with twin

---

[11]In a previous study of these pages of the Madrid Codex (Milbrath 1980b), I suggested this page refers to the summer solstice rather than the second solar zenith, but since the summer solstice may have been celebrated 10 days late (on June 31; see footnote 10), page 16 may refer to both the summer solstice and second solar nadir.

Chacs, a turtle, and an eclipse sign. The dating of page 13 is January 22 to March 16, a few days from the spring equinox, whereas page 17 dates from August 21 to October 12, with the autumn equinox falling at the midpoint. Thus the pages incorporate the solar half-year as well as the eclipse half-year. The two winged *kin* panels probably refer to two successive solar eclipses spaced 177 or 178 days apart. They are placed toward the edge of the page, perhaps indicating that the one on page 13 took place near the spring equinox (for it is at the right edge, or the end of the period designated by the page), whereas the one on page 17 may have taken place in late August (at the left or leading edge of the page), and may have been a partial eclipse since the winged panels are half white, rather than all black as on page 13 (total eclipse?).

The prominence of the turtle here may be a reference to Gemini or Orion, both of which transit the lower meridian during the period specified by page 17. It will be noted that here the serpent is represented only by its head, and the body is cut by the edge of the page. This may be a very graphic image of the Pleiades-Perseus serpent position, for at the autumn equinox the first star of Perseus appears, and by the close of the period on page 17 (October 12) the serpent head represented by Algol has cleared the horizon, well above the Pleiades (serpent tail). The position of the serpent is now rotated, facing left, for Algol is to the west of the Pleiades once again upon rising, but the head is raised above the body, just as Algol is raised above the Pleiades.

MADRID CODEX PAGE EIGHTEEN AND THE END OF THE COUNT

On this final, unfinished page the serpent tail is now shown, indicating that the rest of the Pleiades-Perseus serpent is now visible, except for the rattles formed by the Pleiades which are just on the eastern horizon at dusk. The observers' horizon elevation, and the exact date of the Codex, would affect the date of visibility of the Pleiades. With optimum visibility the Pleiades would have risen on November 1, A.D. 1500 (Colgate University Computer Center, n.d.); in A.D. 1400 this event would occur one day earlier, though still falling in the later part of the period designated by page 18 (October 13 to November 9). Thus the serpent images give a very accurate indication of the position of Perseus and the Pleiades throughout the sequence, and the Pleiades from dusk zenith to dusk rise define a 260-day period corresponding to the fixed agricultural count.

THE GLYPHS

Space does not permit an adequate treatment of the glyphs. As Kelley (1976:293) points out, scholars are able to read only one-tenth of the glyphs with any certainty. However, it will be useful to point out certain correspondences. The fire glyph (T122:563a; Kelley 1976:148) appears three times (pages 13, 15, 16) prior to the first glyphic reference to God B, the presumed rain god corresponding to *Chac* (T668:103; pages 16 and 18). Since the burning precedes the rains, this relationship is appropriate. Nevertheless, these glyphs may not refer to the agricultural cycle, since placing the burn and onset of the rains in July through September (pages 15, 16) would be too late in the year. Since the fire glyph occurs only in the augural band, along with the names of other deities, it may refer to a god of fire or a fire-drill constellation (it appears over a flaming fire-drill bundle on page 13). The *tzab* glyph (T207) has been identified as the Pleiades (Dutting 1979:147), and signifies "owner of the rainy season." Its appearance on page 16 in the upper panel may refer to the lower meridian transit of the Pleiades. The turtle-head glyph (T743) appears frequently on page 17 in the upper panel, during the period Gemini and Orion transit the lower meridian. Lamb (personal communication) suggests the text refers to the burial (*muc* meaning buried or hidden away) of the turtle constellation, an interpretation which seems completely appropriate since this page refers to the lower meridian transit of the turtle constellation.

CONCLUDING REMARKS

The interpretation of these pages of the Madrid Codex may provide a significant contribution to the understanding of Maya codices. If, as proposed, specific astronomical entities are identified as gods, and their movements are recorded throughout the manuscripts, we may have a form of Rosetta Stone for deciphering the meaning of glyphs. It seems clear that the *Chicchan* serpent is the Pleiades, for its rattles and the *tzab* glyph are identical, and the *Chicchan* serpent is usually accompanied by the *tzab* glyph in the manuscripts (see Dresden Codex pages 33-35, 61, 69; Paris Codex page 21). The frequent pairing of God B and the *Chicchan* serpent is appropriate for identifying God B with a constellation close to the Pleiades, most probably Castor and Pollux.

It will be important to examine the complex of Maya serpent imagery to see exactly where the *Chicchan* serpent and the Pleiades fit in. There is certainly some association with the concept of *Itzamna*, the symbol of Maya rulership, since representations of *Itzamna* wear the rattles (for example on the Copan hieroglyphic stairway), and some scholars believe *Itzamna* may actually be the Pleiades (Roys in Tozzer 1941:133; Coe 1975:23). The ruler may have identified himself with the Pleiades, and at important moments in his historical reign (birth, ascension, death) may have taken on the guise appropriate to the position of the Pleiades at that time. Perhaps we should look more closely at the mythological lineages of rulers (Kubler 1974) and see whether they may have claimed descent from the Pleiades at some distant point in the past when the Pleiades marked another season (for example, around 2200 B.C. the Pleiades marked the position of the vernal equinox sun). The ancient Maya rulers may well have based their divine right to rule on celestial ancestry, just as rulers have over the ages (for example, the Pharoah upon death rose to join Orion's belt).

This analysis also clarifies the way in which solar events coordinate with the Yucatec Maya festival calendar, and indicates the importance of the zenith and nadir suns in dividing the solar periods. Both *Yaxkin* and the Aztec equivalent *Panquetzaliztli* may have had special significance, marking the birth of the sun at nadir at the onset of the dry season, as reflected by the midnight zenith of the Pleiades. Thus the important cosmological moments in Mesoamerican thought appear to have a significance linked to the seasons on earth in relation to the sun, moon, and stars.

24

The Use of Native Cosmologies in Archaeoastronomical Studies: The View From South America

Gary Urton

One of the principal goals of archaeoastronomy is the development of hypotheses for explaining cultural behavior associated with celestial phenomena; specifically, the archaeoastronomical studies have traditionally focused on the interest in astronomy evidenced by the incorporation of astronomical alignments in architecture. Since architectural testimony, in the form of ruins, often exists long after a culture has disappeared, archaeoastronomy provides us with a way of saying *something*, however hypothetical, about societies which have long since ceased to exist. In this respect, it is perhaps not surprising that archaeoastronomers often tend to view the discovery of precise alignments in archaeological ruins as something bordering on an explicit statement from the people who constructed the buildings concerning the purpose and meaning of one aspect of their behavior. However, this can never be the case, because archaeoastronomy can tell us only whether or not alignments exist; it can never tell us *why* they exist. Since culture history is composed equally of one part description and one part analysis, archaeoastronomy comprises only one-half of the study of ancient and primitive systems of astronomy. The other half of the task falls within the domain of ethnoastronomy, the study of historical documents, oral tradition, ethnographies, etc., which give testimony from indigenous sources concerning astronomical beliefs and practices. While these sources can seldom provide completely objective information, they do represent our best available evidence for explaining the "why" of archaeoastronomical alignments.

There are not many places left where anthropologists have access to relatively indigenous societies. Where we do find such societies still in existence, we have the opportunity of undertaking ethnoastronomical fieldwork in an effort to understand, and preserve, indigenous calendrical and astronomical concepts. In the Americas, the continent of South America represents an area of profound importance for studying American Indian astronomy in general. While certain areas of South America have undergone tremendous cultural transformations since the Spanish and Portuguese conquests of the early sixteenth century, there still exist a number of remote villages, primarily in the Andes and in the Amazon and Orinoco river basins, where the style of life (i.e., language, dress, economy, architecture, etc.) closely approximates that described in the earliest colonial documents. In light of this fact, it seems apparent that if we want to study the astronomical knowledge of South American Indians, the best place to begin is with the testimony of the indigenous population.

The purpose of this paper, in relation to the above observations, is to discuss selected ethnoastronomical data drawn from several ethnographies of contemporary South American Indians. We will focus on those accounts which tell us something of the larger cosmological systems within which the astronomical beliefs and practices are integrated. In this way, the *indigenous* cosmological models will provide us with a framework for the development of archaeoastronomical hypotheses.[1]

In beginning our description of native systems of astronomy in South America, we should first discuss some characteristics of the celestial sphere as viewed from this part of the New World. If we accept as a point of departure the notion that people everywhere develop cosmologies in order to understand and explain observed natural phenomena, rather than that they are arbitrary systems of order imposed upon the natural world, then it would seem that we must also accept the proposition that people who look at the universe from different vantage points might well develop different systems of cosmology.[2]

More than two-thirds of the continent of South America lies between the tropics, the lower tip of the continent extending down to 55° south latitude. The equator passes through northern Ecuador, southern Columbia, and the north of Brazil (some 3° north of the Amazon). Therefore, the sun stands in the zenith, at different times, over two-thirds of the continent of South America, but perhaps more importantly, the continent contains the equator, the line along which the equinox sunrise and sunset occur on the same day as the passage of the sun through the zenith. We find in South America then the geographic and astronomical bases for nearly symmetrical organizations of time and space.

Along the equator, of course, the paths of the stars are seen as points of light which rise, culminate, and set in a series of parallel lines. When we cross the equator into the southern hemisphere, we find an inversion of northern hemisphere polar-coordinate systems such that polar observations made north of the equator produce a right-to-left pattern of stellar movement, while south of the equator the stellar movement is left-to-right (when facing the south celestial pole). Therefore, we have in South America the potential for opposing systems of cosmology and astronomy based upon movement right and left, along with a resolution of these opposing systems based upon the front-to-back (or back-to-front) movement of the celestial sphere as viewed along the equator.

We will take this principle, the resolution of oppositions, as a theme for our survey of native South American systems of astronomy. We will see three primary ways in which this principle has been described in the cosmological systems of South American Indians: (1) in a "horizontal" view in which intercardinal points and axes are mediated by the east/west line; (2) in a "vertical" view in which the zenith is used as a point of orientation for the non-zenith passages of the celestial bodies; and (3) in an "integrated" view in which each community becomes the center of orientation for the division

---

[1]See Isbell (1976) for a good discussion of the relation between cosmology and the archaeological configurations of Pre-Columbian ceremonial centers.

[2]See the discussion by Aveni (n.d.) on the influence of geography (i.e., tropical vs. non-tropical) on systems of astronomical orientation.

of space into four or eight parts based upon the intersection of the east/west axis by the zenith/nadir axis.

For those acquainted with the anthropological literature on South America, it is apparent that the theme of the resolution of oppositions in native cosmologies is the same as that developed by Lévi-Strauss in his studies of the mythology of the Indians of the Amazonian tropical forest. In his analysis of Amazonian mythology, Lévi-Strauss has shown that the logic used in the mythology is based upon a systematic resolution of oppositions by a mediator, and that while the form of the oppositions and mediators can be transformed from motif to motif, the system of logic--embedded in what Lévi-Strauss refers to as the "architecture of the mind" (Lévi-Strauss 1973:473)--remains the same. In addition, we have evidence that many of the same principles (and in fact similar motifs) found in Amazonian mythology are the basis of the mythological system of the colonial Quechua in the Peruvian Andes (Zuidema 1977:43-48). Since it is not reasonable to suppose that the Indians of South America employ one system of logic in their mythology and a totally different system in their astronomy, the theme which we have chosen allows us to integrate the astronomy within its larger cultural context.

Having appealed to the theoretical framework of Lévi-Strauss' work, we should therefore begin with a brief discussion of the empirical data which he provides for the existence of oppositions and mediators in Amazonian astronomy. In the astronomy of the tropical forest tribes we find a series of calendrical oppositions between stars of the rainy season and those of the dry season; the relationship is most commonly seen in an opposition between Orion and the Pleiades (e.g., Dumont 1976:93-96; Farabee 1918:101-102; Wagley 1940: 256-257; and Lévi-Strauss 1969:226-228). Orion and the Pleiades are opposed not only because they are often related to different seasons but also because they are thought of as brothers who belong to different moieties. The fact that Orion is located just south of the equator (2° south) and the Pleiades to the north of it (24° north) is a first suggestion that two stars or constellations may be placed in a moiety relationship when they are opposed to each other across the line of the equator. Lévi-Strauss suggests that Orion and the Pleiades are consistently opposed because their cycles of visibility and invisibility are closely coincident and because, when both are visible, they form a contrast between the categories of the continuous and the discontinuous; i.e., the line of the belt of Orion as opposed to the cluster of the Pleiades (Lévi-Strauss 1969:227). Thus, Orion and the Pleiades represent opposite, but complementary, cosmological principles.

We have said earlier that we are interested not only in stars or constellations which are in opposition but also in the resolution of that opposition. Lévi-Strauss accomplishes this by showing that the Orion/Pleiades opposition is part of another between Orion and the constellation of Corvus (at 19°-24° south celestial latitude). Since we cannot fully recapitulate Lévi-Strauss' argument here, we need only say that in its opposition to the southern constellation Corvus, Orion's position is inverted with respect to its position in the Pleiades/Orion opposition. Therefore, the complete system of celestial, seasonal oppositions is seen as a triad involving the Pleiades in the north, Orion only a few degrees south of the equator, and Corvus in the south. According to our earlier terminology, Orion is the mediator whose role is reversed depending upon whether it is opposed to a constellation in the north or south (Lévi-Strauss 1969:230-239). With this example in mind, we will examine a number of South American systems of cosmology in order to study the operation of this principle in the astronomy.

In Figure 24.1, we see an example of the circular village plan common among many tropical forest groups. The village is composed of a ring of huts, a central building (the men's hut), and the dividing line of the moieties and that of the up-stream and down-stream groups. In general, most villages of the southern Amazon are divided into four parts by one axis which runs east/west (based upon the east/west line of the sun at the equinox), and another axis running north/south, which is roughly parallel to the south-to-north flow of the southern tributaries of the Amazon (Lévi-Strauss 1948: 369). In their physical lay-out, therefore, tropical forest villages often reflect a principle of order based upon a unification of terrestrial and celestial axes. Among the Tatuyo of the central Amazon, however, the principal cosmological axes (i.e., the lines of the sun and the Amazon) *coincide* rather than intersect; the two east-west pathways are considered to be parallel, one a duplicate of the other, ". . . one below on the earth, the other in the sky precisely at the vertical" (Bidou 1972:59).

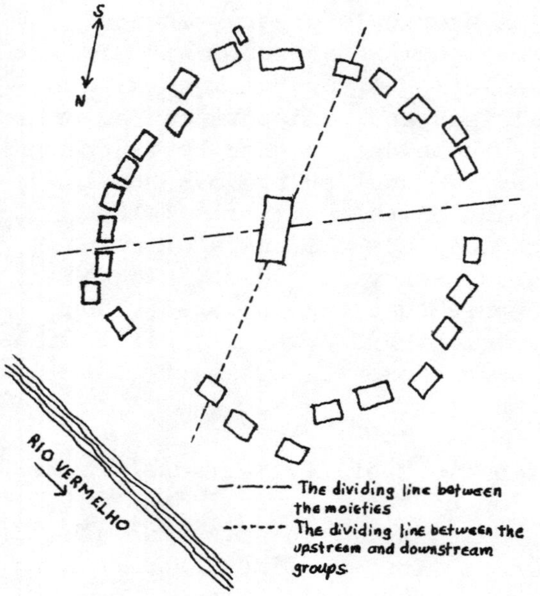

Fig. 24.1 Amazon circular village plan (after Levi-Strauss 1964:203)

That the divisions of celestial space are related to both terrestrial axes and social divisions is further supported in Figure 24.2, which is taken

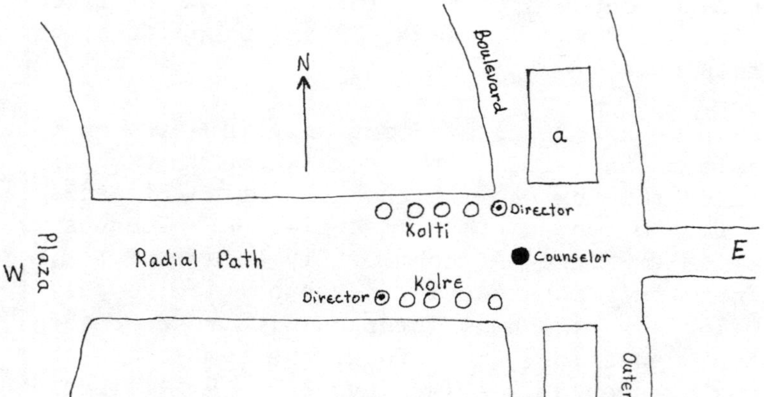

Fig. 24.2 Arrangement of *Kolti* and *Kolre* dancers at the *peny-ta'g* ceremony

from the work of Curt Nimuendajú (1967) on the Apinayé of southeastern Brazil. The complete plan of Apinayé villages is similar to that seen in Figure 24.1, but in this case we see another common characteristic of tropical forest villages: pathways which radiate from the center. The Apinayé example is important for our discussion because it illustrates the principle of mediation in astronomy, village structure, and social organization. In Figure 24.2, we see the arrangement of the dancers of the two moieties, *Kolti* and *Kolre*, on the day of the *peny-ta'g* ceremony. The *peny-ta'g* is a ritual ball game, played with a rubber ball, which takes place roughly every ten years at the time of the September equinox. *Peny-ta'g* is a part of the men's initiation ceremonies (Nimuendajú 1967:37 [compare to the ten-year initiation cycle of the Canella in Izikowitz 1955:204]). The night before the equinox, the dancers line up along the radial pathway leading to the east (see Fig. 24.2). Very early in the morning, just before sunrise, the counselor of the Kolti moiety (an old man who spends the night in the hut marked *a*) leaves the hut and very slowly moves toward the double line of dancers. In Nimuendajú's words:

> When the sun was about to rise, the counselor once more stepped out of the house *a* in the east . . . , holding in his raised right hand the large rubber ball, which now had paty wool stuck on it. Very slowly and almost imperceptibly he advanced by short stages without raising his feet from the ground so long as the *me'anro* dancers were standing still. . . . Thus he gradually got to the opening of the dancers' double line, *while precisely behind him the sun's orb began to rise above the horizon* of the steppe. Now the Kolti director of ceremonies slowly approached the old man and bent down before him. The counselor feigned throwing the ball, but put it into the director's hand, who passed it to his Kolre colleague beside him [Nimuendajú 1967:66; my emphasis].

Among the Apinayé, then, we find that the opposition between the village moieties is related to the opposition of the northern and southern hemispheres, and that the resolution of opposition between the moieties and the hemispheres is accomplished by the equinox sun which moves from the north to the south and by the rubber ball which moves from the Kolti moiety in the north to the Kolre in the south.

A further indication of the correspondence between social and celestial divisions is Nimuendajú's statement that among the eastern Timbira, falling stars are thought to be Timbira men who move from one celestial moiety (= hemisphere) to the other in order to find a mate. "Thus," says Nimuendajú, "they divide the heavens into exogamous moieties like those of their own village circle" (1946:233).

The Warao Indians of Venezuela live between 8° and 10° north latitude. They have been studied by several anthropologists, but we will discuss here primarily the work of Johannes Wilbert (1975) on Warao cosmology (see also Roth 1908-1909:254-270). In Figure 24.3, we see Wilbert's drawing of the vertical, three-dimensional structure of the universe as conceived by the Warao.

The vertical cosmos of the Warao consists of five stratified levels: (1) the flat surface of the earth; (2) the cardinal and intercardinal gods of the world mountains at the horizontal ends of the world; (3) the gods residing on the imaginary belt mid-high around the bell-shaped cosmic vault; (4) the gods and spirits of the zenith; and (5) the dark, featureless realm of the

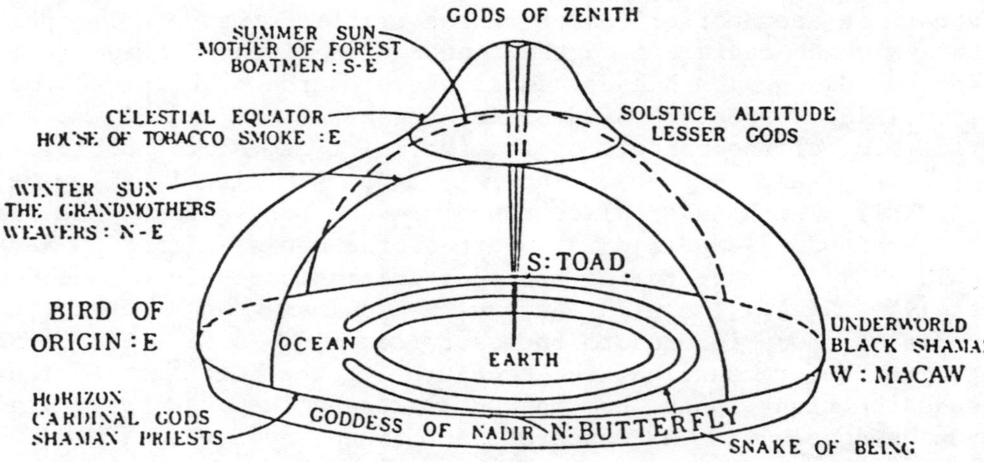

Fig. 24.3 The Warao Universe
(From Wilbert 1975:165)

goddess of the nadir (Wilbert 1975:168). It is obvious from Figure 24.3 that there are a great many symbolic relationships which could be discussed concerning Warao cosmology, but we will only point out those features which are particularly relevant to our general discussion here. We see in Figure 24.3 (compare to Fig. 24.4) the opposition of the two solstices (at a lower

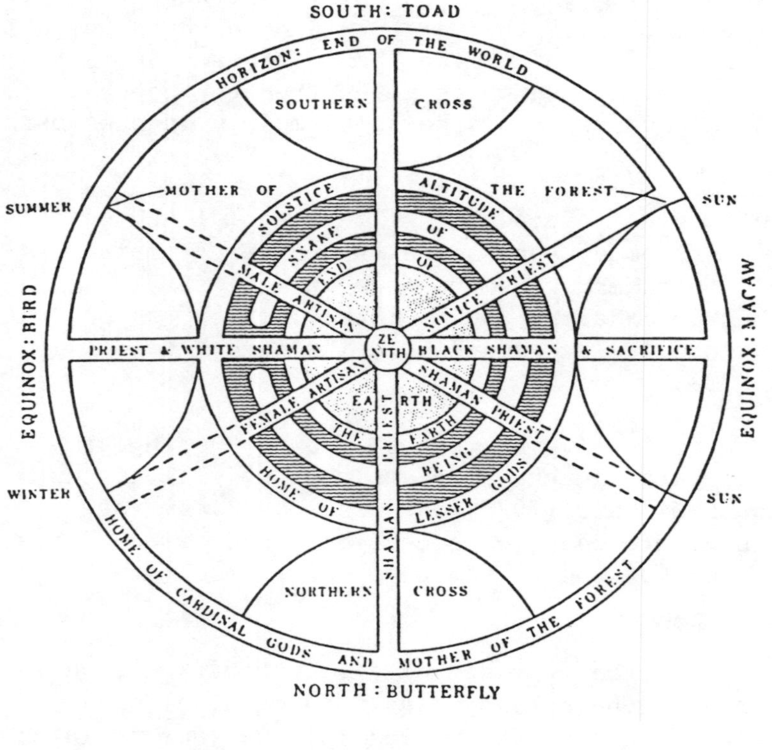

Fig. 24.4 Top view of the Warao universe (From Wilbert 1975:164)

altitude) mediated by the equinox line which, in this idealized view, intersects the zenith/nadir axis.[3] The zenith/nadir axis passes through the center of the earth, which becomes the center point of orientation for the beings, both natural and supernatural, who reside on the different horizontal and vertical levels. This concept is well illustrated in a top view of the system (Fig. 24.4) in which we see clearly the inter-relationship of opposing points (or domains of the Warao universe) which are joined by several cardinal and intercardinal cosmic axes.

We should note especially the diagonal (NE/SW and NW/SE) axes which are the result of relating the rise of one solstice to the set of the other, since these axes will be crucial in our later discussion of Kogi cosmology. In addition, we should note the symmetrical, semi-circular arcs or pathways formed by the rising and setting of the Northern and Southern crosses. The only *visible* lines in the sky, the alternating axes of the Milky Way when they cross the zenith, are not included in Wilbert's diagram, and one wonders if they perhaps could be included as perpendiculars to the diagonal axes of the solstices (cf. Urton 1978b, and below; see also Lipkind 1940:249 on the Southern Cross and the Milky Way). Another important point which should be made before leaving the Warao is that each community considers the center of the universe to be the dancing platform in the middle of its own central plaza (Wilbert 1975:163). Therefore, the system of cosmology described by the Warao, with its stratified levels and cosmic axes, results in a kind of portable coordinate system which can be used to orient any ego (whether an individual or community) through which the zenith/nadir axis is considered to pass.

That the zenith/nadir axis of tropical forest cosmological systems may, in a number of cases, be a moveable axis is well illustrated in Nimuendajú's description (1942) of the mechanism for communication between the terrestrial and celestial spheres among the Sherente of eastern Brazil. When the Sherente, for instance, experience an especially long drought and want to communicate with the celestial powers, young penitants who are drawn from the two Sherente moieties gather at a clearing outside the village. At night, the elders erect within the clearing a 10-meter-high *pindahyba* pole; the pole is called the "road to the sky." Representatives of the two moieties climb the pole to ask for rain and to communicate with the spirits. At the top of the pole the moiety director carries on discussions with K Orionis: ". . . after him [the moiety director] others climb up in order to learn how long they are to live, for at the top, the souls of their dead kin appear and answer questions" (Nimuendajú 1942:97). In both its function and structure, therefore, the *pindahyba* pole of the Sherente is similar to the zenith/nadir axis of cosmological orientations within the various Warao communities.

When we move now westward from the Warao Indians of the Rio Orinoco to the Kogi Indians of the Sierra Nevada de Santa Marta, we remain at approximately the same latitude but have now entered the relatively low Columbian Andes. The Kogi Indians of the Sierra Nevada have been studied for many years by the Colombia anthropologist Gerardo Reichel-Dolmatoff (1950; 1951; 1967;

---

[3]The drawing is "idealized" because the zenith sun and the equinox sun occur on the same day *only* along the equator. The Warao, as mentioned above, live between 8° and 10° north latitude and the zenith sun will therefore rise 8°-10° north of the east/west line.

1975; 1978). Reichel-Dolmatoff has given us one of our best accounts of the incorporation of celestial phenomena in the belief system of a South American Indian group and has provided us as well with a systematic description of the relation of astronomy and sacred architecture.

Figure 24.5 is a drawing of the Kogi conception of the cosmos, which is thought of as two cones joined at the base. The juncture of the cones is at the level of the earth; above and below the earth are four stratified levels, resulting in a nine-layer cosmos. Each of the nine layers is related to one of the nine daughters of the Earth Mother who created the world, to a different type of soil on the earth and to one month in the nine-month period of human gestation (Reichel-Dolmatoff 1975:204-205).

The earth, which is the fifth or middle plane of the universe, is considered by the Kogi to be the spindle whorl of an immense spindle which the Universal Mother works incessantly to produce yarn for weaving. The *axis mundi* of the Kogi universe is the shaft of the spindle:

> "In the beginning of time, the Mother Goddess took a spindle and pushed it upright into the newly created and still soft earth, right in the center of the snowpeaks of the Sierra Nevada, saying: 'This is the central post . . .!' And then, picking from the top of the spindle a length of yarn, she drew with it a circle around the spindle-whorl and said: 'This shall be the land of my children'" [Reichel-Dolmatoff 1978:13].

The intercardinal points marked on the drawing are ideally related to the points at which the solstices rise and set along the horizon. These four points are referred to as the "corners of the earth," and they are the places where four sons of the Great Mother reside. The fifth and favorite son of the Earth Mother, Mulkuéxe, resides at the center of the universe--at the point of intersection of the two intercardinal axes, and the point at which the zenith/nadir spindle shaft enters the earth/spindle whorl (Reichel-Dolmatoff 1975:206). We should also note that as shown in Figure 24.5, the cosmic spindle is right-spun, or spun clockwise, not left-spun. This conforms with the spinning techniques of most South American Indians in most (though not all) circumstances.

In Figure 24.6 are two drawings of a Kogi temple. Each conical-shaped temple is constructed by a Kogi shaman (*mama*). In both their external and internal features, Kogi temples are models of the cosmos; the conical structure which rises above the earth shares its base with an inverted conical temple which is conceived of as extending below the earth. When we note, in Figure 24.6b, that the internal framework of the temple is composed of a series of four quadrilateral frames rising above the fifth foundation frame, and that the four upper frames are mirrored in four subterranean ones, we see that the Kogi temples reflect not only the basic geometric form of the cosmos but its structural segmentation as well.

In a top view of the floor plan of a Kogi temple (Fig. 24.7), we find a structure similar to the top view of the Warao universe and of the village plans of the southern Amazon as described earlier. When a *mama*, a Kogi shaman, lays out this ground plan for the temple, he proceeds as follows. He first determines (on the basis of divination and practical considerations) where he wishes the center of the temple to be, and at that place he drives a small

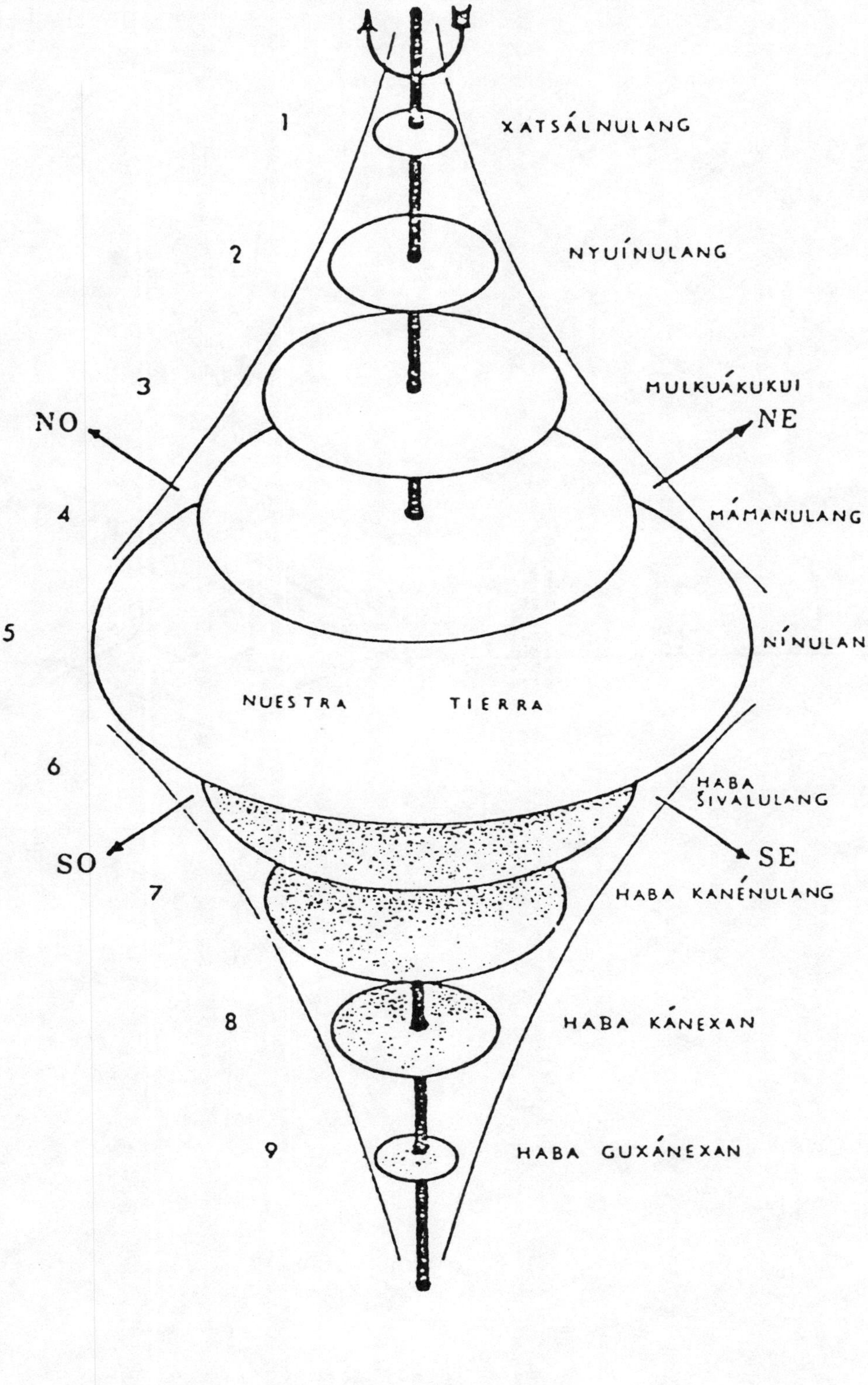

Fig. 24.5 The Kogi Universe
(From Reichel-Dolmatoff 1975:235)

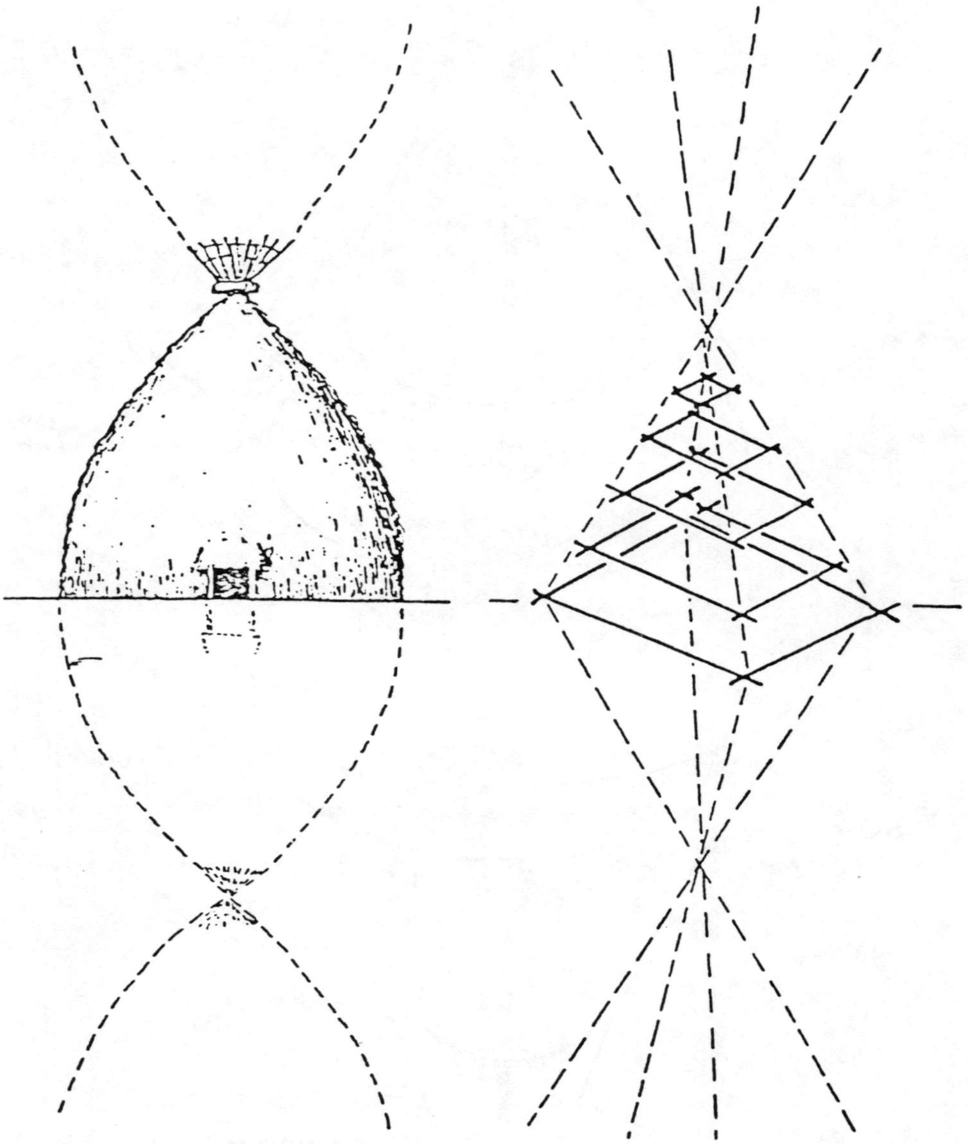

(a) (b)

Fig. 24.6 A Kogi temple
(From Reichel-Dolmatoff
1975:236)

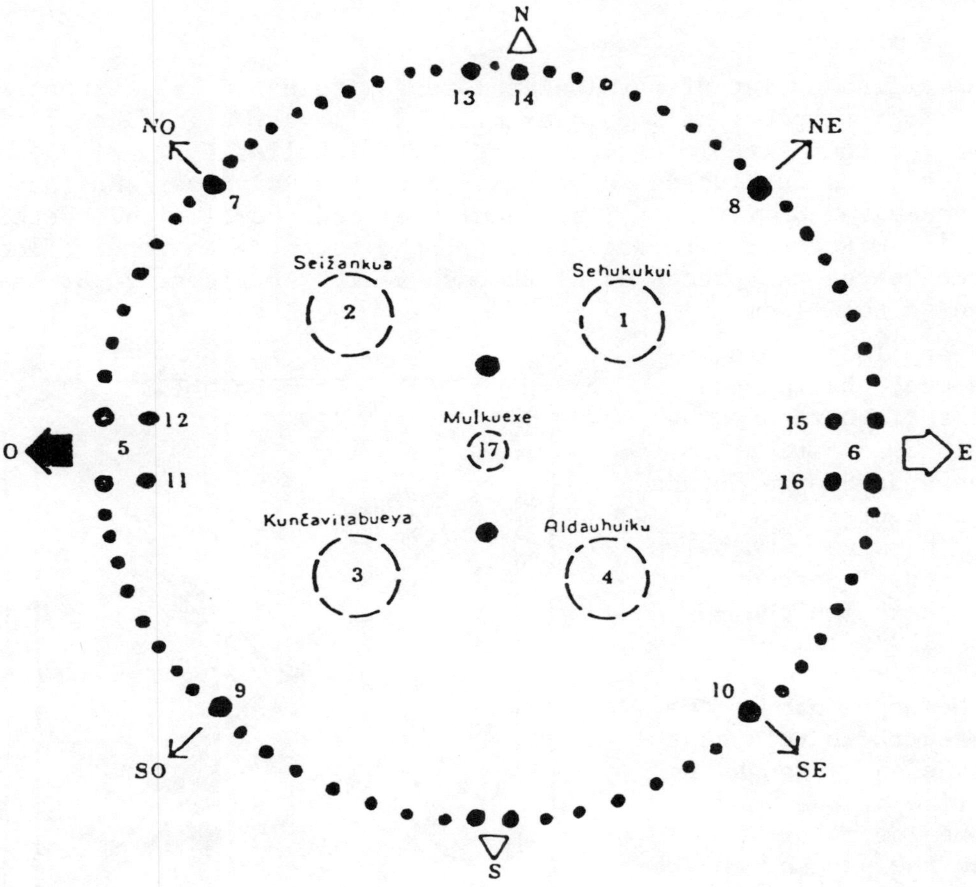

Fig. 24.7 Floor plan of a Kogi temple (From Reichel-Dolmatoff 1975:237)

stake into the ground. He then fastens a cord at the base of the stake. The cord is made by the wife of the *mama* and has a series of knots tied into it which guide the *mama* in determining the proportions of the temple. Although Reichel-Dolmatoff does not tell us this explicitly, I assume that one knot gives the radius from the stake to the two center posts, another knot gives the radius to the "circle" of four fireplaces, and a third gives the radius to the peripheral ring of the temple (Reichel-Dolmatoff 1975:207). In this way, the shaman defines the space of the temple in precisely the same way that the Mother Goddess defined the circular space of the Kogi universe--with a fixed post and string.

The *mama* determines the position of the large peripheral posts along the circumference by sighting the June solstice sunrise from the center stake. This point is marked on the peripheral ring in the northeast and a line is then drawn back over the center stake toward the southwest where it intersects the circle at the setting point of the December solstice sun. The June solstice sunset is then observed, marked on the peripheral ring, and a line is extended to the December solstice sunrise point (Reichel-Dolmatoff 1975:207). It will be seen that these "solstitial" posts, marked 7, 8, 9, and 10 in Figure 24.7, are not at the solstice points but rather at the true intercardinals. This "squaring off" of the posts is done, Reichel-Dolmatoff tells us (1975:207), as an adaptation of the posts to a desired quadrangular form.

As a final point of particular interest to us in the astronomy and cosmology of Kogi temples, we should examine briefly the significance of the placement of the four fireplaces in Figure 24.7 (labelled 1, 2, 3, 4). Just as the temple itself is considered to be a model of the universe, the four fireplaces are conceived of within a larger cultural and cosmological context--that of weaving. According to Kogi beliefs, the earth is an immense loom on which the sun weaves two pieces of cloth each year. In Figure 24.8, we see an example of a Kogi loom.

> A loom--any loom--thus becomes a map; the four corners of the square are the four Colombian cities that encompass the Sierra Nevada . . . the center of the square, where the two poles form a cross, represents the snowpeaks [Reichel-Dolmatoff 1978:11].

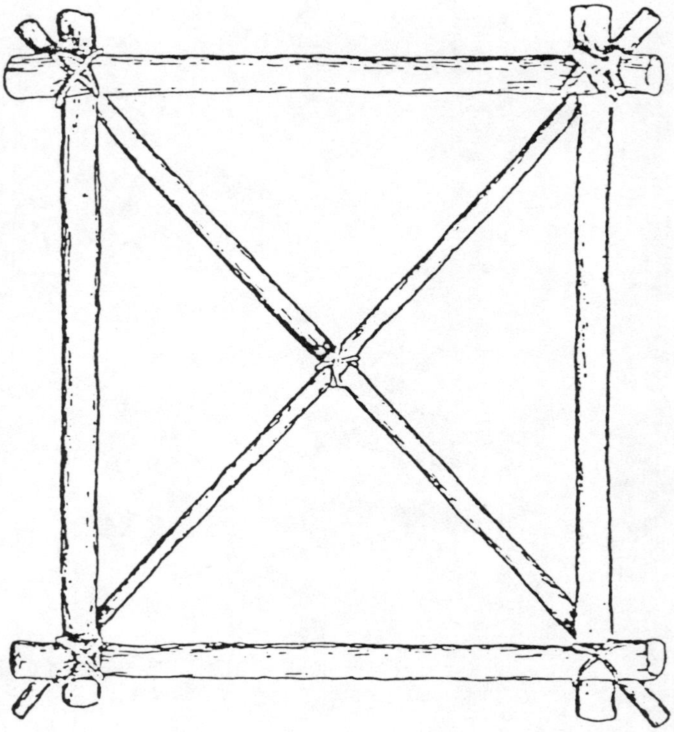

Fig. 24.8 A Kogi loom (From Reichel-Dolmatoff 1975:240)

> The upper bar of the loom of the *earth* is related to the path of the sun through the sky at the time of the June solstice; the lower bar is equated with the path of the December solstice sun. The cross at the center marks the idealized point of intersection of the diagonal axes of the four solstice points with the equinox/zenith sun (Reichel-Dolmatoff 1978:15-16).

If we return now to Figure 24.7, we find that the four fireplaces are arranged like the corners of a Kogi loom. The center of the temple loom (number 17 in the drawing) is the focal point of the loom, since this is the point through which the zenith/nadir spindle-shaft penetrates the earth.

The temple loom is operated by the sun. At the apex of the conical temple is a small hole which is usually covered by a pot-sherd. On certain occasions, the shaman of the temple removes the pot-sherd and allows sunlight to enter the temple. At the beginning of the Kogi year, on the morning of the June solstice, the pot-sherd is removed at 9:00 a.m. and sunlight strikes the fireplace in the southwest; during the day, the shaft of light moves along the temple floor from west to the east until, at 3:00 p.m. it rests in the southeastern fireplace. On the morning of the December solstice, the shaft of sunlight makes a line on the ground from the fireplace in the northwest to the fireplace in the northeast. The Kogi say that in this way the sun is weaving a fabric on the loom of the temple (Reichel-Dolmatoff 1978:18). The warp threads of the fabric, running north-south, are laid down by the Earth Mother, and the sun completes the textile by weaving in the east-west threads of the weft. Thus, during the daytime throughout the year, the sun weaves a white fabric. During the night, however, the sun weaves another fabric, but this fabric is black.

The weft of the black fabric is woven during the night when the light of the night sun (what the Kogi refer to as the "black sun") passes through the apex of the inverted subterranean temple. The night sun, however, lays down the black weft of the temple fabric in lines moving from right to left, as opposed to the left-to-right pattern of the white weft of the daytime fabric. Throughout the year, then, the sun weaves one black and one white fabric by the alternation, or opposition, of the left-to-right moving rays of the daytime sun and the right-to-left moving rays of the nighttime sun (Reichel-Dolmatoff 1978:18-19).

In this and the earlier examples of the structure and symbolism of Kogi temples, we find that the theme of the resolution of oppositions (via a process of inversion in the example of the cosmic loom), is reflected in virtually every aspect of Kogi life and thought.

> The concept of inversion, which is so important in the Kogi world view, is obviously related to the ray of light falling through a focal point which is the articulation of two cones joined by their tips, as exemplified by the chain of inverted temples [Reichel-Dolmatoff 1978:18-19].

As a final ethnographic example of an indigenous cosmological system which may aid in formulating hypotheses for archaeoastronomical studies, I will discuss my own field research in the community of Misminay, near Cuzco, Peru (Urton 1978a; 1978b; 1979; 1980; n.d.). Misminay is a small community (pop. 350) located about 13,250 ft above sea level in the southern Peruvian Andes. The community is composed of several named house clusters linked together by a complex network of footpaths. In addition to the several small inter-community footpaths, there are two major arteries which connect Misminay with the surrounding communities. The principal footpath, *Chaupin Calle* ("Middle Road"), runs from the southeast to the northwest and divides Misminay roughly into equal halves. The other major footpath, *Hatun Raki Calle* ("the great road of the four-part division"), runs southwest-northeast and intersects *Chaupin Calle* at a chapel called *Crucero* ("cross"). These two axial footpaths are also the routes of the principal irrigation canals of the community (Fig. 24.9).

The quadripartition of Misminay space does not, however, end at the boundaries of the community land. Instead, the two intercardinal axes are conceptually extended to the horizon, and all mountains and villages visible (and in some cases invisible, being sub-horizon) in the distance are named and related to the quarter of the community within which they are located. Therefore, we find in Misminay an intercardinal quadripartition of terrestrial space which is structurally similar to the quadripartite division of tropical forest villages, and it is a division very much like that of the Kogi in which the delineation of (sacred) space comes about by the intersection of two intercardinal axes.

In addition to the intercardinal quadripartitioning of terrestrial space, the people of Misminay also partition the celestial sphere into quarters, the boundaries of which are defined by two intercardinal, intersecting celestial axes. The principal feature by which the Quechua-speakers

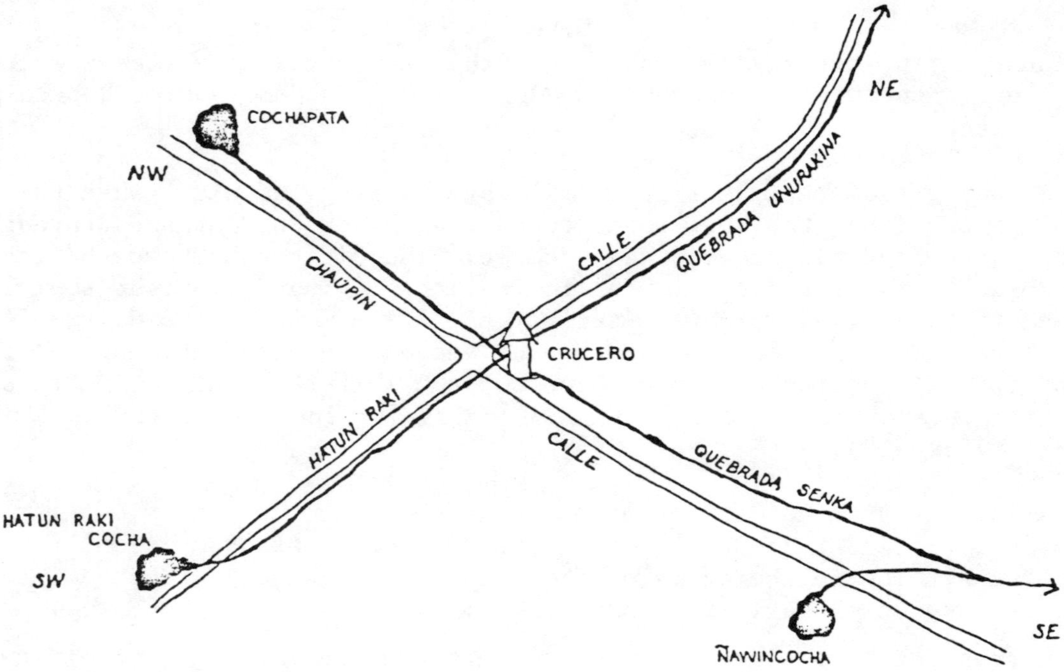

Fig. 24.9 Axial footpaths of the Misminay
(From Urton 1979:69)

of Misminay orient the celestial sphere is the Milky Way. Within the southern hemisphere, the Milky Way appears as a bright band of light which cuts the celestial sphere into roughly equal hemispheres. The ring of the Milky Way is not oriented along the line of the earth's rotational axis but rather is inclined to that axis by some 27° in the south and 30° in the north. Therefore, as the Milky Way revolves (apparently) around the earth, it forms two alternating intercardinal axes: one axis passes from the southwest through the zenith to the northeast; twelve hours later, when the other side of the Milky Way crosses the zenith, a line is formed in the sky from the southeast to the northwest (see Fig. 24.10). The point of intersection of the two celestial axes, located at the zenith, is referred to (as in the case of the point of intersection of the terrestrial axes) as a "cross" (*Calvario*).

By a complex method of relating points on the horizon to stars which divide the ring of the Milky Way into four equal sections (Urton 1979:227-230), the two intercardinal axes of the Milky Way are related to the two intercardinal axes formed by the four solstice points. The "quarters" of the Milky Way are shown in Figure 24.11.

The translation of the center points of the quarters of the Milky Way to the horizon of the community of Misminay is illustrated in Figure 24.12. The planes, points, and axes of terrestrial and celestial space are not, however, a series of unrelated phenomena in the Quechua mind. Instead, they are merged into a model of the universe in which a point or axis on the earth is correlated with a point or axis on the celestial sphere (Urton 1979; Earls 1972). The result is an integrated system of astronomy and cosmology in which the convergence of the terrestrial and celestial spheres occurs along the plane of the horizon. A point on the horizon serves not only for orienting space, but also (since the celestial bodies which form the background of the ring of

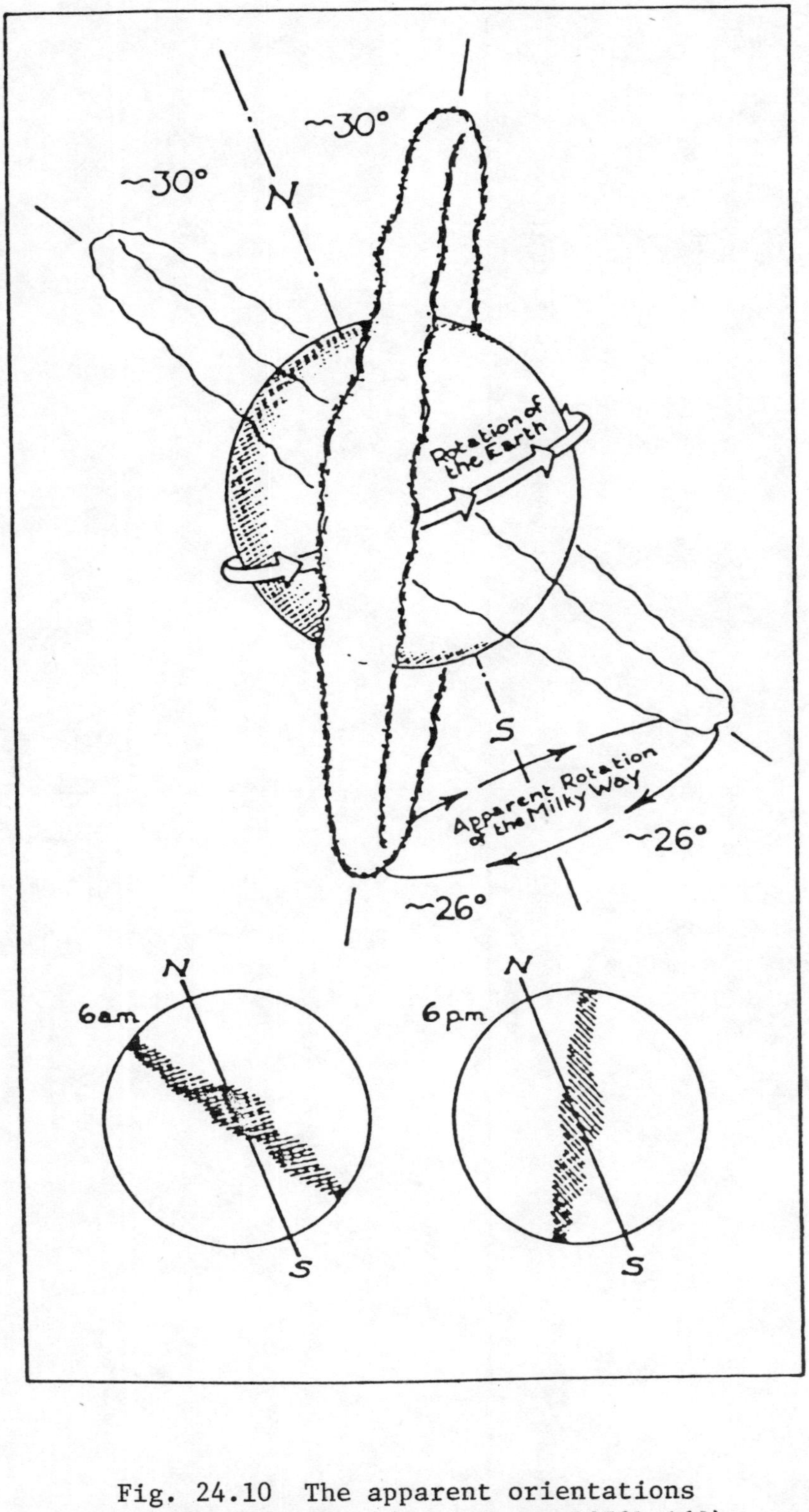

Fig. 24.10 The apparent orientations of the Milky Way (From Urton 1978b:163)

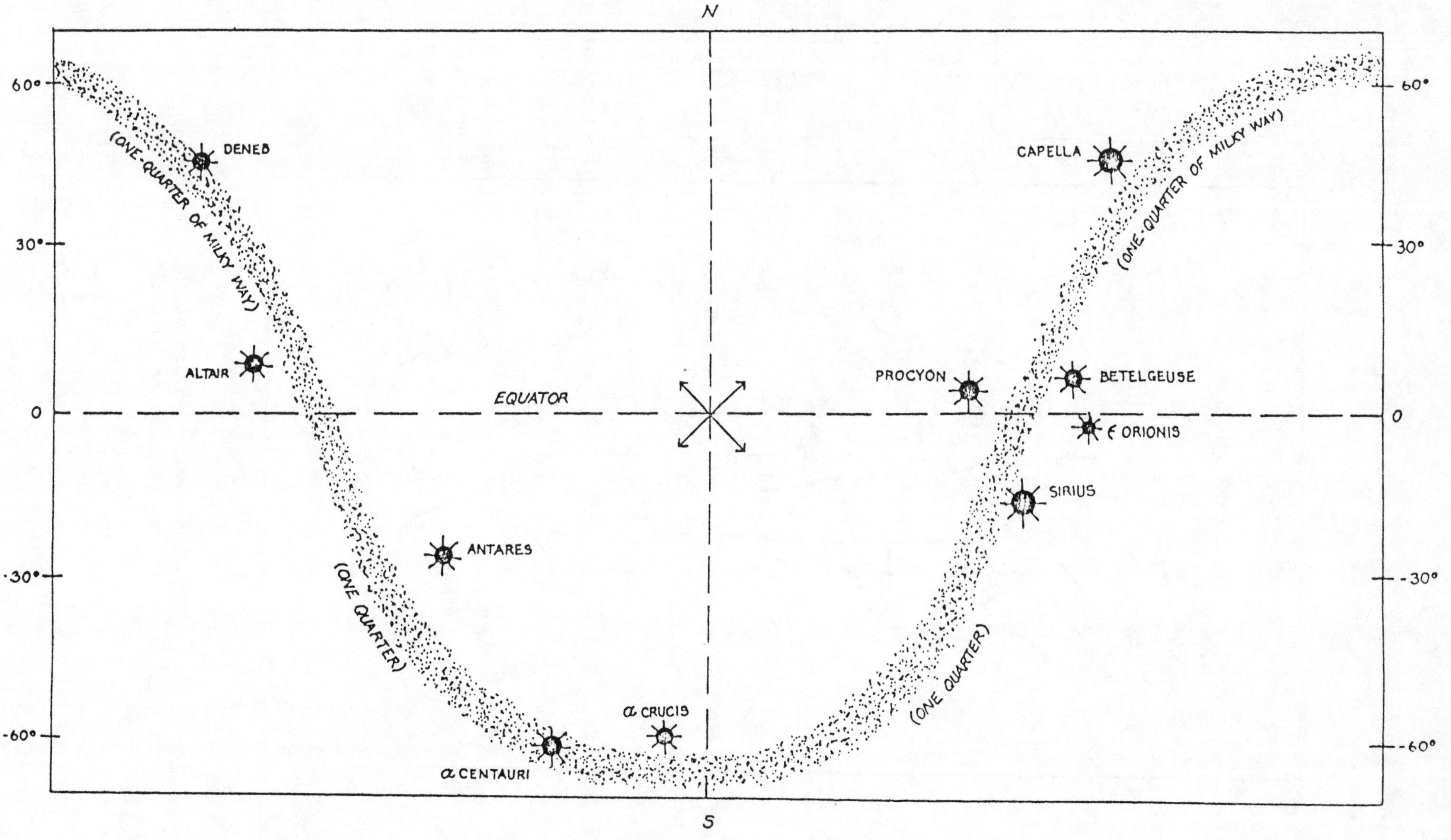

Fig. 24.11 The quarters of the Milky Way
(From Urton 1979:226)

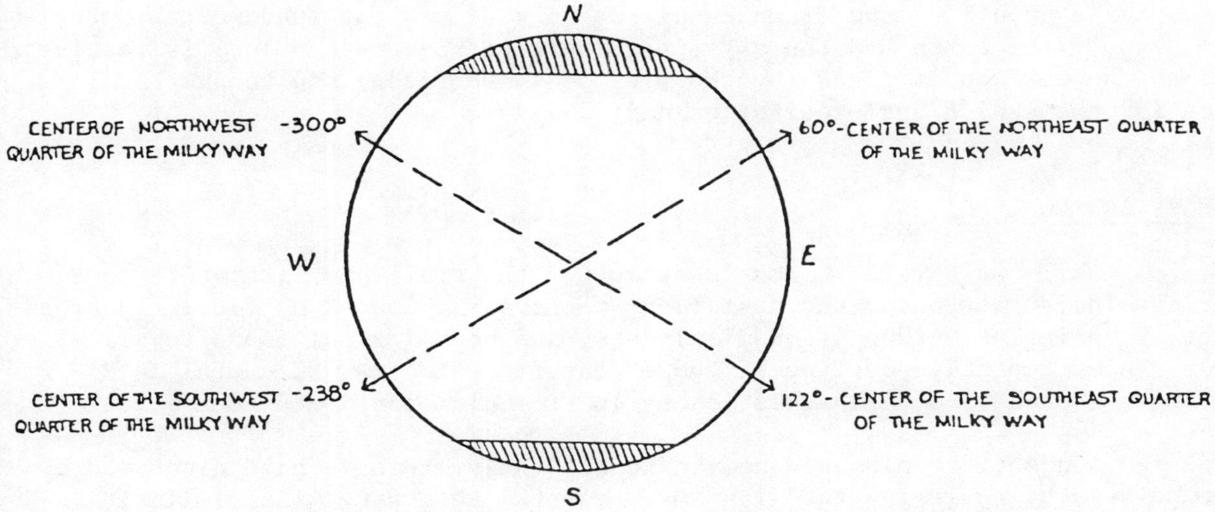

Fig. 24.12 The quadrants of the Milky Way from Misminay
(From Urton 1979:229)

the horizon are constantly changing) for temporal orientation as well (see Fig. 24.13).

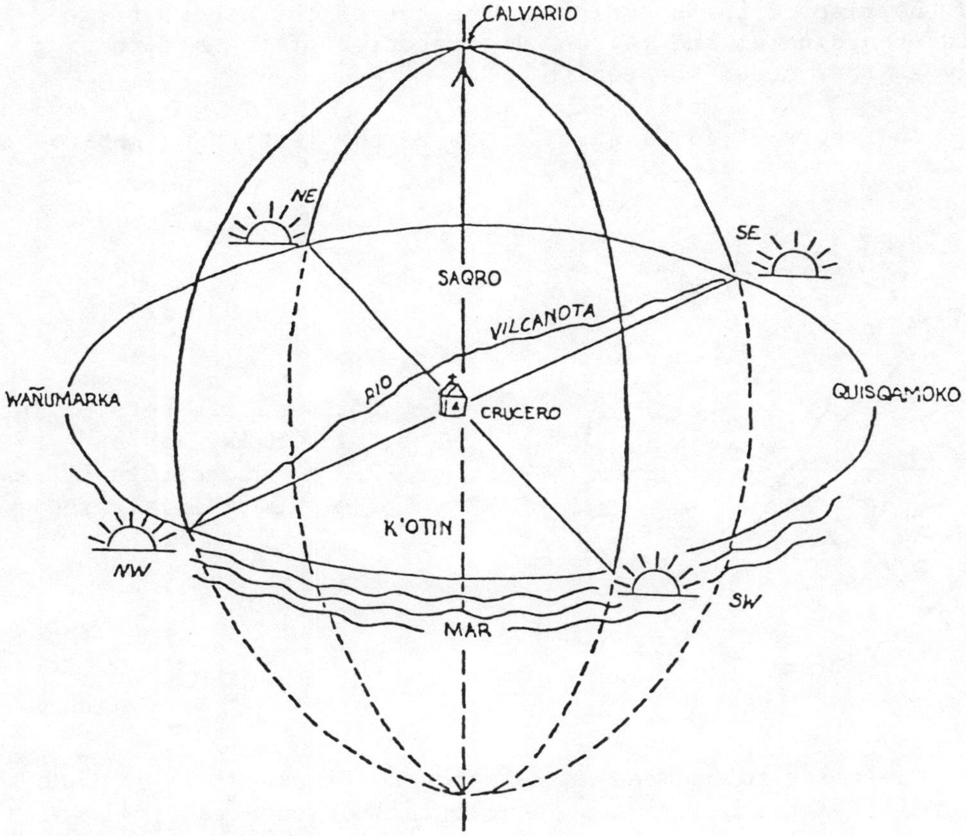

Fig. 24.13 A Quechua model of the universe
(From Urton 1979:95)

As in all of the examples discussed earlier, the fundamental opposition between the earth and the sky (i.e., between above and below) is resolved on both the horizontal plane (by the ring of the horizon) and the vertical plane (by the zenith/anti-zenith, or nadir, axis).

SUMMARY AND CONCLUSIONS

While we have by no means exhausted the available literature on South American Indian astronomy and cosmology, we have examined a sufficiently broad range of societies, ethnoastronomical data, and cosmological ideas to allow us to draw up a tentative outline of those features which may be useful in directing our studies of archaeoastronomy in Pre-Columbian South America.

A number of elements common to all the systems we have discussed can best be seen by reviewing the diagrams presented thus far. The similarities are especially obvious when we note, for instance, the structure of a Kogi temple in relation to the ground plan of a Borroro village, or the similarity of the cosmological diagram of the village of Misminay with that of the Warao. The element among the several diagrams which appears most critical is the organization of space around a center point defined by the intersection of one or more of the following axes: (1) the zenith/nadir axis intersected by the east/west (equinoctial) axis; (2) the intersection of the two axes formed by relating the rise of one solstice to the set of the other; (3) the intersection of the intercardinals; and (4) the intersection of the alternating axes of the Milky Way as they cross the zenith.

In Figure 24.14, I have diagramed the last three sets of intersecting axes as described above:

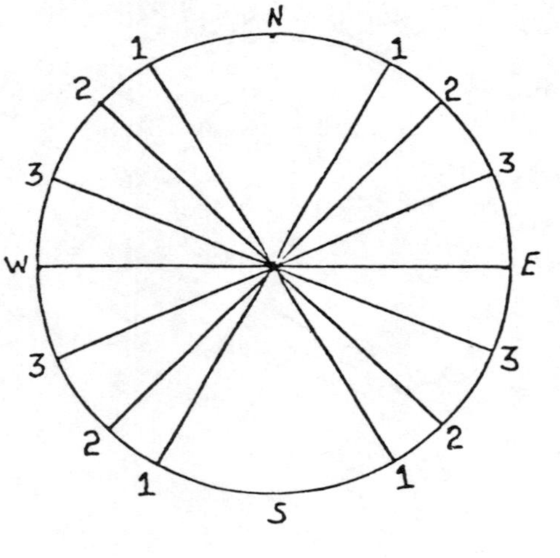

Fig. 24.14 Three sets of intersecting axes: 1 = Milky Way axes
2 = intercardinal axes
3 = solstitial axes

The point to be made in relation to Figure 24.14 is that the systems of terrestrial, astronomical, and cosmological orientations all focus at a point *on* the earth (usually at the center of the village), not on an *external* point. This observation is especially important when we note that coordinate systems centering on a point on the earth are exactly the opposite of most coordinate systems which developed in the northern hemisphere. The latter were almost without exception based on an external point of orientation (e.g., the pole star). In addition, the principal axis of cardinal orientations of the systems we have described does not appear to be the north/south line as is again true in the northern hemisphere, but rather the east/west line. Therefore, we may summarize these observations by stating that space is defined in relation to a terrestrial center which serves as the point of intersection of several diagonal (intercardinal) axes with the east/west line.

This summary, however, applies only to the definition of space as viewed from above, as though space were only two-dimensional. If we now shift our focus and view these systems in three dimensions, we find that in all cases vertical space is oriented around a zenith/nadir axis which passes through the same terrestrial center point as defined in Figure 24.14. In fact, just as there is a consistent tendency in these diagrams to place the center point along the axis of the equator, the zenith/nadir axis can itself be conceptually rotated through space to intersect the horizon at the equinox sunrise and sunset points (Fig. 24.15).

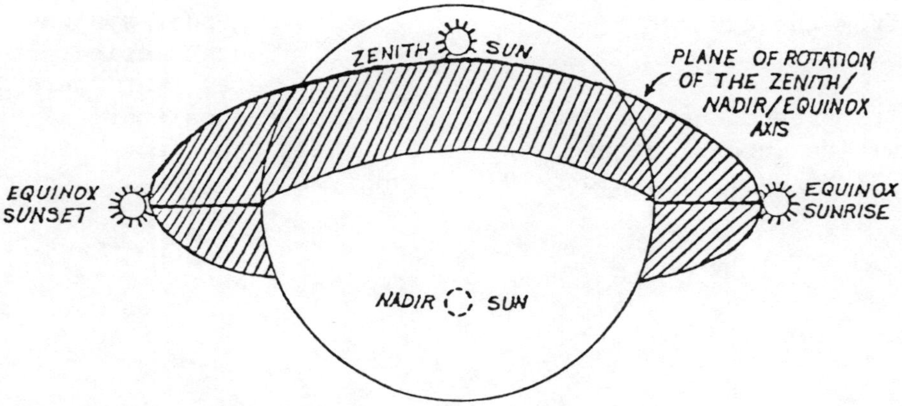

Fig. 24.15 Plane of rotation of the zenith/nadir/equinox axis

In effect, the vertical zenith/nadir axis and the horizontal east/west axis can be joined in a plane which encircles the earth, dividing terrestrial and celestial space into two equal spheres. As we move north or south away from the equator, however, the zenith/nadir and east/west axes no longer coincide with the plane of the earth's rotation. Thus, if a community is located at 10° south latitude, the zenith sun will rise at an azimuth of approximately 100° (= 90° + 10°) and set at an approximate azimuth of 260° (= 270° - 10°); however, the nadir (anti-zenith) sun will rise 10° north of true east and set 10° north of true west (i.e., at azimuths of 80° and 280°). Thus, as we move away from the equator, the zenith/anti-zenith sunrise and sunset points form not one plane, but rather two diagonal axes which intersect at the center point (Fig. 24.16).

In archaeoastronomical studies which have been carried out in Cuzco, Peru (latitude -13°30') by Zuidema (this volume) and Aveni (this volume), we find in fact that the zenith/anti-zenith points were of central importance in the Inca calendrical and cosmological system.

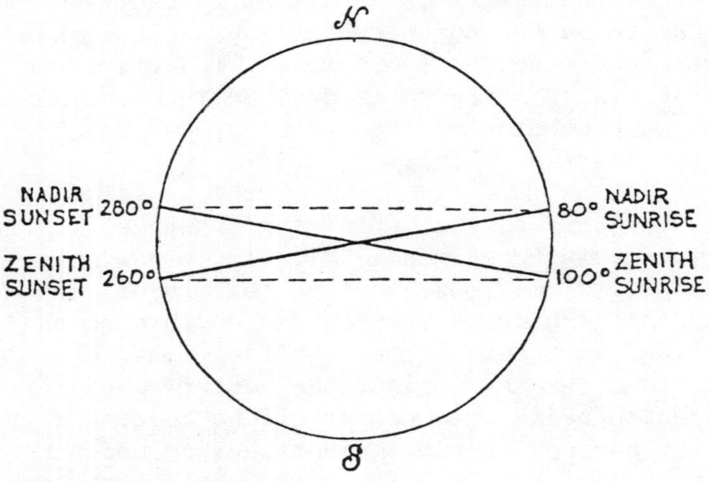

Fig. 24.16 Zenith/anti-zenith sunrise/sunset points away from the equator

We can see in the several points which have been made in this summary that the principles of opposition and mediation are of fundamental importance in native South American astronomy and cosmology. This is especially evident in the definition of vertical and horizontal space by points in opposition which are conceptually joined by axes intersecting at a central point on the earth. While we do not have space here to discuss the principles of opposition and mediation in indigenous calendar systems throughout South America, it should be noted that there is a large body of data to extend these principles as well to the calendrical organization of activities in agriculture, economy, ritual, etc. (e.g., Vallee 1972; Brownrigg 1973; Morey 1971; Dumont 1976; Maybury-Lewis 1971).

25 Horizon Astronomy in Incaic Cuzco[1]

Anthony F. Aveni

INTRODUCTION

It is often casually stated that the Inca of Peru could scarcely be expected to have developed a sophisticated system of astronomy like that of the Maya because of their inability to produce a written record. One tends to think of them as New World Romans--too occupied with war and conquest to turn to matters as esoteric as star-gazing. But their establishment of an empire stretching over nearly 30° of geographic latitude gave them reason enough to engage in calendrical matters, which can be anything but superfluous when one is faced with the task of regulating civil, agricultural, and religious dates on a national basis. When we assemble all the data from the post-Conquest chronicles alluding to the astronomical methods of the Inca, we are staggered by both the quantity and quality of sky-watching which took place in and around Cuzco, the Inca capital. We are even more astonished by Incaic astronomical accomplishments when we realize both that the Inca empire began less than three centuries before Pizarro's entry into Cuzco, and that the Spaniards who wrote the history of the proceedings (like those who produced the "true" story of the conquest of Mexico) were not particularly disposed to believe that their subjects could have been sophisticated enough to pay much attention to the heavens. But we should realize that a long tradition of Andean cosmological thought had its roots in the earlier Andean cultures of Huari, Nasca, and Chavin. In this paper we employ the rich historical record left for us by the Spanish chroniclers, together with three seasons of archaeoastronomical field work in Cuzco, to derive specific astronomical sight-lines relating to solstitial and zenith-antizenith solar observations. The underlying framework for our study is the arrangement of *ceque* lines and *huacas* which united many aspects of life in the ancient capital.

THE *CEQUE* SYSTEM

As interpreted by Zuidema (1977a), the *ceque* system was an organization of 41 radial lines (or *rayas* as the Spanish called them), which emanated from the Temple of the Sun (Coricancha) in the center of Cuzco and crossed the

---

[1]We are indebted to the organization EARTHWATCH (Belmont, Massachusetts) for supporting this research and to Stephen Fabian, Gary Urton, and William and Billie Jean Isbell for useful discussions relating to this problem. We thank Peter Dunham for doing the illustrations.

landscape. Used for different purposes related to social organization (Zuidema 1964), irrigation (Zuidema 1978a), wind-directions (Zuidema 1977-8), and astronomy (Zuidema and Urton 1976), these imaginary lines were traced by a number of *huacas* or sacred places (of which there were 329 in all, counting the Temple of the Sun). Many of these *huacas* were rocks, springs, valleys, or mountains. Bernabé Cobo (1956:167-186) gives a full description of the location of the *huacas* and the situation of the *ceques* within the four principal quarters of the city. A large part of our work derives from our success in locating these *huacas*, tracing the *ceque* lines, and reconstructing the astronomical arguments given by the chroniclers based upon this system.

The chronicler F. Guaman Poma (1936:353-354) suggested that the Inca "had observatories with windows" to catch the first and last rays of the sun at the horizon. These observations were employed to know when to shear the llamas, sheep, and alpacas, and when to sow and harvest crops--events occurring at different seasons of the year. In other passages (Poma 1936:883-884) where this chronicler describes the duties in office of the state astrologer, he refers quite explicitly to the changing aspect of the sun in the local environment. The sun "sits in his chair one day and rules from that principal degree (of the December solstice). Then he sits in another chair where he rests and rules from that degree (of the other solstice)." From one seat to the other "he moves each day without resting"--a reference to the successive rising and setting points on the horizon. During the solstices "he rests for more than a day in his chair"--on those days the motion of the sun from day to day on the horizon becomes imperceptible.

Guaman Poma's statements, though brief, serve to establish both a purpose and a methodology in Inca observational astronomy; they also imply that the Inca had developed a sophisticated system of positional astronomy based upon accurate horizon observations. We shall see that evidence from other chroniclers expands and elaborates upon that of Guaman Poma. We will use this evidence together with other statements from the historical record and basic data acquired in the region about Cuzco, to deduce the extent and location of Inca astronomical observatories in the environment of the holy city. The system, as it is revealed to us, turns out to have been mostly practical and functional, and may not have constituted an astronomy as ritualistically-based as that practiced in Pre-Columbian Mesoamerica.

While Guaman Poma mentions the use of windows, there is even better evidence relating to the use of pillars to mark the solar course. These pillars, called *mojones* (singular *mojon*) by the Spaniards, were used as border markers (called *sayhua* in Quechua) and high points of Inca mountain roads; a *ceque* line often terminated with a *mojon*. In Guaman Poma's pictorial treatise, pillars appear explicitly in drawings relating the history of the Inca (1936: 352, 354). In the first picture (Fig. 25.1a) we see a *mededor* carefully measuring off blocks, which he supplies to the *mojonador* or pillar builder; both craftsmen work with great care as they fashion a *mojon* together. A completed model, taller than a man and with a rounded top, appears in the background. In the second picture (Fig. 25.1b), three *mojones* adorn the scenery; one of them, apparently an important landmark, is next to a labeled road. A painting in the Church of Santo Domingo on the present site of the Temple of the Sun depicts a number of worshippers in front of a cathedral (Fig. 25.1c). The painting is in the Flemish style of the late sixteenth century, but the environment is clearly that of Cuzco, although the horizon is somewhat distorted in

Fig. 25.1 *Mojones* or pillars in the environment of Cuzco:
(top left) Inca tower builders. Example of their completed work in background (F. Guaman Poma, f. 352); (top right) towers mark important points in environment of ancient Cuzco (F. Guaman Poma, f. 354); (bottom) towers adorn background scenery of Cuzco in 16th century painting in Church of Santo Domingo in Cuzco.

comparison with the view which actually exists. If one looks to the left of the church, the careful viewer can see two square pillars appearing like towers perched on the Cuzco skyline; from all indications such structures must have been both prominent and abundant in the landscape.

The chronicler Bernabé Cobo (1956) is our principal source for much of our information on these towers. Cobo employs several pages to give readers a meticulous word-map of the *ceques* and *huacas* comprising the four quarters of the city.[2] We have extracted two passages buried in his ordered description which allude to the use of these *mojones*:

> Chinchaysuyu--*Ceque* 6 (Zuidema Number I 2a), *Huaca* 9: "A hill called Quiangalla that is on the road to Yucay where there were two monuments or pillars that they had for signs and when the sun arrived there it was the beginning of summer" [p. 172].
>
> Cuntisuyu--*Ceque* 13 (Zuidema Number V A 1b), *Huaca* 3: "Chinchincalla is a large hill where there were two monuments at which, when the sun arrived, it was time to sow" [p. 185].

Let us take Cobo's second statement first because it is simpler to analyze. As Figure 25.2 illustrates, the Chinchincalla twin pillars probably were fashioned to frame the setting sun at the December solstice (which is the time to *finish* planting). The most likely observation point would have been the Coricancha; the supporting evidence for this argument is quite strong. The line from the Coricancha to Chinchincalla can be drawn accurately from Cobo's description of the *huacas* on three closely-spaced *ceques* in this region of Cuntisuyu quadrant. The line thus drawn is found to coincide precisely with the direction of the setting sun at the December solstice (the first day of summer in the southern hemisphere), as Figure 25.2 shows. Chinchincalla is delimited on the north by our precise determination from local informants of the location of Pantanayoc, a *huaca* on the adjacent fourteenth *ceque* to the north which Cobo describes thus: "It is a large hill parted in the middle that divides the roads of Chincha and Condesuyu" (1956:185). Ravaypampa, the next *huaca* of *Ceque* 14 of Cuntisuyu in the direction of Cuzco, is described as a "terrace on the slope of the hill of Chinchincalla," while the *huaca* Puquincancha limits our angular view on the next *ceque* (No. 12) to the south. Now a straight line from Pantanayoc to the Temple of the Sun passes immediately over a steep slope which must be the northern flank of Chinchincalla hill. The pillars must have been located higher up on that same slope. Of greatest astronomical significance is our discovery, from measurements with the surveyor's transit, that the December solstice alignment as measured from the roof of the Coricancha lies precisely in this direction on *Ceque* 13, between *ceques* 10 and 14. As is the case for June solstice observations, only a pair of pillars is needed to frame the sun at the horizon, when it comes to rest before turning around. Nevertheless, it would be useful to look for signs of rocks, pillars, temples,

---

[2]The city was divided into quadrants, two of which belonged to the upper moiety of Hanan Cuzco (Chinchaysuyu, the northwest, and Antisuyu, the northeast) and two to the lower moiety of Hurin Cuzco (Cuntisuyu, the southwest, and Collasuyu, the southeast). Curiously, the division between the two southern quadrants occurred at azimuth 146° and not 180°. The Temple of the Sun served as the center of coordinates, having been situated at the confluence of the two rivers which flowed through the valley of Cuzco.

or other markers on adjacent *ceques* to the north of Chinchincalla. Such archaeological remains might originally have served as devices to mark the solar slowdown.

With regard to Cobo's other reference to horizon pillars, we find that the hill called Quiangalla can be located fairly accurately on Cobo's *Ceque* 6 of Chinchaysuyu, on which it is the ninth *huaca* as one counts outward from the Temple of the Sun (see Fig. 25.3). But neither the Coricancha (Sun Temple) nor any other monument near it could have served as the observing station, because not all parts of the hill Quiangalla are visible from the center of the city. By meticulously tracing the June solstice-sunset direction from Quiangalla back to Cuzco using the surveyor's transit, we found that it passed about 2 km north of the center of the city. If long-distance June solstice-sunset observations were marked by a pair of pillars on Quiangalla, the most likely observation point from which to observe the event would be 5 km to the southwest in the area of the *huaca* Chuquimarca (III 1c, No. 4). This was called a temple of the Sun; it was the one where the Inca spent his time celebrating the June solstice rituals. Chuquimarca was on or near a mountain called Manturcalla (III 1c, No. 6). Although neither name is used any more, by locating the surrounding *huacas* and using colonial documents that mention Manturcalla in the context of other place-names there, Zuidema has suggested that the most logical candidate for Chuquimarca is a complex today called Lacco. It consists of a sculptured rock with caves and ruins of buildings, terraces, stairways, and a canalized stream. If we suppose Lacco to have been the place, and then trace the June solstice-sunset line more exactly from it, we find that the astronomical sight-line crosses *Ceque* I 2a at a point that falls on a straight line passing through at least three other *huacas* of that *ceque*; i.e., we find that *Ceque* I 2a (*Ceque* No. 6 of Chinchaysuyu) is a straight line, and the Quiangalla pillars lay on it. Though the towers are long since destroyed, an archaeological probing for some remains might prove fruitful. The search could be limited to the 400 m^2 area within which our astronomical-ethnohistorical argument suggests the pillar remains must lie.

Having established a technique that illustrates how we can use astronomical measurements and calculations in conjunction with information on the placement of towers to mark the winter and summer solstices, we turn next to a more complex argument in which we show that the Inca devised a far more basic astronomical line that defined the zenith sunrise-antizenith sunset axis, an alignment that actually cut across the *ceque* system and incorporated several observing stations and a gnomon. The basic reasoning behind the establishment of this line, its connection with the aforementioned solstice lines, and the place of such linear schemes in the general context of Andean cosmological thought is developed further in a separate paper by Zuidema in this volume. Here we elaborate only the technical argument.

We begin with another of Cobo's descriptions, in which he states that the seventh *huaca* of the eighth *ceque* of Chinchaysuyu (Zuidema Number I 1b), the northwest quadrant of Cuzco,

> . . . was called Sucanca. It was a hill where the irrigation canal of Chinchero comes. On it were two pillars or monuments to signal that when the sun arrived there it was time to begin to plant maize. The sacrifice made there was addressed to the sun and they asked that it would arrive on time so that they would have good reason to plant . . . [1956:173-174].

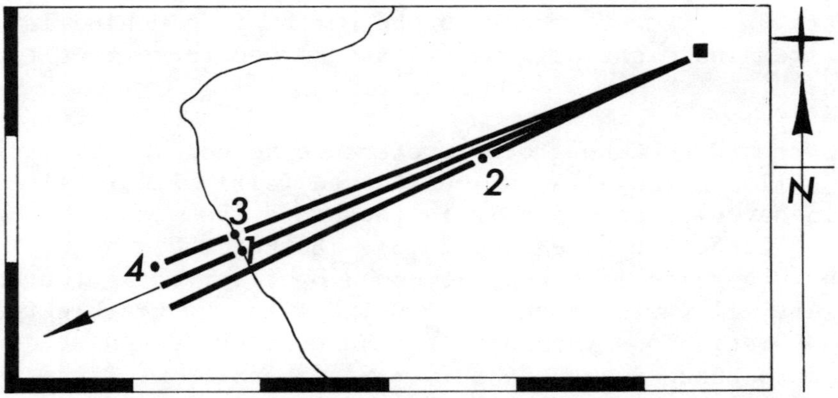

Fig. 25.2 December solstice sunset alignment (*ceque* lines in dark ink). Curved line is the visible horizon as seen from the Coricancha (filled square), the observation point. Bars on the rim of the figure represent 1 km. (1) Location of pillars at Chinchincalla on Cobo's *ceque* 13 of Cuntisuyu; (2) Puquincancha (a *huaca* on *ceque* 10); (3) Ravaypampa (a *huaca* on *ceque* 14); (4) Pantanayoc (a *huaca* on *ceque* 14).

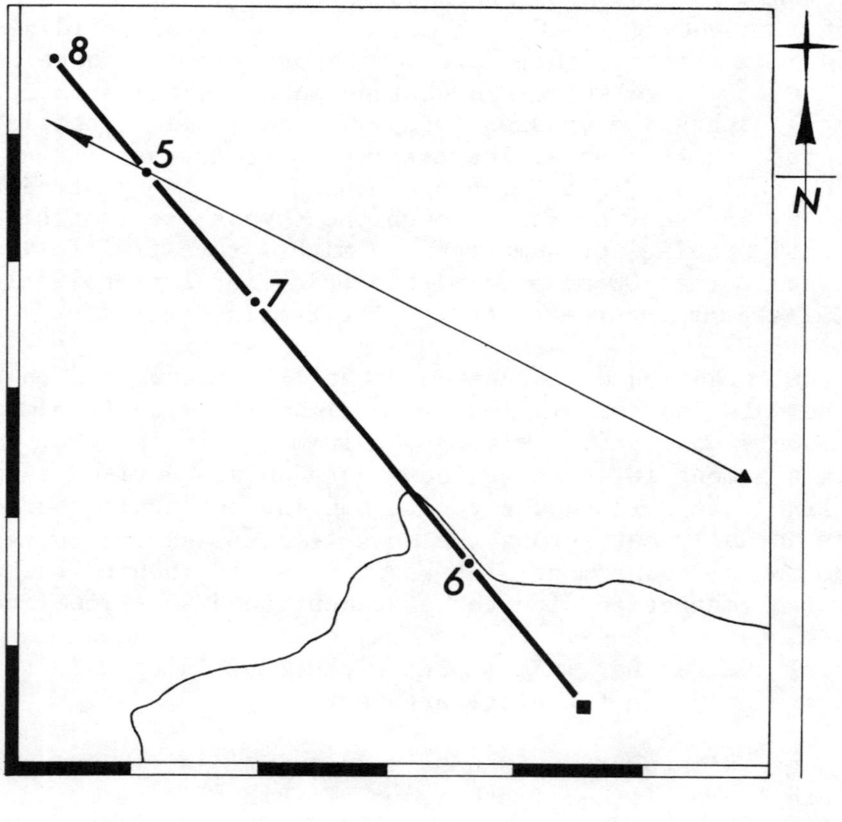

Fig. 25.3 June solstice sunset alignment viewed from Lacco (filled triangle) to a pair of pillars on Quiangalla (5), a *huaca* on *ceque* 6 of Chinchaysuyu. Coricancha, visible horizon, and 1 km lengths as in Fig. 25.2. Other *huacas*: (6) Ticcicocha; (7) Quisco; (8) Guargua Illapuquiu.

The earlier Anonymous Chronicler (probably from Maurtua, ca. 1570: 150-152) mentioned four little pillars serving the same function, located in the same general direction. These were visible at about 2 to 3 leagues from the city and were used to regulate the sun. The chronicler says they were "high on a hill overlooking Cuzco from the west":

> When the sun passed the first pillar they prepared themselves for planting in the higher altitudes, as ripening takes longer.
>
> When the sun entered the space between the two pillars in the middle it became the general time to plant in Cuzco; this was always the month of August.
>
> And when the sun stood fitting in the middle between the two pillars they had another pillar in the middle of the plaza, a pillar of well worked stone about one estado high, called the *Ushnu*, from which they viewed it. This was the general time to plant in the valley of Cuzco and surroundings [author's italics].

The northernmost pillar (see Fig. 25.8a) evidently served as a device to warn that the planting season in the Cuzco valley was approaching. People who cultivated crops at higher altitudes, where growth occurred at a slower pace, would be allowed sufficient additional time to sow their seeds before planting commenced in the valley. Though the chroniclers provide no evidence, the southernmost pillar was perhaps used as a similar warning device during the harvest season when the sun, moving from left to right, passed it in April.

There is no doubt that these particular pillars resided on the slope of Cerro Picchu above the Carmenga district (located in the western section of the modern city and still retaining its old name). Another chronicler, P. Cieza de León, establishes this for us when he states that

> on the hill of *Carmenga* [i.e., Cerro Picchu] they have at *definite intervals*, small towers, which serve to keep track of the movement of the sun, which they regard as important . . . and they had a plaza where they say a long time ago there was a swamp or lake at which the founders smoothed over the mortar and stone. . . . From this plaza the four royal roads go out [1973:214].

Cobo tells us that *Ceque* 8, radiating from the Temple of the Sun, passed over the hill of Carmenga and terminated at Sicllabamba. Different *huacas* on that *ceque* can be pinpointed exactly (Fig. 25.4) because today half of them can still be located and many still retain their old names. Urcoscalla was a place "where those who travel to Chinchaysuyu lose sight of Cuzco" (Cobo 1956: 173-174). On the basis of sightings of Cuzco made from the hills west of Cerro Picchu, we equate the area where the city disappears from view with a place bearing the modern name of Arco Punco. Radially outward from the Coricancha along the same *ceque* (*Huaca* 12) we find Poroypuquio ("the well of Poroy"); Cobo says that the Spanish built a watermill there, and the ruins of the mill can still be seen near the village of Poroy. Collanasayba ("the principal *mojon*"), located 13 km radially outward from the Temple of the Sun, is a rock outcrop about an acre in extent located at a prominent bend in the river coming from Poroy. Original Inca walls were found when the modern road to Chinchero was constructed, and the hacienda there still bears the original name Sicllabamba; the hill which belonged to Sicllabamba is also called

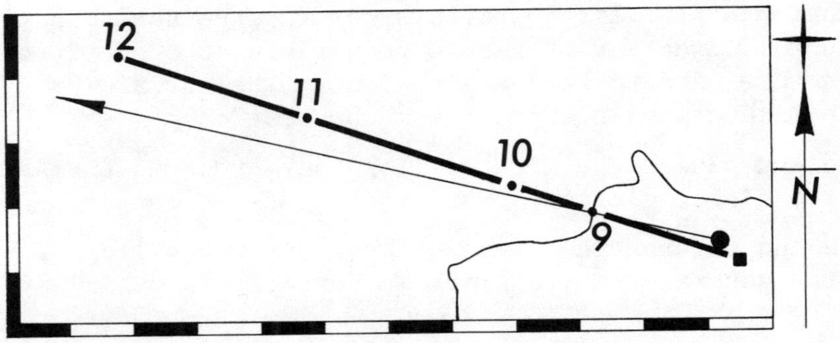

Fig. 25.4 18 August antizenith sunset line viewed from the Ushnu (filled circle) to the 4 *sucanca* or pillars on Cerro Picchu (9) which represent one *huaca* on the 8th *ceque* of Chinchaysuyu. Other *huacas*: (10) Urcoscalla; (11) Poroypuquio; (12) Collanasayba.

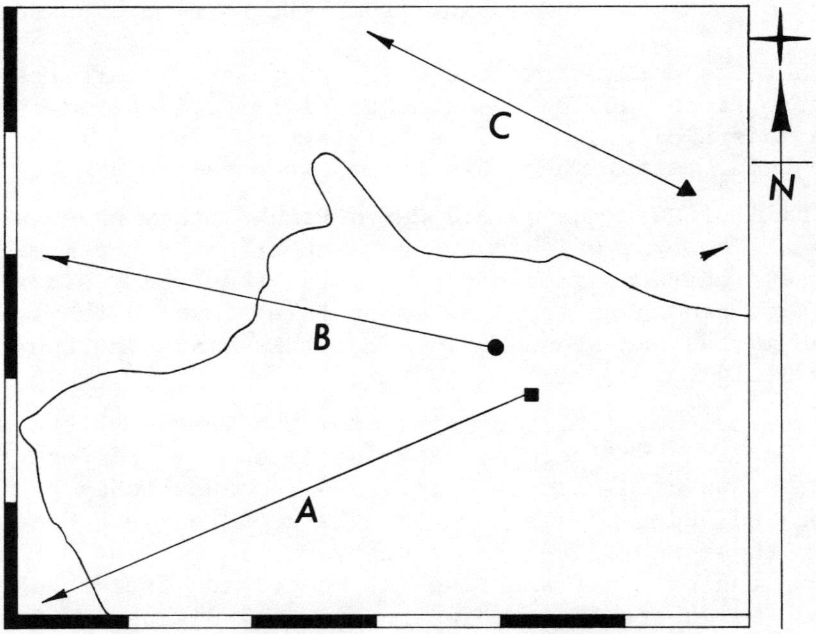

Fig. 25.5 Summary schematic showing all 3 alignments derived for Cuzco: (A) December solstice sunset; (B) June solstice sunset; (C) Antizenith sunset.

Collanasayba. The house of that hacienda still stands adjacent to this, the last *huaca* of its *ceque*. The place must have been ideal for the ritual purpose of throwing offerings (brought from Cuzco) into the river at the end of the *ceque*, a practice referred to often by the chroniclers.

Using the surveyor's transit, we traced the *ceque* as a straight line over the mountain between Collanasayba and the Coricancha in Cuzco. The goal of this exercise was to approximate the position on Cerro Picchu where the straight line would cross. Since Sucanca lay on the *ceque* line but was astronomically observed from the *Ushnu*, we saw a means of fixing the important sunset line and the corresponding date in the agricultural calendar, thereby locating the site of the *Ushnu* (known to have been somewhere in Hanan Haucaypata--the present Plaza de Armas) as well. Other chroniclers such as Betanzos (1968) and Sarmiento de Gamboa (1942) discuss these same pillars, and Garcilaso de la Vega (1969:116-118) also mentions having seen them standing in 1560, though he could not be sure of the place from which they were supposed to be observed when the sun went down.

The position of the Picchu Sucanca thus having been located (we erected our own Sucanca on the spot by piling up stones), we set up the transit there and shot a series of lines into the Plaza de Armas. We sought to determine for what dates sunset could be viewed over the Sucanca from various points in the plaza. One line to which we paid particular attention was that corresponding to the 18 August sunset date (the broad arrow in the enlargement of the Plaza in Fig. 25.7), for it coincides with the sunset point on the day when the sun passes through the nadir (the point opposite the zenith). The importance of the passage of the sun through the zenith is mentioned frequently in the chronicles (e.g., see Garcilaso de la Vega 1969). The zenith-antizenith direction accords very well with the vertical symbolism so prominent in Andean thought (Zuidema 1977-8; Isbell 1978), and with statements made by Guaman Poma (1936:883-884) that the earth opens up in February and August, the zenith-antizenith sun dates. On the antizenith sunset day the sun disappears over the western horizon at a point about 180° opposite the sunrise point on the day of the zenith passage. (Actual differential horizon elevations cause the line to deviate by as much as 0.5° from linearity, but the azimuth is, nevertheless, about 283°.)

By a fortunate circumstance, our assumptions and calculations were further aided in the summer of 1978 as a result of our examination of a 1653 painting by the artist Estrada Monroy (Fig. 25.6), located in the cathedral of Cuzco. It shows the city in the aftermath of the earthquake of 31 March 1651, barely a century after the conquest. The view shown is from the top of the cathedral looking toward the southwest (the same as the orientation of the map). Near the center of the picture (enlarged in the inset) we see a vertical column supporting a cross. This is the *rollo* (or "*picota*") which Pizarro was said to have placed on the site of the *Ushnu* "in the middle of the plaza" shortly after he conquered the city (Betanzos 1968:70). (The *Ushnu* itself is a very elusive concept; its relation to verticality in the Inca cosmos has been fully developed by Zuidema [1977-8], who has demonstrated that it was manifested in a multitude of material objects including an altar of sacrifice, a platform of stones, a *mojon*, and even a hole in the earth.) Note that if we join the Plaza de Armas (the old Hurin-Haucaypata) with the Plaza de Regocijo (the old Cusipata) to the southwest to form a single plaza, the stone

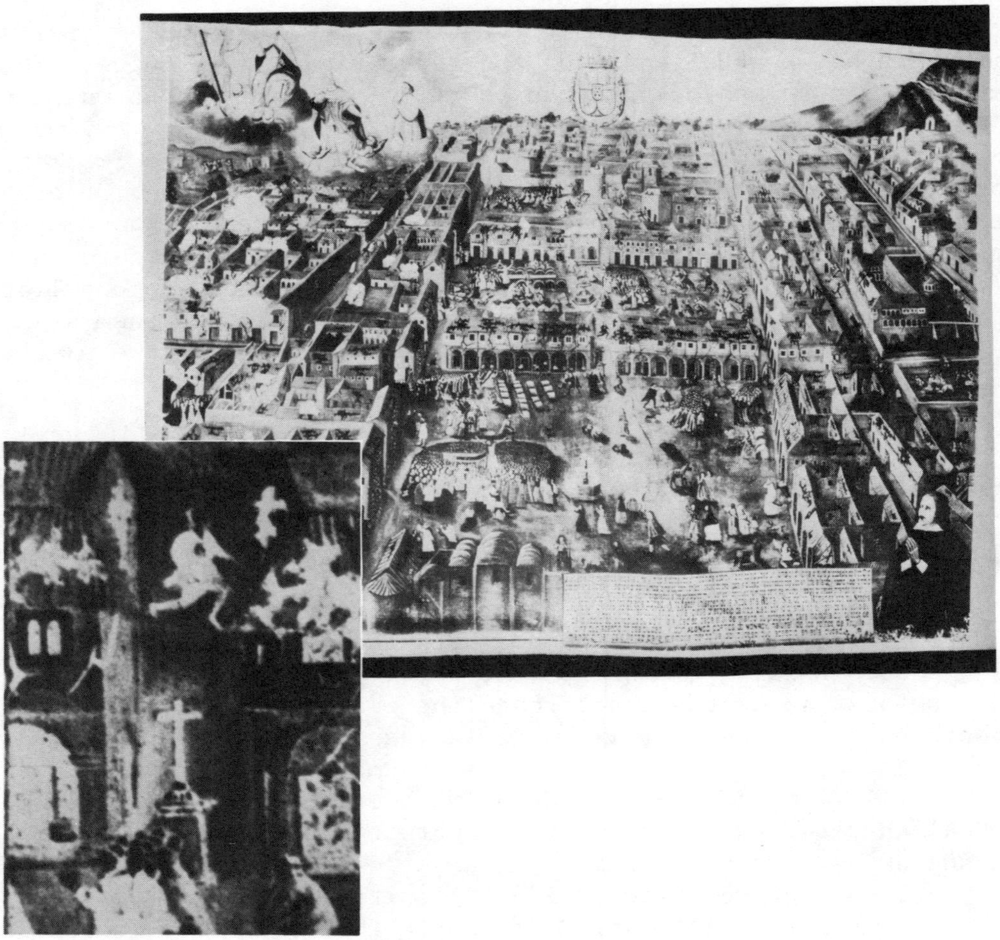

Fig. 25.6 Monroy (1653) painting of the Plaza Hanan-Haucaypata of Cuzco. *Rollo* (at lower left), which Pizarro placed on the site of the Ushnu, once marked the center of a much larger plaza which can be visualized by removing the row of buildings passing horizontally across the picture and including the *rollo*. The plaza boundary also can be traced roughly in Fig. 25.7 if the Ushnu (B) is imagined as the center. For orientation, the cathedral just to the left of bottom center is at position C in Fig. 25.7.

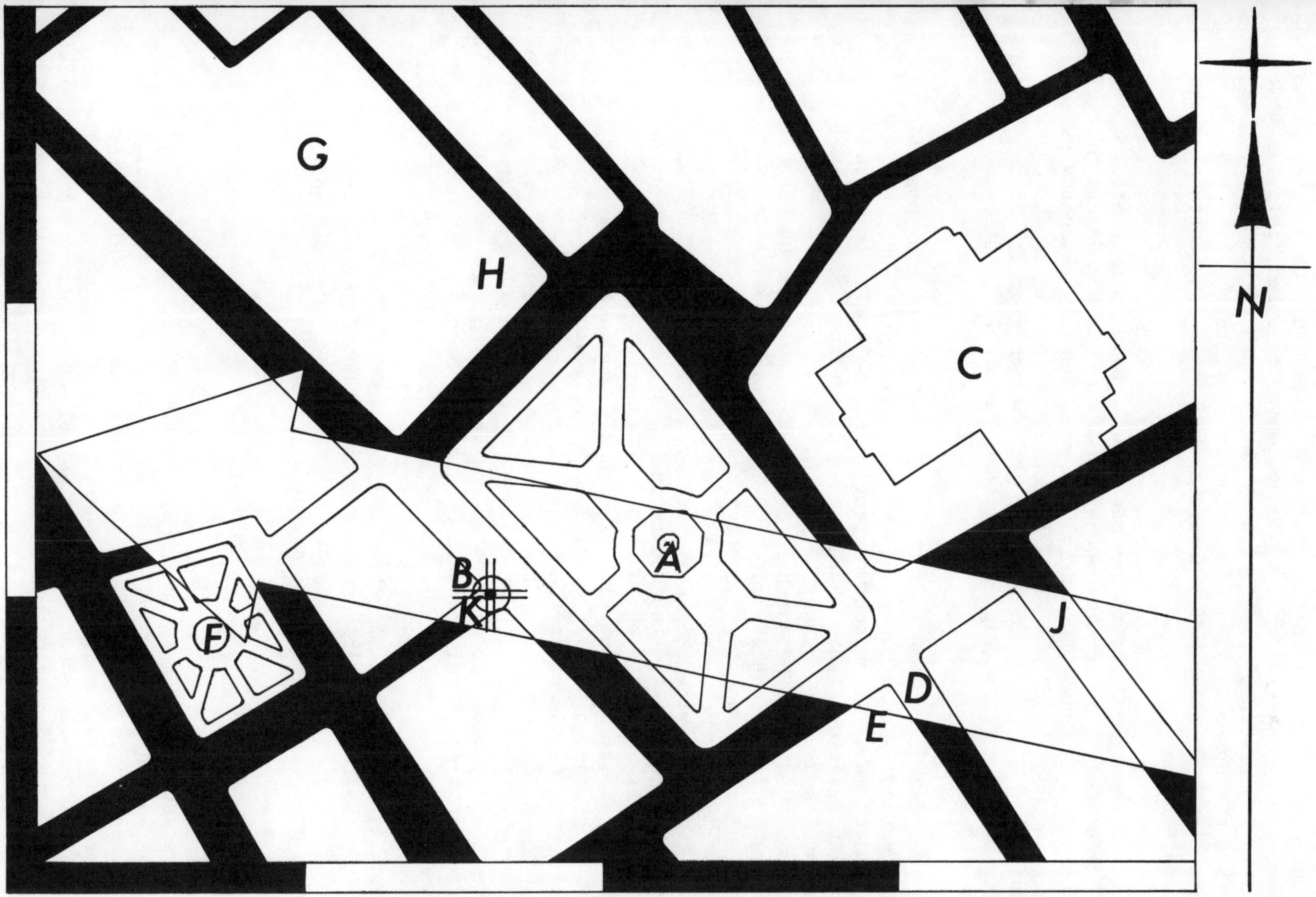

Fig. 25.7 18 August antizenith line (arrow) traced backwards from the known position of the *sucanca* across the Plaza de Armas, the center of which today is marked by a fountain (A). The *rollo* or *picota* discovered in the Monroy painting (Fig. 25.6) was located at B. It was placed on the site of the Ushnu by Pizarro and is seen to fall within the predicted observational zone. Bars bordering the figure represent 100 meter lengths. Other features identifiable in Cuzco today: (C) Cathedral; (D) Calle Loreto; (E) Compania; (F) Plaza Regocijo; (G) Garden of Ticcicocha; (H) Casana Palace; (J) Calle Sta. Catalina Angosta; (K) Calle Medio.

would have lain near the center; the old layout can be imagined in Figure 25.7. The chronicler Santacruz Pachacuti Yamqui talks about the center of the double plaza Haucaypata-Cusipata, so we now understand that this was what other chroniclers meant when they referred to the "middle of the plaza." We note further that the *Ushnu* is situated comfortably within the 18 August viewing band as calculated from our measurements of the *ceque* line containing the Sucanca on Cerro Picchu.

While the Anonymous Chronicler mentions a chain of four pillars, presumably running perpendicular to the *ceque* lines, why does the description of Cobo include only two of them? Evidently Cobo mentions only those towers located on the *ceque* whose *huacas* he was delineating, as he did in the case of the solstice pillars. In Figure 25.8 we see the horizon geometry of the Picchu scheme and attach a photograph (Fig. 25.8b) that shows the situation at present as viewed from the plaza.

The line from the *Ushnu* to the pillars on Cerro Picchu turns out to be the tip of a rather large iceberg, of which I shall reveal only the basic outline. As Zuidema shows in an accompanying paper, this line turns out to be derived from observations taken in the other direction, that of the rising sun at the day of its passage across the zenith. This line passes from the *Ushnu* of Hanan-Haucaypata of the northern moiety through a second *Ushnu*, that of Hurin-Haucaypata of the southern moiety. The latter *Ushnu* is located on the site of the plaza Limapampa. This observational line is extended to point to the eastern horizon, where sun temples were situated. The concept of reversing the sight-line to view the antizenith event (which coincides with both the time of planting and the time the earth opens) became a guiding principle for the Inca, a brilliant stroke which also embodied the dualism of the vertical-antivertical axis so central to Inca and even contemporary Andean cosmic thought (e.g., see Isbell 1978:197-220). Lying along the same line (which crosses from one side of the Inca capital to the other) is the Sunturhuasi building and the two *Ushnu*, functioning as gnomon used to make measurements of the sun when it migrated to the overhead position. The sighting of the moon at the zenith, when the sun lay at the antizenith, may also have been involved. These ideas are merely presented here without proof, in order to indicate the direction of our research. The entire argument, together with the solstice-watching scheme, is summarized in Figure 25.5, which provides strong evidence concerning not only the position and function of the Cerro Picchu pillars, but also the reality, in some instances, of Cobo's straight-line description of the *ceques*. Archaeologists are encouraged to begin explorations of the predicted *Ushnu* site as well as of the sites of the pillars.

CONCLUSIONS

We have attempted to explicate here both the exactitude and the complexity of the most basic types of Incaic astronomical orientations. In this task we have employed ethnohistorical evidence, astronomical arguments, and architectural and topographic data gathered through appropriate instrumentation in the Cuzco environment. We have established beyond doubt the existence of three astronomical sighting directions which emanated from three different points in the Cuzco area: (1) a pair of towers which mark the June solstice-sunset point as viewed from Lacco, a complex of rock carvings on a

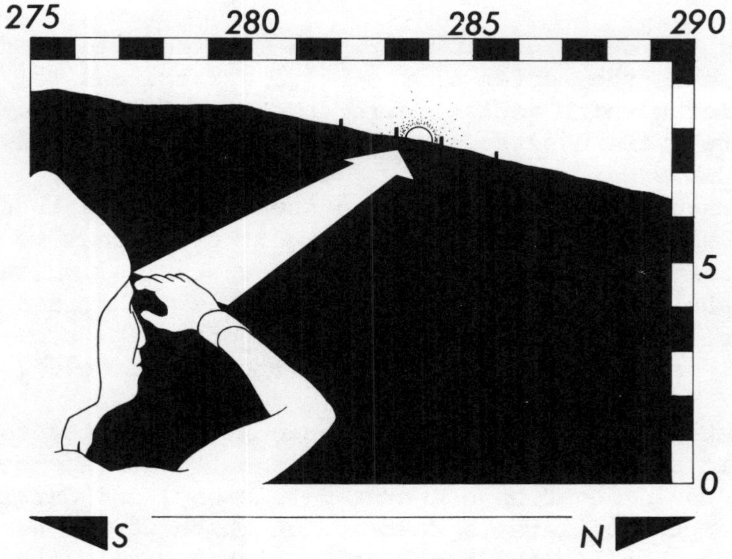

Fig. 25.8a Observing the 18 August antizenith sunset on the western horizon of Cuzco. Observer is at the Ushnu 2 km from the pillars. The horizontal scale tabulates azimuth, the vertical scale, altitude, both in degrees. The size of the sun is exaggerated.

Fig. 25.8b View as it appears today from close to the site of the Ushnu. (Rooftop blocks the view from the exact site today.) Antizenith sunset occurs behind the light pole, on left (see arrow sight-line of Fig. 25.7).

hill north of Cuzco; (2) a pair of towers which mark the December solstices as seen from Coricancha, the center of the *ceque* system; and (3) four towers situated on Cerro Picchu which mark time for the planting season, centered about the place where the sun sets on the day of passage through the antizenith.

We have discussed only formative concepts relating to the last sightline, but have done so in enough detail to make it clear that it was of great importance, serving as an embodiment of the up-down dualism which enters so frequently into Andean thought. The scheme of employing horizon observations to determine planting dates is similar to that reported by Urton (1978b) in his study of present-day Misminay, a village northwest of Cuzco.

We find a tricentric concept with unequal numbers of towers at only one horizon difficult to deal with because it is not what we expected to find. Western astronomers would never have operated this way, and Garcilaso's confused statement about the pillars (a description which we can no longer take too seriously) reveals how shallow the penetration of the Renaissance mind into Incaic astronomy was. To add to the complications, we are confronted with a basic astronomical concept--the zenith-antizenith passage of the sun--which has no analog outside the Tropics. Near the equator, vertical motion on the celestial sphere is dominant. By contrast, circular orbits pivoted about the celestial pole characterize the principal aspect of the heavens as seen in the temperate latitudes that served as the environmental backdrop for the more familiar astronomy of the Classical world. It will be interesting to see how this principle holds in other tropical zone astronomies, such as those of Polynesia, Java, Central Africa, and Mexico.

In spite of perceptual difficulties, this early foray into the fundamentals of Inca astronomy raises questions not only for astronomers, but also for agriculturists, ethnologists, and archaeologists. There are many hypotheses to test, and what once appeared (astronomically speaking) to be an unknown world sealed by the absence of the written word has opened into a rich field for future research.

26

Inca Observations of the Solar and Lunar Passages Through Zenith and Anti-Zenith at Cuzco

R. T. Zuidema

INTRODUCTION

Development of the Research

The research on Inca astronomy, carried out by Aveni and me during the last four years in Cuzco,[1] has focused on checking out the empirical data contained in the "Relación de las huacas" preserved in the chronical of Bernabé Cobo. His manuscript refers to observations of three sunset points and to the calendar dates on which these observations occurred.

Two of these observations appeared to deal with the solstices. Aveni (this volume) describes our determination of not only two sunset points, but also of the places from which these were observed. Only in the case of the December solstice sunset observation (DSSS) did the temple of Coricancha, the center of the *ceque* system, prove to be an observatory as well. The June solstice sunset (JSSS) was observed from another temple of the sun, Chuquimarca, which was located northeast of Cuzco. Chuquimarca was the site where the king and his court went during the Incaic month which included the date of the June solstice. Analysis of the JSSS observation made us aware that in addition to Coricancha, which was the central temple of the sun in Cuzco, there were other temples of the sun as well, each used for a specific and distinct observation. Moreover, a horizon point could derive its astronomical importance not only from its direction and *ceque* orientation as seen from Coricancha, but also from another place in or around Cuzco.

Defining the JSSS and DSSS observations helped us to interpret their astronomical associations. Quiancalla, the JSSS point, was a hill that had "two markers or pillars which they regarded as an indication that when the sun reached there, it was the beginning of the summer." "Beginning of the summer," therefore, referred to that period as understood by an educated person in the northern hemisphere. The DSSS point, Chinchincalla, was "a large hill where there were two markers; when the sun reached them, it was time to plant." We

---

[1]I want to express my gratitude to Aveni for his collaboration in the fieldwork and the analysis of the data that we collected (although the mistakes in the latter are my own). Further, I thank the following institutions that, for the last seven years, have supported this research: National Science Foundation, Social Science Research Council, Earthwatch, and the Center for Latin American Studies of the University of Illinois.

must conclude that this description does not refer to the beginning of planting but to its end. Guaman Poma (1584:1164 [1154])[2] observes that December was the last month for the planting of most crops, since after that month the rains would become too heavy.

A far more complicated problem was to understand the significance of the third sunset point. The Relación describes it as

> . . . *Sucanca*. It was a hill by way of which the water channel from Chinchero comes. On it there were two markers as an indication that when the sun arrived there, they had to begin to plant the maize. The sacrifice which was made there was directed to the sun, asking him to arrive there at the time which would be appropriate for planting, and they sacrificed to him sheep, clothing and small miniature lambs of gold and silver.

The agricultural interest in the timing is clear, but no calendrical date is mentioned. Moreover, we must allow for the possibility that, as in the case of the DSSS, the observation was defined by way of an astronomical argument.

Although the term *sucanca* was an Inca word generally used for a sunrise or sunset point, the Relación recognizes only one such place. Chinchincalla and Quiancalla are not mentioned by other chroniclers; Guaman Poma (1584:316) does refer to Quiancalla in a drawing, but without specifying its meaning. However, the pillars of Sucanca were singled out by various chroniclers for their astronomical importance. Cieza de Leon (1553:Ch. 92) observed the pillars when he described the city as a traveler to Cuzco.

Our best data for understanding the importance of Sucanca in Inca astronomy come from an anonymous author, probably writing about 1580; his central theme was the correlation of the lunar and the solar calendars. When the Incas, according to him, became aware of the regular yearly movement of the sun, "they observed carefully the high places on the western horizon where the sun sets" (Anonymous n.d.:151). Among these natural markers was one area in which man-made towers were constructed:

> . . . in the highest mountains to the West, as seen from Cuzco, they made four pillars in the form of small towers that can be distinguished from two and three leagues, with a distance from the first to the last of 200 *pasos*, and between the middle ones of 50 *pasos*, and those at the ends distributed with their interests in mind; with the result that when the sun passed the first pillar that then they became ready for sowing in general, and they started to sow in the higher places as the plants are later here, and when the sun entered the central pillars, it was generally time for sowing in Cuzco, and this was always in the month of August.

Later, in a list of the twelve Inca lunar months, we read:

> The moon of the month of August was called *Tarpuyquilla* ("Moon of sowing"). In this month they would only sow, the poor as well as the rich, everybody helping each other. And during this month of August the sun entered in between the two small towers of the four that were indicated there by the Incas, as has been dealt with before [Anonymous n.d.:158].

The sun was observed between the pillars from another pillar in the plaza:

---

[2]Guaman made some mistakes in his own pagination. The number in brackets is the correct number.

> . . . a pillar of well-dressed stone, one *estado* high, in the middle of the plaza, in a place that was singled out for that purpose, that they called *Osno*, and from there they measured the sun in between the two pillars, and when it was exactly there, it was the general time for sowing in and around the valley of Cuzco [Anonymous n.d.:151].

This description is repeated when Anonymous (n.d.:158) refers not to the Incaic month of August, but to that of September. We may assume, therefore, that the Ushnu (= Osno), as an observatory, was used in defining the Incaic month of September as well as that of August. The reference to sowing allows us to identify the central two pillars with those of Sucanca which are mentioned in the Relación.

Without considering the locations mentioned, we can still draw certain conclusions. In his calendar, the chronicler first refers to movable lunar months. He speaks of the "moon of the month of . . . ," and mentions that in the month of September men would come out on the plaza with torches during the full moon. However, in referring to agricultural activities and describing the four pillars, he correlates a fixed period within the solar year to the Spanish month of August; the period was defined by the position of the outer two pillars of the four pillars described. When the sun passed the northern pillar, it was time "to sow in the higher places, as they are later"--that is, they needed a longer growing season. The same argument is given in the description of the month of July. Therefore, the sun passing the outer pillar did not indicate the time when people *started* to sow later crops, but rather when they *ended*; this coincided with the time when normal sowing would begin. The full moon which occurred at the time of sunset between the two outer pillars fell within the Incaic month of Tarpuyquilla, the "moon of planting." As the day when the sun set between the two central pillars had to be included in this month (it was the only setting point that would be included at all times), the whole range of setting dates between the outer pillars defines a period equal to that of a synodical month of 29½ days. The day the sun set between the central pillars had to fall in the middle of that period.

Similar data from Guaman Poma help us to time the first passage of the sun through the zone of the pillars. Poma's Incaic month for sowing corn in Cuzco (1584:1152 [1142]) started on the day of Santiago, July 25, a day still very important in the ritual calendar of many Andean peoples; July 25, the Julian calendar used by Guaman Poma, corresponds to August 3 in the Gregorian calendar. On this day, the king himself would ritually sow a special field, and Guaman Poma represents the Inca thus in two drawings (1584:250, 1153 [1143]). Today, August 1 is celebrated as a New Year in a way which suggests its pre-Spanish origin. The Spanish interest in Santiago, with his association with warfare and thunder, led to a rapid integration with indigenous cultural elements with comparable associations. We should, therefore, attempt to determine the original Andean reason for selecting Santiago as the first day of August.

The dates indicated by the pillars of Sucanca are (or are close to) August 3, August 18, and September 2. Even though the Inca king began ritual plowing on or near August 3, non-official plowing had already been carried out before. This initial plowing was determined by altitude, soil conditions, etc., as we are told by the Anonymous Chronicler. Any date around August 3 could have served equally well as the official date; therefore, additional reasons why the Incas in Cuzco were interested in August 3 and August 18 had to exist. Our effort to understand the importance of these dates led us to formulate five hypotheses. All were confirmed in terms of Inca astronomical theory, but we had to revise our ideas about how the observations were actually carried out. In the following section, I will define the five hypotheses and present the results of our fieldwork concerning Inca astronomical practices. I will then discuss the ethnohistorical and archaeological data supporting our conclusions.

First Hypothesis

August 18 or 19 and April 26 or 27 are the two occasions when the sun at Cuzco (latitude -13°30') goes through nadir (Fig. 26.1). We know that the Incas used the central pillars of Sucanca for defining the first date in relation to the opening of the agricultural season. We also know that they used the same towers to end this period officially (Sarmiento 1572:Ch. 30), when the Inca harvested the same field that he had ritually plowed at the beginning of the planting season. Our hypothesis is that the Incas were capable of, and interested in, defining these two dates (without implying here how they understood this practice theoretically).

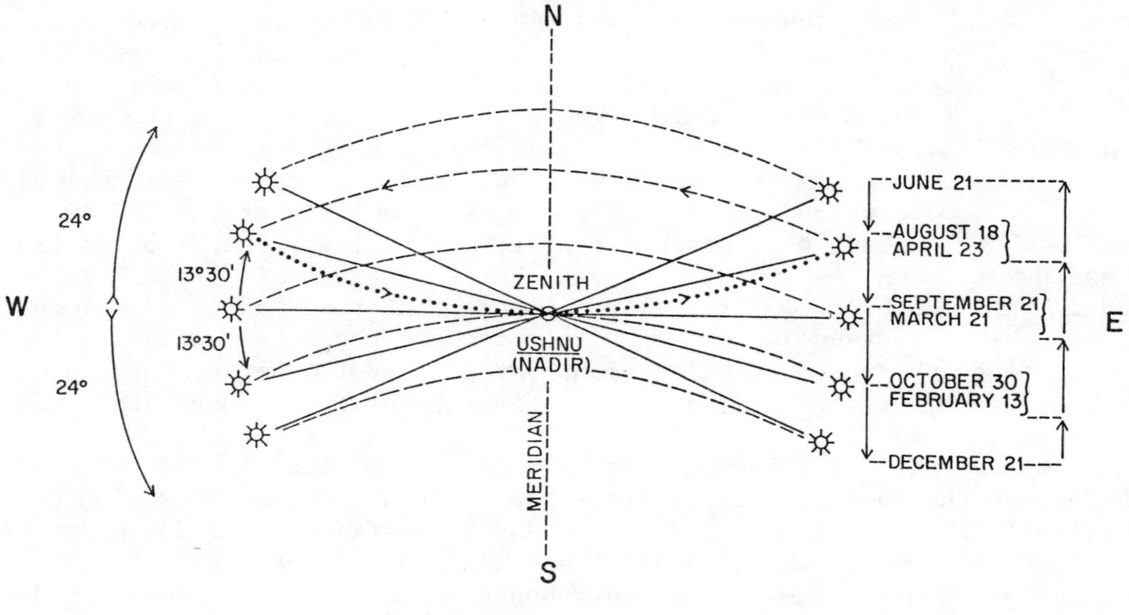

Fig. 26.1 Zenith and anti-zenith

Second Hypothesis

The only way in which the Incas could have determined the dates of August 18 and April 26 would be to note the days when the sun goes through the zenith. Reversing the direction of sunrise on those days would have given them the days of antizenith (= nadir) sunset. One of our most important chroniclers, Polo de Ondegardo, mentions the existence of not one *sucanca* but two, indicating two beginnings of the year. I will analyze his data later and will come to the conclusion that one *sucanca* refers to sunrise on the days of zenith passage (ZSR) and that the central pillars of the other *sucanca* (already known to us from the accounts of the Anonymous and other chroniclers) were used for measuring the antizenith sunset (AZSS). Both places were on mountaintops some

26 km apart and functioned reciprocally as observatories and horizon points. This double line of observation was horizontal, and the dates indicated were exactly those of zenith passage (October 30 and February 13) and antizenith passage (August 18 and April 26).

Third Hypothesis

The passage of the sun through zenith is observed when a vertical pole or gnomon does not cast a shadow at noon. We know of at least three stone pillars in Cuzco which were used on the dates of ZSR and AZSS or for rituals in relation to them. However, there was one tall tower, the Sunturhuasi, which was erected on the central plaza. The Sunturhuasi is described as a temple of the sun and it was one of the places from which sunrise and sunset observations were made. Its location could have served the purpose of observing AZSS as delimited by the pillars of Sucanca. According to the Anonymous Chronicler, this observation was made from the Ushnu in the plaza Haucaypata, close to the Sunturhuasi (Fig. 26.2). The central pillars of Sucanca, the Ushnu, and the Sunturhuasi were probably aligned with each other.

Fourth Hypothesis

There are two difficulties in deriving the date indicated by the northern pillar of Sucanca from a locally observed AZSS position. First, although today the date of August 16 rather than the original date of August 18 or 19 is still important in the indigenous calendar, its significance stems from the use of the period August 1-16 in predicting the agricultural year to come; August 16 is the derived date, not August 1. Second, the date of August 1 (originally August 3) has ritual importance not only in Cuzco, whose latitude (-13°30') can explain a relationship to the date of AZSS (August 18), but in all the Andes, from Ecuador (Fock 1977) to Chile (Barthel 1959). Inca influence does not easily explain this fact, since the Incas occupied these outlying areas of their empire for only a short time. Explaining it in terms of a similar timing for the opening of the agricultural year is not probable either, since this date loses much of its practical value for that purpose elsewhere and especially in Ecuador.

Fig. 26.2 A man observing the Sunturhuasi (F. Guaman Poma, p. 329)

August 3 is close to one of the four dates (August 5, November 5, February 4, and May 6) that, together with the dates of the solstices and equinoxes, divides the year into eight nearly-equal time intervals. The hypothesis suggests that August 5 was the date for the Incaic New Year, and that Andean peoples in general were interested in such a division of the year. Guaman Poma's approximation of this date is the day of Santiago, because of its importance in Spanish mythology in relation to thunder, weather, and agriculture in general. The northern pillar of Sucanca might therefore indicate the date of August 5 rather than August 3.

Fifth Hypothesis

Only when the sun goes through the nadir, on August 18 and April 26, is the full moon close to the zenith. At this time the full moon can best be observed by way of a gnomon. A full-moon zenith passage can occur in Cuzco from August 3 to September 2 and from April 11 to May 11. The practical astronomical importance of beginning the year in Cuzco on August 3 or 5 might also stem from this observation of the moon.

Although the five hypotheses could be confirmed, our research in the field changed our understanding of the logical order in which the Incas arrived at the dates of AZSS. These changes also have implications for the order in which the ethnohistorical data are presented. Originally, we first located the central pillars of the *huaca* called Sucanca, based on the data in the Relación. Assuming that these pillars indicated AZSS, we calculated the possible location of the Ushnu on the plaza Haucaypata from which the August observations were said to have been made. Arguing from a prior zenith observation of the sun, I came up with the following suggested logical sequence:

(1) The Sunturhuasi was used for observing the sun passing through zenith.

(2) The dates of zenith passage were established by observing sunrise from within the Sunturhuasi itself.

(3) Reversing the direction of ZSR, the high tower of the Sunturhuasi itself would indicate the dates of antizenith sunset (AZSS).

(4) In addition to the Ushnu of the plaza Haucaypata in Hanan-Cuzco in the northwestern part of town, there was another Ushnu on a second plaza (also called Haucaypata) in Hurincuzco, in the southeastern part of Cuzco. Both Ushnus can be located approximately. I suggested that they were used in establishing the zenith and antizenith axis, as mentioned in the third hypothesis.

A flaw in this theory on the sequence of observations was exposed when we established the zenith/antizenith axis in between the *sucancas* on the two mountaintops. The location of the Sunturhuasi, as calculated from ethnohistorical data, was some 3° south of this axis as viewed from the AZSS point. Such a shift results from the 9° vertical angle of the sight-line between the Sunturhuasi and the central pillars on the mountain at the AZSS point (Cerro Picchu) as compared to the level sight-line between the two *sucancas*. This means that there is no reason to locate the Sunturhuasi as a gnomon on the ZSR-AZSS axis itself. However, its astronomical importance still remains, since it was essential for observing the sunset on the dates of August 3 or 5 and of September 2 or 4. The fact that the Incas did use the same Cerro Picchu both for establishing the ZSR-AZSS axis and for observing sunset points from August 3 or 5 to September 2 or 4 was the result of practicality on their part rather than astronomical necessity. In each case, Cerro Picchu served a different purpose: in the first instance, it is the only part of Cuzco visible from the eastern *sucanca*, and it is covered completely by the disk of the setting sun; in the second instance, it is the only mountain in the appropriate direction which is close to Cuzco, and which has a gentle slope that could be used to set up a system of four pillars. Another difficulty with our initial theory is that both Hanan- and Hurin-Ushnu could be located only approximately; however, as their exact location will not be necessary for our present astronomical argument, I will leave it for future research.

In the following sections in which I analyze the ethnohistorical data, I will first discuss the empirical data confirming both an Incan and a more general Andean interest in measuring exactly the passage of the sun through zenith; these data are independent of those referred to so far. Next I will discuss the data on the ZSR-AZSS axis between the two *sucancas* mentioned by Polo de Ondegardo. The third section will discuss the Sunturhuasi and the measurement of the four pillars on Cerro Picchu.

This paper is primarily concerned with an identification and interpretation of Inca places and placenames, and their relationship to our present understanding of the *ceque* system described in the "Relación de huacas." The *ceques* were 41 sight-lines which radiated outward in all directions from Coricancha, the central temple of the Sun in Cuzco. A distinctive feature in the landscape was taken to define a particular direction; depending on the distance to the horizon, this feature was either on the horizon itself or just beyond or before it. The *ceque* system also incorporated other natural or man-made points in the landscape that were close enough to a direction of interest to make them usable. In this way, a mental and visual map was made of the area around Cuzco. Because of their ritual and religious uses, these points within the *ceque* system were considered as *huaca*, "sacred." The *huacas* of the *ceque* system were neither the only nor the most important ones around Cuzco; they served this system and they could be important for other reasons.

In addition to the local *ceques* of Cuzco, there were also interlocal ones which extended far beyond the horizon as viewed from Coricancha. High points close to the city, like Cerro Picchu, allowed much further visibility. The eastern *sucanca* of the two referred to by Polo de Ondegardo was established by way of such an interlocal system of directions.

This paper encountered one dilemma that may be with us for quite some time. My interest here is to indicate the importance in Inca astronomical theory and practice of certain objects like the *sucancas*, the Sunturhuasi, and the Ushnus; however, the importance of these objects to the Incas was not simply astronomical, and it is useless to discuss their first or foremost *raison d'être*. Moreover, the Incas did not explain their interests in ways which might be convenient for a modern armchair astronomer. We will discover data of astronomical interest only if we first try to understand other data as they come to us and as the Incas, through their Spanish interpreters, wanted to present them. Astronomy might be only a minor part of their significance. On the other hand, astronomy can establish connections between objects that otherwise might not be clear to us. In this discussion I will confine myself to the study of the astronomical use of objects like the Sunturhuasi and the *sucancas*. In other articles, I will document and analyze in detail their use in other contexts.

DATA SUPPORTING AN ANDEAN CALENDRICAL INTEREST IN OBSERVING THE SUN AT ZENITH

The principal aim of this paper is to establish Inca interest in observing the sun at zenith as that observation was used for calendrical purposes. This interest can be established unambiguously by two previously unmentioned pieces of evidence; the first of these involves the alignment of a particular building in Cuzco toward ZSR and AZSS.

In addition to the Sunturhuasi, which stood in the central plaza in Cuzco, there was another important round building called Muyucmarca, which still stands within the fortress of Sacsayhuaman above Cuzco. The building and the fortress were considered a "temple of the Sun of War." When Santacruz Pachacuti drew the temple of the Sun of Coricancha, he placed it on the hill Collcampata of Sacsayhuaman, making one temple equivalent, to a certain extent, to the other. Muyucmarca was integrated into a complex of rectangular buildings, most of the foundations of which have survived (Fig. 26.3). The space between the three concentric walls of Muyucmarca (*Muyu* = "round") is divided into sections by other walls and by three gutters which pass from the inside to the outside of the round structure. The central gutter, the niche opposite it, and the straight walls extending to the east are all very close to the ZSR-AZSS direction. As Muyucmarca is on a hill above Cuzco, and therefore has a more level view of the horizon than one obtains from the town, we conclude that the round tower plus the extension of the rectangular buildings next to it could have served the double purpose of observing the sun at zenith and at ZSR and AZSS. In this context I will later analyze other ethnohistorical data referring to Muyucmarca (Zuidema n.d.b).

The other piece of evidence concerns two Incaic temples of the Sun, one on the Island of the Sun in Lake Titicaca, and the other in Copacabana on the coast opposite the island. (According to the Incas, the Sun was born in Titicaca, or was created by the god *Viracocha* there.) Along with the temples of Cuzco and of Pachacamac on the central coast of Peru, these two temples were among the most sacred ones in the Inca empire. Our interest in Titicaca resulted from the following question: If the Incas in Cuzco were interested in the date of August 3 not only because it indicated the lunar month of AZSS at their latitude but also because they sought to divide the year into eight equal time intervals, where in the Andes would observations of the zenith and antizenith sun generate such a division of the year? The answer is at the latitude of -16°30', which is the latitude of both the Island of the Sun and of Copacabana. Moreover, when the Indians of Copacabana established the cult of the Virgin there in 1582 initially against the wishes of the Catholic church, they first chose February 2 as the day of her worship (the "Virgen de la Candelaria"). They later changed this to August 5. Both dates (of zenith passage and of antizenith passage, respectively) are related to the division of the year into eight equal parts. An original pan-Andean interest in these dates might have supported the religious importance of the Island of the Sun as well as of Copacabana, its port of access (Fig. 26.4).

Incaic myths such as the one of the god Viracocha, who traveled from the Island of the Sun through Cuzco to Ecuador, and rituals such as the one in which water was brought from the Island of the Sun to Cuzco and poured into a hole that was part of the Ushnu, are an attempt to relate the Cuzco dates for observing the sun and the moon in zenith and antizenith to those at Titicaca, the Inca place of origin.

POCCOY *SUCANCA* AND CHIRAO *SUCANCA*

Many modern ethnographies of the Andes mention the notion that the Earth-mother, *Pachamama*, "opens" on the first of August. On this day she "rests" and no agricultural activities are allowed. Afterwards she will be impregnated by the seeds that are sown. She remains open during the whole

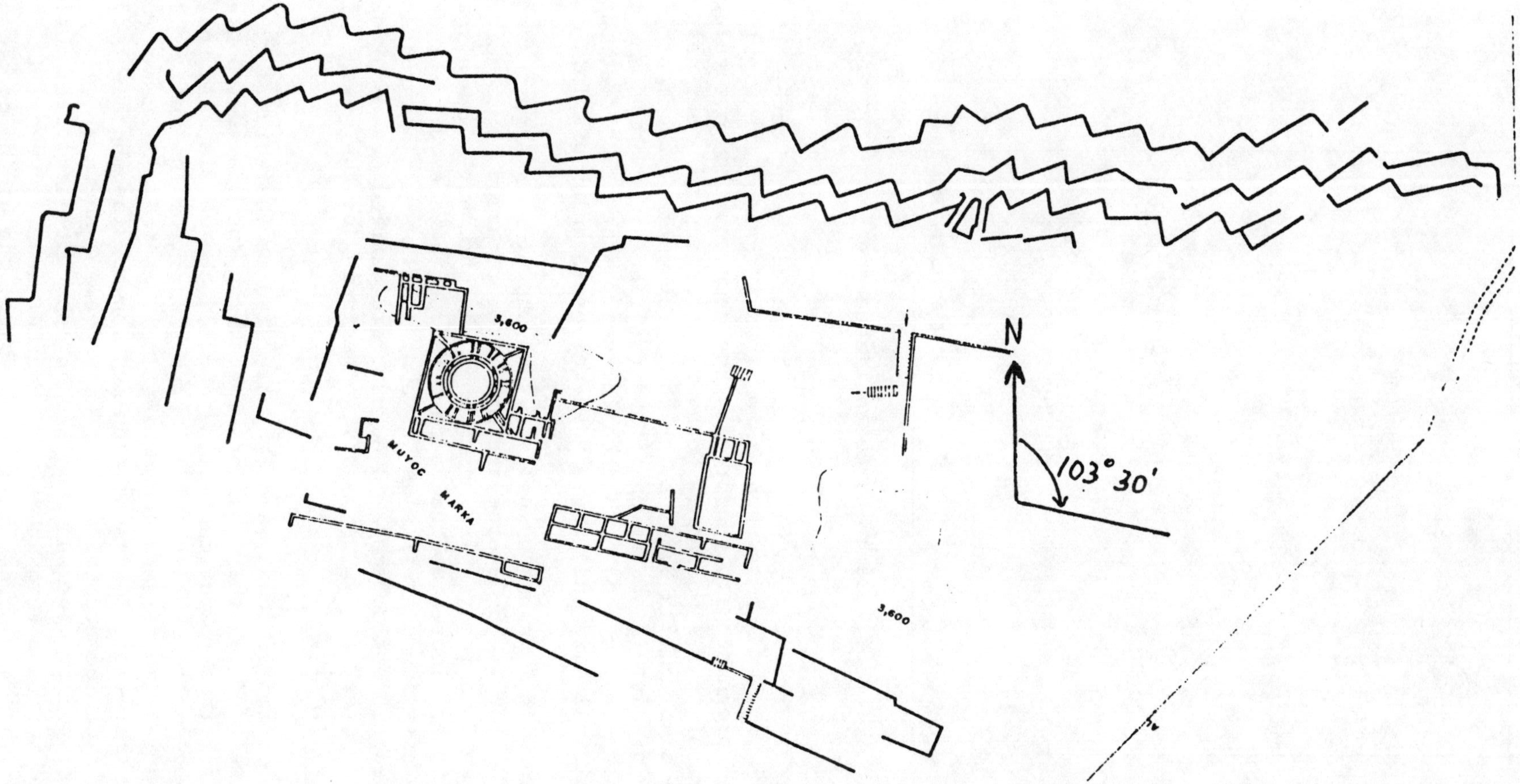

Fig. 26.3 *Muyucmarca* on *Sacsayhuaman*

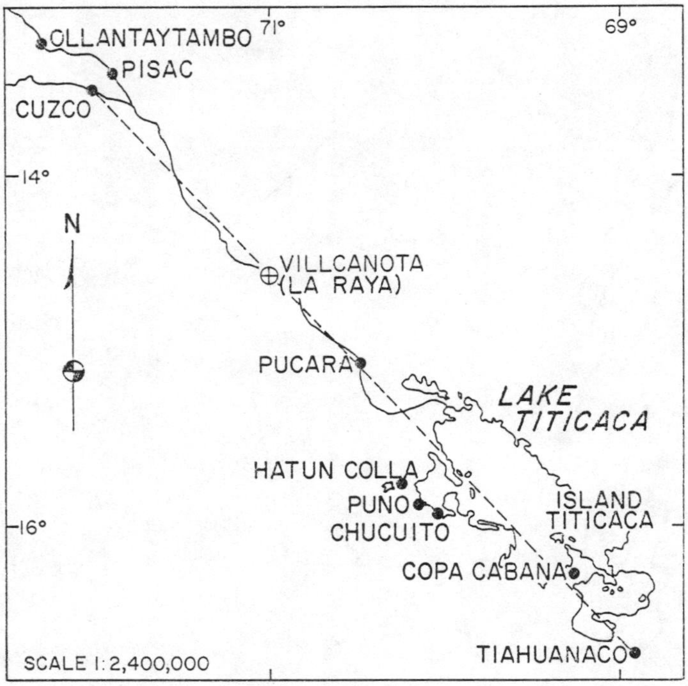

Fig. 26.4 Cuzco to Tiahuanaco

month of August. The first days of August are called the "cabañuelas," as the weather on each of these days is thought to predict that of the eight months when the Earth is pregnant with the crops. The name and the particular form of the custom are of Spanish origin. However, according to other descriptions, either the first sixteen (2 x 8) days, or the first half of the month, are used in order to predict (from the phase and position of the moon and from other natural phenomena) a good or bad coming year. These observations seem to be of a more empirical kind and possibly are of pre-Spanish origin.

The earth is "opened" sexually by men who plow in August after the first irrigation in July; the custom and its interpretation are modern as well as Incaic. Guaman Poma reports its occurrence not once but twice a year. The heavy rains in January loosen up the earth. February was used to break virgin soil in order to distribute new fields to young families. Similar customs, related to the "opening" of *Pachamama* in August, are carried out today during the feast of "compadres" in February, at the beginning of the movable feast of Carnavâl (Lira 1946:18-19, 86-91).

These customs of August and February, two months which are half a year apart, are related to other opinions concerning the sun and moon. Today in the village of Sarhua (Department of Ayacucho), the sun is said to "stand still" from the December solstice until February and from the June solstice until August (John Earls, personal communication). Data from Guaman Poma (1584:894 [884]) that support this opinion mention how the sun has two "seats" where he "rests." One occurs when, after his movement southward along the sunrise and sunset points, the sun comes to an apparent standstill during the December solstice and then turns north. The other seat, however, is not the June solstice but (as Guaman Poma insists on twice) the beginning of August. We have noticed elsewhere Poma's claim (1584:1152 [1142]) that the Incas began plowing ("opening the earth") in August, starting on the day of Santiago.

However, in his discussion of the northern seat of the sun in August, Poma relates this to St. John the Baptist (whose day is June 24). We can clarify this confusion if we realize that for Guaman Poma, or his informants in the sixteenth century, the sun was "standing still" or "sitting" from the June solstice to the beginning of August.

We come now to the information of Polo de Ondegardo that refers to the two systems called *sucanca*. He gives them seasonal names which, as I will explain, refer to the months of February and August; he also refers to a beginning of the Incaic year in February. I will combine the two types of information. Another chronicler, Cristobal de Molina "el Cuzqueño," writing at the same time as Polo, mentions two *sucancas*, just beyond the reach of the *ceque* system and east of Cuzco. These two *sucancas* indicate the JSSR and the ZSR as seen from Cuzco. I will conclude that one of them is the Poccoy *sucanca* (ZSR) referred to by Polo.

There is good reason to combine the data of Polo and Molina and to integrate them with those on the *ceque* system. Each chronicler knew the *ceque* system and reported on it, although both reports are now lost. Even if we are not certain that Cobo copied his "Relación" from either of these two, or from another similar source, we know he employed them extensively in his discussion of the calendar and calendrical rituals in Cuzco. The "informes" of Polo which have survived and the description by Molina of Inca rituals coincide well with each other. Therefore, we are able to use Polo, Molina, and Cobo together as one corpus of data.

Polo de Ondegardo wrote various "informes" on Inca culture from 1559 to 1571. Although the data we are interested in are somewhat confusing, we can draw conclusions from them concerning three problems that are of interest to us.

The first problem involves the beginning of the year. Most chroniclers agree that the two months of *Capac raymi* ("royal solar feast") and *Capac raymi Camay quilla* ("royal solar feast, the moon or month of Camay") were related to the December solstice, and that they coincided best with the months of December and January, respectively. While all the other Incaic months have a lunar name (*quilla*, "moon") that refers to an agricultural activity, the name *Camay* lacks such an agricultural reference and the month name *Capac raymi* is never paired with a lunar name. Elsewhere (Zuidema and Urton 1976) I came to the preliminary conclusion that *Capac raymi* was the only non-solar month and that it ended on the December solstice. Polo, who should be our best source on these matters, says in his oldest "informe" (from 1559 or 1560) that the Inca year originally began in January but that one king moved the beginning date back to the solstice in December. Assuming that the shift back was of one month (as later chroniclers say), the two dates in the Julian calendar would be January 10 and December 11. Part of the discrepancy between Polo and these later chroniclers--who followed him in their descriptions but who used the Gregorian calendar--might be explained by the fact that the first *Capac raymi*, which fell after the solstice, was included mostly in December, while for the others, most of the month of December came before the solstice. In his second "informe" from about the same time (1561), Polo says that the year started with February. In his two later "informes" (from 1571), he says that the feast of *Raymi*, which presumably is *Capac raymi*, fell in February. Even if Polo meant *Capac raymi Camay quilla* with his reference to "raymi," the time indicated would still be one month later than what we read in most other chronicles. Many of them agree that the Incas originally had a calendar which began one month later. On the basis of their data, I would conclude that the year originally started on the December solstice and was then moved back in time. Whichever theory we accept, it is clear from Polo's data that the beginning of February (January 21 in the Gregorian calendar) was important as a Spanish approximation of an Incaic calendrical interest.

The second problem concerns Polo's use of the concept of *sucanca*. He distinguishes it from that of *sayhua*, a word which he translates in Spanish as "mojon" (landmark) or "pilar" (pillar). He also refers to *sayhua* with the Quechua word *topo*, which is a land unit of variable size or a unit of distance (Rostworowski 1978:8). The *sayhuas* indicated the beginnings of the Incaic months. There were probably twelve *sayhuas* and two *sucancas*; Bernabé Cobo copied and interpreted these data in this way. We have no more data on the system of *sayhuas*, but we can conjecture that the two outer pillars of the *sucanca* on Cerro Picchu were part of it.

The third problem concerns the names of the two *sucancas*, Poccoy *sucanca* and Chirao *sucanca*; the names refer to seasons and we have Spanish translations for them. The difficulties are that (1) the times indicated do not exactly correspond to the Spanish seasons; (2) the concept of a season is different in the two cultures; (3) the Spaniards in Peru (i.e., in the southern hemisphere and in an area where the climate in the mountains and on the coast are very different from each other) used the terms "summer" and "winter" for the same periods of the year as in Spain; and (4) the dictionaries distinguish between the Spanish words "estío" and "verano," which both today mean "summer," but in some of the texts "verano" sometimes has preserved its old meaning of "spring." However, from an analysis of these data we can conclude that Poccoy refers to the time when fruits ripen (from *poccoy*, "to ripen") during the latter part of the rainy season. Most chroniclers include the term *poccoy* in the names of the months of February, *Hatun poccoy quilla* (*hatun*, "big, great") and March, *Pacha poccoy quilla* (*pacha*, "earth"). *Chirao* refers to the time when the first changes in the weather occur after the dry season; this was probably one reason why the Spaniards called it "spring." Today the term *chirao* is still used in this sense. Therefore, it is most probable that Chirao *sucanca* can be identified with the *sucanca* on Cerro Picchu, which refers to a date in August, and that Poccoy *sucanca* refers to a date in February. Polo does not locate the *sucancas* nor describe their appearance. As we need a ZSR point in order to define AZSS which occurs in August, we will look for the first of the *sucancas* on the eastern horizon of Cuzco and consider now the data of Molina.

Molina refers to two *sucancas* in his description of an elaborate ritual, an understanding of which is essential for a general comprehension of the Inca calendar and astronomy. During the month when the king celebrated the June solstice at a temple north of Cuzco, priests went to Huanacauri, which was the most sacred mountain southeast of the city. At the same time, other priests went to a temple of the Sun far beyond Huanacauri to celebrate the "birth of the Sun" there. The place is now called in Spanish La Raya, the "dividing line," but in Incaic times it was known by the Aymara name of Villcanota. This name is translated by Bertonio in his Aymara dictionary as "House (*nota*) of the Sun (*villca*)"; i.e., "House where the Sun is, or was, born." The river which originates there and flows toward Cuzco has preserved this name (see Fig. 26.4). Apparently, the reason for associating the mythological event with this place was its direction from Cuzco, which it had in common with both Huanacauri and the Island of the Sun. When the priests pursued a straight course toward Villcanota, they passed 21 named stations. They

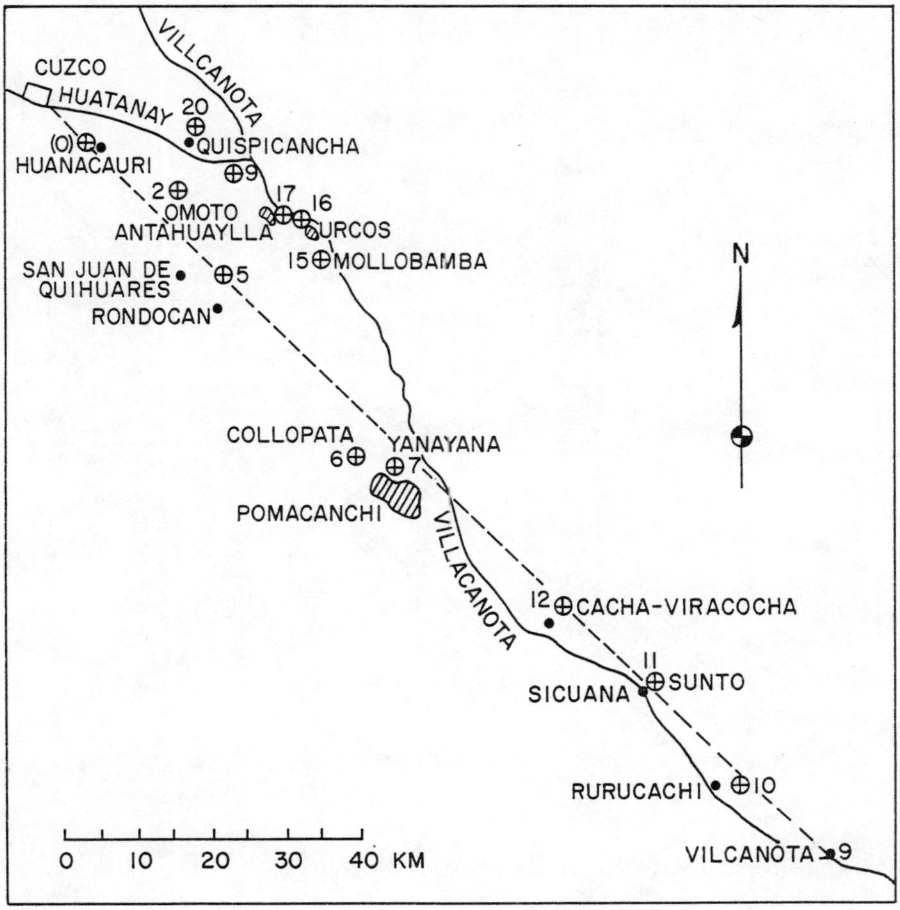

Fig. 26.5 Cuzco to Vilcanota

returned to Cuzco by way of the river (Fig. 26.5). The first and the last stations are called *sucanca*. The first *sucanca* was on, or close to, the second of these stations and was called Omotoyanacauri. This mountain, which is known today as Mutu, is also mentioned in the *ceque* system (the last and tenth *huaca*, fifth *ceque*, Collasuyu; II 2 b): "a small mountain called Omoto *urco* ("mountain Omoto"), opposite Quispicanche in the *puna* or *paramo* (high grass plain)." Mutu is visible from Cuzco. Its location and the data on other *huacas* in the same direction indicate that this was the *sucanca* of the DSSR point.

The other *sucanca* was on, or close to, the next to the last station, "a mountain in Quispicancha." Modern Quispicanchis is a small village in a large province of the same name. The Spaniards moved the village to the bottom of the valley; however, higher up in the hills are the ruins of an important Inca city which today is known as Tipon (Fig. 26.6). Cieza (Cronica, Ch. 97), in 1551, described the Inca buildings ("aposentos") of Quispicanche on the left side of the road from Cuzco to Muyna. Quispicancha was probably the original name of Tipon. The change in the ending of the name from -cancha to -canche to -canchis may have been induced by the proximity of the province Quispicancha to its neighbor, the province of Canchis. The name *quispicancha* might mean "crystal (*quispi*) enclosure (*cancha*, Gonzalez Holguin [1608]; *kancha*, Lira [1946])." But, given the importance of the place in Inca

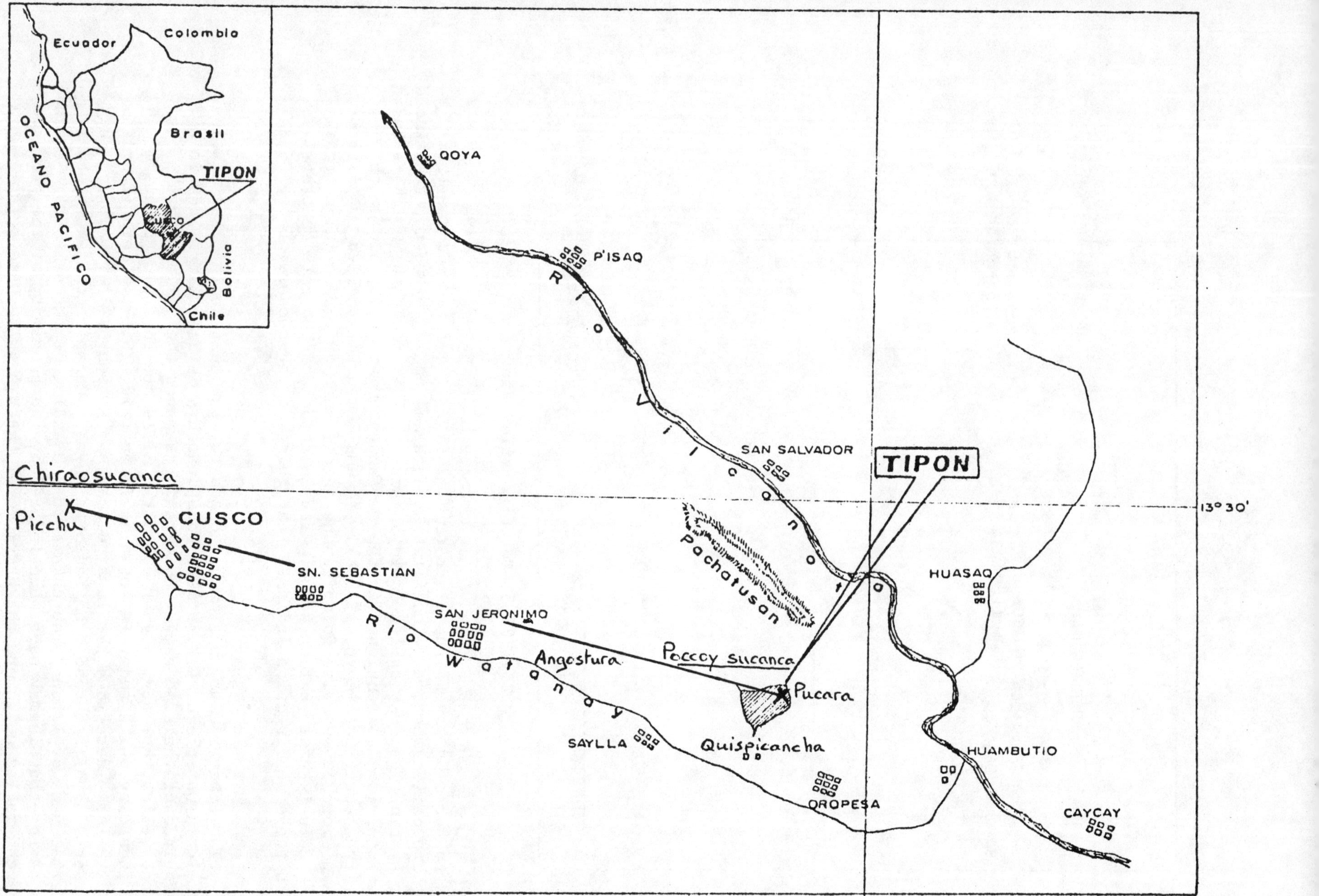

Fig. 26.6 *Poccoysucanca* and *Chiraosucanca*
(adapted from the map by Fidel Ramos)

astronomy, the translation "crystal light" or "first rays of crystal light" seems more appropriate. The following words, given by Gonzalez Holguin (1608) and Lira (1946), support this interpretation. Gonzalez Holguin (under *Pasca*, "clearness of the clouded moon") presents the following: "*Canchariy* (or) *illariy*--This is a ray or light of the clear (or full) moon." Lira gives the following information: "*K'ancha*--Luminousness, brightness, splendour or glitter. Luminary or skylight." He goes on to give these examples: "*K'anchariy*--Action of shining a little; beginning to shine when there is little light; becoming day with a diffuse light; giving someone light in heavy darkness, to light with a weak light or from afar."

Molina describes the mountain as part of (the area of) Quispicancha; knowing the local situation, this is understandable. The ruins of Tipon consist of two parts. The lower part has elaborate terraces of large, well-fitting stones, and the "aposentos" that Cieza talks about are also of well-built Inca masonry. This section is separated by a ridge from an extensive, second urban area. On the ridge is an irrigation canal coming down from a mountain slope to the north. The latter is covered with terraces and a road which is carved, in part, from the rock. Going up, one discovers that the mountain is a promontory from the *puna* (grass plains). One reason for the importance of Quispicancha must have been its access to an abundant supply of irrigation water. Two cyclopean walls were built all around the mountain to defend the canal and the town. These extend from the top to the lower and flatter parts where the urban complex is bordered on both sides by two small rivers (Fig. 26.7). For this reason, the mountain is known as Pucarca, "fortress." An Inca synonym of this word is *Quespina*, which has the same root as *Quispicancha*; I will return to this point later.

Quispicancha is part of the valley system of the river Huatanay that starts just above Cuzco; it belongs, however, to a second section. This section is separated from the first by a gorge called today Angostura, "the narrow part." From two promontories in Tipon one can see the upper part of Cuzco, which lies at a distance of about 26 km to the northwest; the names of these promontories first attracted our attention to Tipon.

The first promontory, Intihuatana, "where the sun is bound or measured," is at the lower end of the ridge separating the two parts of town. It contains a complex of buildings on and next to a rock. The ruins do not resemble those at Pisac and Machu Picchu which are today also called Intihuatana, except for a similar location on a promontory. The name at Tipon may be of a genuine local tradition.

The other place is the very top of the mountain Pucara and is called Limaqawarina. Gonzalez Holguin calls *kahuana* or *cahuarina* an "outlook" (*mirador*, in Spanish) or "a high place to overlook and observe the country and seacoast from a watchtower or eminence (*atalayar*). The Spanish words *otear* ("to observe, to examine, to pry into, to discover by artifice; to inspect" or "look from a high place") and *otero* are translated respectively as *kahuarimuni* and *kahuana urcu* (*urcu*, "mountain"). We know of various Inca and modern place-names referred to as Cahuana or Cahuarina. The chroniclers describe certain places as "mirador"; and Limaqawarina has all the necessary characteristics of a look-out. It is remarkable, however, that it is not called *Cuzco qawarina*. A possible explanation is that Cerro Picchu, with snowcapped mountains in the far distance behind it, is the most visible part of Cuzco. The city of Lima is

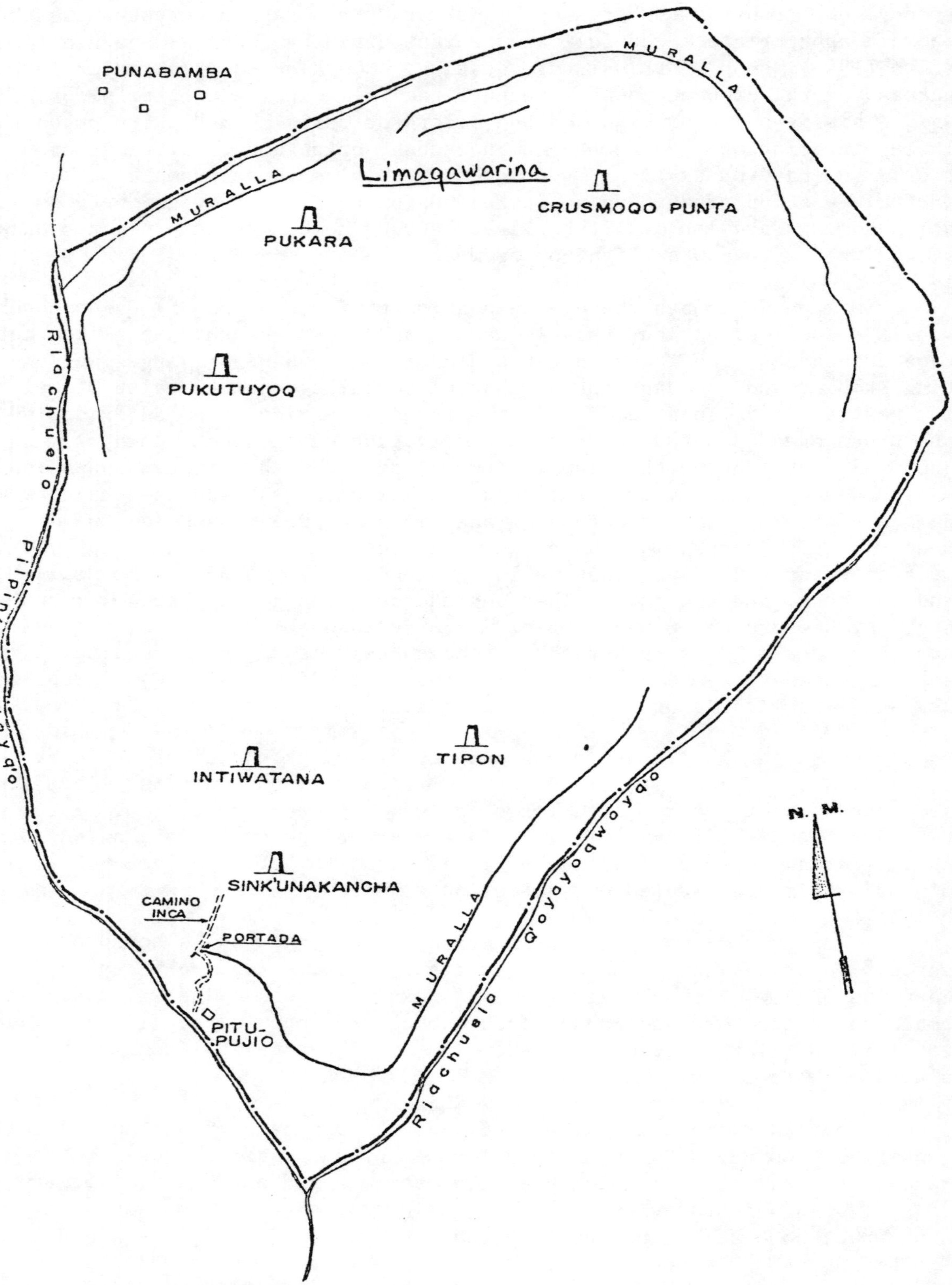

Fig. 26.7 Quispicancha (Tipon)
(Map by Fidel Ramos. Scale 1:10000)

in the same direction, and the road from Cuzco to Lima starts on the slope of Picchu. It passes by an Inca town some 70 km from Cuzco which is still called Lima tambo, "the waystation to Lima."

Limaqawarina is located exactly on the direction of ZSR as seen from the *sucanca* on Picchu. I will now discuss the various data and arguments that help to identify this place as one of the *sucancas* described by Molina and as the Poccoy *sucanca* mentioned by Polo de Ondegardo.

First, in addition to the fact that the whole mountain slope was part of the town of Quispicancha, the Incas marked the top in a special way. One of the rocks there has a petroglyph consisting of pecked holes in roughly the form of two circles. One reason for marking the top in this way could have been because of its use as a look-out for military and astronomical purposes. (See Aveni et al. 1978 for similar data in Mexico.)

Second, the *sucanca* mentioned here by Molina was of astronomical significance not only because of the technical meaning of the term *sucanca*, but also because of the ritual and astronomical context in which it was mentioned. The pilgrimage was carried out during, and in relation to, the June solstice. The starting point of the pilgrimage was related to the DSSR. The goal of the pilgrimage was one of the three sanctuaries (all in the same direction) that were related to "the birth of the Sun." One of these, the Island of the Sun (together with Copacabana), celebrated the days when the sun locally went through zenith and antizenith (February 2 and August 5). These are the dates which were, and still are, considered as "New Years" in Cuzco. The ritual expressed a progression in time from the December solstice to February 2 and February 13 and, by implication, from the June solstice to August 5 and August 18.

A third argument involves the term *quispi* and its ritual, mythological, and calendrical use. *Quespi* (or *quispi*), according to Gonzalez Holguin (1608), means "a translucent thing like glass, the heavens or crystal; any precious stone." A second meaning is expressed by the following words:

Quespiccuni--to free oneself from work; to escape, to evade; to save, free oneself from danger.
Quespichik--liberator, saviour.
Quespircun--to trespass something from one part to the other, and to begin to appear, to become visible that which passes over (or crosses, or trespasses) on the other side.
Quespircuni--passing something with difficulty and getting out of it or because of some difficulty or obstacle.
Quespina--shelter or defensive place.
Quespi quespilla--to go safe and free from danger.
Puccara or *Quespina*--castle or fortress
Qquespicupuni--to become better or healthy
Caci qquespilla, Cussi quespilla--healthy and safe.

Although the connection between "crystal," on the one hand, and meanings like "to become free," "to pass something with difficulty," or "fortress (as a place where one is safe)," on the other, may not be clear to us, certain myths do make this connection within an astronomical context. For instance, when the future king Pachacuti Inca, on the eve of an attack by the Chanca people, plans to defend Cuzco, he first has an encounter with his "father," the Sun god, who rises out of a well of water into which Pachacuti has thrown a crystal. This well, called Susurpuquiu (Molina 1573:20) or "the well of Susunmarca" (eighth *huaca*, fifth *ceque*, Anhisuya; III 2 b) is located in the direction of the June solstice sunrise as seen from Coricancha. The name Susunmarca still belongs to a former hacienda there. The myth relates the concept of "crystal" to sunrise at a time when Pachacuti and Cuzco were in dire need and when the city had to be defended. The Sun god announced that the Incas would get through the trouble and foretold a glorious future for them.

Another mythical example concerns the Island of the Sun in Lake Titicaca. The king who invented the ritual of "drinking to the Sun" in Villcanota (La Raya) because "the Sun was born there" was also the first to bring sacred water from a well on top of the mountain on the Island of the Sun. In Cuzco, he poured the water out into a basin (a part of the Ushnu) at the birth of the crown prince. The water was called or described as *Capacchana quispisutoc uno*, "the royal and victorious water that drips like crystals." The term *suttuni*, "to drip or distill of any liquid" (Gonzalez Holguin 1608), was used also for a "house that drips," *Suttucun huaci*. We can also interpret the particular reference of *quispi* to the temple of the Sun in that sense: from its roof hung crystals. The Sunturhuasi, the round temple used as a gnomon and as an observatory, also had such a roof (see below). The concept of "crystal," in the case of the "water like crystal," connects the birth of a crown prince to that of the birth of the Sun; the importance of the well on the Island of the Sun for the birth of the latter is transferred to the Ushnu by way of the water. Both places were used for a sunset observation on August 3 or 5, and the Ushnu was also used for such an observation on August 18. The term *quispi* therefore supports the interpretation that an AZSS observation was transferred from Titicaca (carried out on August 5) to Quispicancha (carried out on August 18).

The last example of the symbolic use of *quispi*, "crystal," refers directly to the month of August. During the month of Tarpuyquilla, corresponding to our August, Molina (1573:29) says that the priests of the Sun or *tarpuntaes* prayed to the Sun for heat for the food to grow; other priests prayed to the Thunder for water and no hail. For this reason Molina claims that the month was also called Quispi. The Inca observation of ZSR, as argued here, establishes a direct relationship between the month Quispi and the place name Quispicancha.

A fourth argument relates Quispicancha to the rising sun by means of an important myth about Huayna Capac, the last king to rule and die before the Spanish conquest. The association of a solar myth to king Huayna Capac was facilitated by his very name, "the royal (*capac*) young (*huayna*) person or king"--whatever the reason was for giving him this name. Today the name is used in southern Peru in reference to the sun. In Inca times, the sun of the June solstice was distinguished from the one of the December solstice by the use of the name Huayna Inti, "the young reborn Sun." We have the myth of Huayna Capac in two versions: that of Sarmiento de Gamboa (1572:Ch. 55-57) and that of Santacruz Pachacuti Yamqui (1613:254-257).

In Sarmiento's story, two factions of nobels were competing for the crown to go to their claimant. The winning claimant was hiding in Quispicancha. His followers brought him to the temple of the Sun and crowned him. Then they led him to Huanacauri, the mountain close to Cuzco, which is associated with the direction of the birth of the Sun. They took him back to

Cuzco by the same road that Manco Capac, the first king, had traveled on his way from Titicaca and Huanacauri. After presenting him to his new subjects, they gave him the name of Huayna Capac. His first act of state was to go to the temple of the Sun. He took for himself the stewardship from the person who held it--until then it was a separate function from that of king--and called himself "Pastor of the Sun."

In the version of Santacruz Pachacuti, Huayna Capac needed a regent at the time of the death of his father because of his young age. The regent wanted to rebel and take power for himself. He also invented the cult of the Sun, Moon, and Lightning by worshipping them from the mountains. However, a bastard uncle of Huayna Capac dreamed early in the morning that the city was surrounded by armies and that the regent pointed with bow and arrows to the young king. Thus advised, the latter was able to nip the rebellion in the bud; he was crowned and officially married. Coricancha has a central role in these stories, but the temple is said to have been dedicated to the Creator and not to the Sun. The clue that interests me in the second version is that the rebellion was discovered by a person who was dreaming early in the morning. Dream knowledge is used here in an empirical situation, an actual instance of rebellion, since the dreaming is carried over into the day. The situation inverts a normal one, where a lord would receive knowledge from a guard on vigil at night, when other people slept. Sarmiento, moreover, called the faithful uncle Guaman Achachi or "Falcon Grandfather," a name that suggests the position of shaman, with which the act of dreaming best agrees. Dreaming into the day, rebellion or the rejection of rebellion, and assuming the throne in circumstances that are contested all belong to a category that the Quechua language defines as *pachacuti*, "to reverse the time or the world." Guaman Poma, who also described a person sleeping until after sunrise, gave him the name of Pachucuti for that same reason. However, regular reversals of time or of the earth occurred after the time of travel, trade, and warfare, when the agricultural season opened in August and when it closed in April. We can therefore read a special significance into the fact that Huayna Capac, the young king (as a rising Sun) came out of hiding in Quispicancha. Seeing the sun set from here meant in annual terms the time of planting or harvesting, but in heroic terms it meant the rejection of rebellion and the ascendance to the throne of a new king, called the "royal young (king or Sun)."

Two other myths that I will not discuss here relate Huayna Capac first to Titicaca and the appearance there of a light kept in a box, and second to a building like the Sunturhuasi in the Ecuadorian town of Tumibamba (modern Cuenca). The building was dedicated to a golden statue of Huayna Capac's mother, who kept his afterbirth in her (dark) belly. In this way, his birth in Tumibamba and his ascendancy like a "young Sun" were honored.

All these data support the conclusion that Quispicancha had a particular astronomical significance to the Incas in Cuzco, and that the best translation of the name is not simply "the crystal enclosure" but "the place where a new light comes out of the darkness (*cancha*) and where it frees itself from an obstacle [in terms of the horizon as a place and in terms of the time of the year] (*quispi*) that had retained and sheltered it so far (*Quespina*, *Pucarca*)."

CHIRAO *SUCANCA*, THE SUNTURHUASI, AND THE USHNU

The *Sucanca* on Cerro Picchu

In the foregoing section I established the importance of the zenith/antizenith observations between Cerro Picchu and Cerro Pucara of Quispicancha. As the setting sun will frame the whole of Picchu, no importance can be attached to the pillars in this observation. The Sunturhuasi was not located on this line of observation, but it was probably used for determining the passage of the sun through zenith at noon. The outer pillars of Chirao *sucanca* served to frame the fixed period of the year corresponding to the lunar month of the antizenith passage. The pillars could have been set up at any convenient place, but in this case Picchu served two different purposes. The pillars were observed either from Hanan-Ushnu or from the Sunturhuasi or from both; we know their general locations in the plaza of Hanan-Haucaypata and we know that they were close to each other (Fig. 26.8). Our main interest is the distance between the outer pillars as given by the Anonymous Chronicler, given their distance from the Ushnu or Sunturhuasi; the small difference in distance from either of them to the pillars does not influence our conclusions. Again, since we do not need to know the exact location of the pillars in order to interpret the measurements (the zenith/antizenith observation was not decided by either the inner or the outer pillars), I will not enter here in detail into the problem of their location. A problem of greater interest is why the Incas apparently used two constructions, Sunturhuasi and Ushnu, to observe the outer pillars. The distinction was more hierarchical, and thus social, than anything else.

In addition to our anonymous source, three other chroniclers--Betanzos, Sarmiento, and Garcilaso--provide measurements of the distance between the pillars; of these chroniclers, probably only Betanzos is an original source (Zuidema 1977:253, 254). The other three sources are not of practical use for present purposes. First, they all confuse two systems of measurement: one involving the setting up of wooden stakes, observed from a short distance, and the other involving a system of pillars on the horizon. Second, none of the three chroniclers gives distances from the observer to either stakes or pillars.

According to the Anonymous Chronicler, the outer pillars were 200 *pasos* apart and the inner ones 50 *pasos* apart. Although there are modern *pasos* of 0.33 m, 0.65 m, and 0.75 m, the *paso* used in the sixteenth century was the "paso geográfico" of 5 *pies* or 1.392 m. (This is the old "Roman step" where two "steps" are taken as one "paso.") The distances between the pillars are 278.4 m for the outer and 69.6 m for the inner pillars, respectively. In 1978, we calculated the location of the inner pillars by tracing the *ceque* to which the *sucanca* belonged. Since all the identifiable *huacas* of this *ceque* were located very close to a straight line as traced over the horizon, we are confident of the location of the *sucanca* and of the inner pillars. On the basis of these data we "paced" the distances to the outer pillars. The location of the southern pillar came out very close to the top of Cerro Picchu. Thus it is very possible that the Incas took the top as their point of reference and paced from there the location of the other pillars toward the north. This gentle slope (its name, *pihchu*, "the coca in the mouth that makes the cheek bulge," explains its form) offers the best means for observing the progressive movement of sunsets during this time of the year.

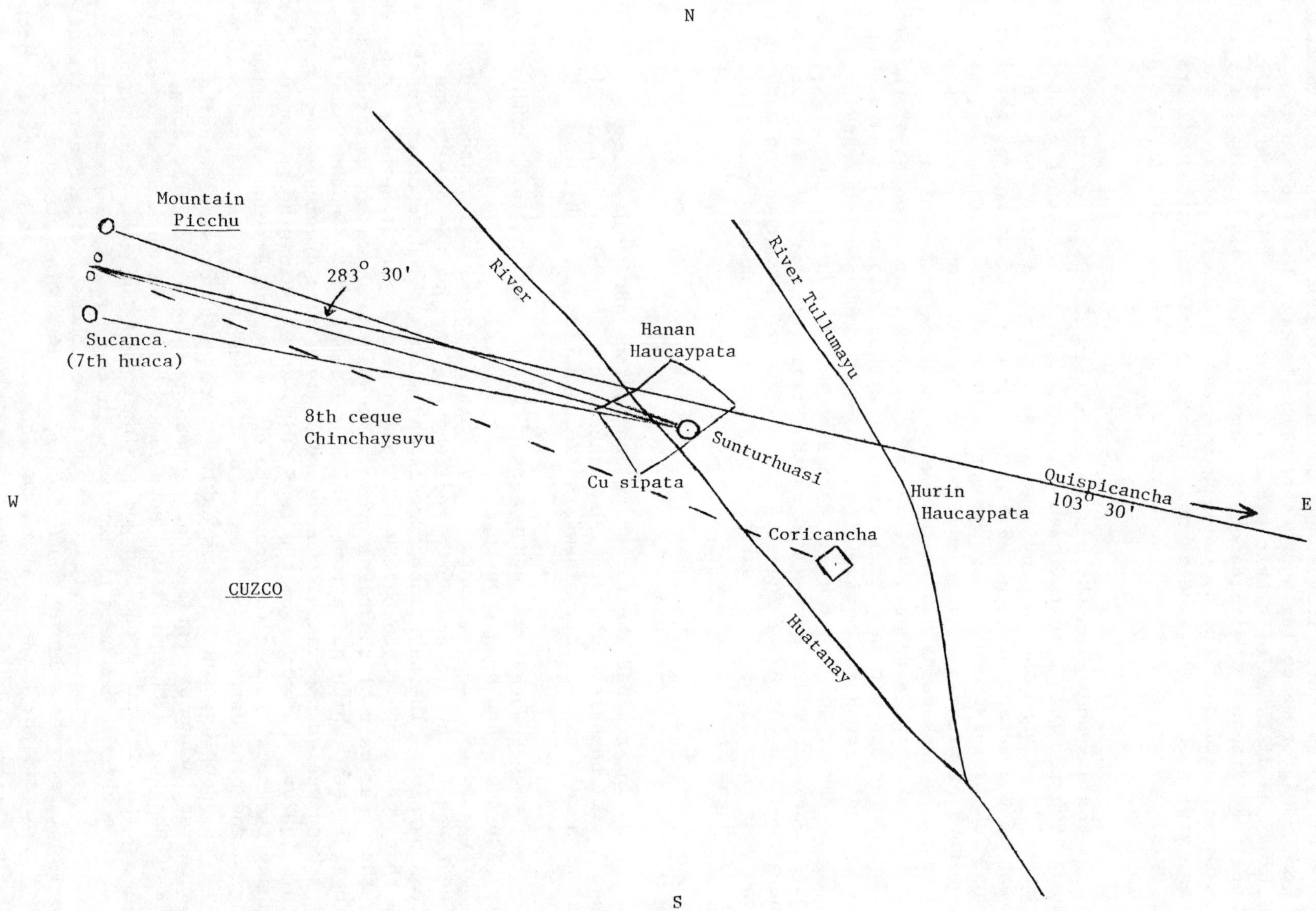

Fig. 26.8 Cuzco

Our chronicler says that the pillars "could be observed from a distance of two and three leagues" (± 11 km or 16 km). He also claims that they were used for measuring sunsets from the Ushnu. Was there a reason for mentioning the distance of either two or three leagues? We might think of the distance from Picchu to Quispicancha. Sarmiento, when discussing the myth of Huayna Capac, said that the distance from Cuzco to Quispicancha was three leagues, giving the same value as one of the two from which the pillars could be observed. However, the real distance to Quispicancha (26 km) is closer to five leagues.

The distance from the pillars to either the Ushnu or the Sunturhuasi is about 1900 m. This results in an angular distance between the outer pillars of some 8.5°. Around the date of August 18, this gives us a period of the movement of the sunset of about 29.5 days. However, if we accept the idea that the sun disk was observed during the period when it was within the outer pillars, we have to deal with an angular movement of the sun along the horizon of 8°, with the sun's disk being 0.5°. This will account for a period of about 27.5 days. Without being more precise, we can conclude that the data of the Anonymous Chronicler support an Incaic period of either 29.5 or 27.5 days. This period was calculated respectively from August 3 to September 2 and from April 11 to May 11, or from August 4 to September 1 and from April 12 to May 10. The first estimate would correspond to the period of a synodical lunar month; in the second, the dates of August 4 and May 10 are closer to the dates dividing the year into eight equal parts (which could be measured with zenith/anti-zenith passages in Titicaca.

The Sunturhuasi and the Ushnu

Having reached the foregoing conclusions about the pillars on Cerro Picchu, I will be brief about the Sunturhuasi and the Ushnu, especially since I deal with them extensively elsewhere (Zuidema n.d.a, n.d.b).

The Sunturhuasi or "house *suntur*" was a round building with windows, a very high roof, and a mast on top (see Fig. 26.2). The term *suntur* is also used for the royal scepter, *sunturpaucar*, "the multicolored *suntur*," a stick with a crown of feathers and three red feathers on top; the object was similar to a parasol. The name *sunturpaucar* was also applied to other royal objects that were compared to the Sunturhuasi or *sunturpaucar*: the parasol itself; the upper part of the royal crown which had an imitation parasol; and the royal litter, which had a roof similar to a parasol. Another name for the royal litter was *quispe rampa*. Guaman Poma gives us a drawing of it, with crystals (*quispe*) hanging from the roof (Fig. 26.9). This helps us to understand how the royal litter (*quispe rampa* or *suntur paucar*) and the Sunturhuasi belong to the same category of "temple of the Sun" as Coricancha and Muyucmarca. Guaman Poma describes the first as having a roof with crystals hanging from it, and Santacruz Pachacuti mentions Collcampata as a base for Coricancha, the temple of the Sun, although it was actually the base of Muyucmarca on Sacsayhuaman. Molina describes the Sun god rising out of the well of Susurpuquiu, into which a crystal had been thrown, with three rays rising from behind his head, rays like the three red feathers on top of the *sunturpaucar* in its different meanings. The scepter *sunturpaucar*, the Inca king with the *sunturpaucar* in his crown, and the house in which the Inca officiated when he observed the rising and setting sun (be this house Coricancha, Sunturhuasi, Muyucmarca, or the royal litter) all represented the idea of *Axis Mundi*. This notion stresses the time when the sun goes through zenith. The Spaniards who conquered Cuzco

Fig. 26.9 The crystal litter of the king (F. Guaman Poma)

were very impressed by the Sunturhuasi as an expression of Inca royalty; for this reason, we have rather good data on its location. It stood on the east side of the plaza Hanan-Haucaypata, in front of the present Jesuit church of the Compañia. Its location according to the ethnohistorical sources conforms closely to the location that we could reconstruct from the position of the four pillars of Chirao *sucanca*.

Locating the Ushnu presents greater difficulty, even though the first Spaniards who arrived in Cuzco still saw it ritually used. One problem is that the term *ushnu* referred to at least three structures: (1) a basin; (2) a stone pillar; and (3) a throne. The Anonymous Chronicler mentions all three as part of the Ushnu. The basin, which had a low, gold-covered wall and a roof like that of the Sunturhuasi which was removed during the daytime, was connected to a subterranean sewer into which libations were poured to the Sun god and the ancestors. The sacred water from the Island of the Sun that was compared to *quispe*, "crystal," was poured out in the middle of the double plaza Haucaypata-Cusipata, which was divided by the river Huatanay. The basin of the Ushnu was probably used for this purpose. Such a location of the basin would generally coincide well with the direction of Sunturhuasi toward Chirao *sucanca*. But we do not know where the pillar and the throne (which was probably wooden) were located. We can at least say that the data on the Ushnu are not in conflict with our conclusions.

CONCLUSIONS

In this paper I have considered two problems related to the astronomical interests of the Incas. The first of these involved determining their method of defining the dates of the sun's passage through zenith and antizenith. The second concerned the calculation of the first month in their agricultural calendar and its relation to the antizenith passage. Since the first problem was solved by observations between two mountains, Picchu and Pucara of Quispicancha, it was important to bring together and interpret the ethnohistorical data on these mountains, especially those referring to Quispicancha. The concept of *quispi* or "crystal," which here became of central interest, symbolized the idea of "sunrise."

Historical and modern ethnographic descriptions of Andean culture abound with data on the religious and ritual importance of mountains and on the relationships between mountains, which are described mostly in kinship or other social and ritual terms. The data on Picchu and Quispicancha may be crucial to an understanding of problems of Andean political structure. Here, for the first time, precise technical data of calendrical content can be offered that document the political use of a visual connection between two cities (in this case, Cuzco and Quispicancha) by way of two mountains that rise above them. Elsewhere, I have analyzed the references to another, similar example (Zuidema 1978b).

The astronomical and calendrical importance of Cerro Picchu was further clarified by the location of the four pillars on its slope as viewed from Cuzco itself. The observation of AZSS was not only placed in the context of the lunar and agricultural calendar--in a political sense, Cuzco itself was connected locally to Cerro Picchu and its significance for observations from afar as well. The observation was done from the Ushnu and the Sunturhuasi. In this context we discover the importance of a remark made by Santacruz Pachacuti Yamqui (1613:219). When the Incas conquered a new province and established their government there, their first act was to build an Ushnu in the newly integrated valley; Cuzco provides us with an example of this practice.

I have argued that the same astronomical and calendrical observations were made from the Sunturhuasi and the Ushnu; I should now clarify this. The king, accompanied by the high priests (Guaman Poma 1584:262, 263), made his observations from the Sunturhuasi, a symbol of royalty and the *Axis Mundi*. On his throne, the Ushnu, he was visible to and surrounded by his non-royal Inca and non-Inca subjects (Guaman Poma 1584:384, 398). Here he communicated his decisions, one being the opening of the agricultural year. We now understand why the Anonymous Chronicler, in discussing the planting time of the non-Inca subjects living on the mountain slopes around Cuzco, mentioned the Ushnu when he described the calendrical importance of the four pillars of the *Sucanca* on Cerro Picchu.

27

Irish Henge Orientation: Preliminary Results and Some Problems

Ronald Hicks

In previous work on archaeoastronomy in the British Isles, the greatest effort has been devoted to the study of stone circles, of which over 900 examples are known. This report will present the orientations of 42 Irish henges and examine their implications for early Irish astronomy. The term "henge" is a back formation from *Stonehenge*, in which the Saxon word "henge" referred to the "hanging" stone lintels; in archaeology, however, it has come to mean Neolithic and later monuments of the British Isles resembling the very first phase at Stonehenge: the circular earthen bank constructed centuries before the famous stone circles were built. Most henge monuments have an internal ditch separating the bank from the central platform. The largest may range up to nearly a mile in circumference, like Avebury, while the smallest are quite small indeed, only 10 to 12 m in diameter.

Although henges and stone circles are clearly related in some way, relatively few of the stone circles have surrounding henge enclosures; in addition, most of the 175 known henges lack stone circles, although a number are known to have once contained circles built of massive wooden posts. What I have chosen to study, therefore, is not the alignments provided by these rather precise stone or wooden markers, but rather the orientation of the earthen henge entrances. By definition, henges have entrances, which sets them apart from such undoubtedly related monuments as embanked stone circles, which have none (Atkinson 1951:82). In most cases henges have a single entrance or two more-or-less opposed entrances, marked by a causeway across the ditch and a break in the bank. A few henges, specifically some of the largest monuments like Avebury, have more entrances.

This brief report will not include a comprehensive history of henge studies, which are discussed in detail elsewhere (Hicks n.d.). It is important to mention here, however, that some henges do not strictly match the specifications I have given. There may be a question as to whether these should all be considered henges, but they resemble henges more than anything else. For example, a few sites in the west of England, Wales, and Ireland lack a ditch entirely, have the whole central area quarried down to a level lower than the surrounding land, or have causeways but no breaks in the bank or vice versa.

---

[1]The research on which this paper is based was supported in part by a Fulbright-Hays fellowship, a doctoral research grant from the Department of Anthropology of the University of Pennsylvania, and a grant-in-aid from the American Council of Learned Societies. It would also have been impossible without the assistance of a number of my friends and colleagues, both in Ireland and the U.S.

There are a number of other variants, including a couple of henges in the northwest part of the island of Ireland (one in the Republic and one in the North proper) that have slanting entrances which from the center of the enclosure appear only as notches in the bank. In two or three other cases the entrances are narrow and lined with stone slabs, and a few henges have only dips rather than complete breaks in the bank.

A further complication is that it is relatively difficult to distinguish an original entrance from a later addition. Only 42 of the 55 probable henges identified so far in Ireland have entrances that I am relatively confident are part of the original design.[2] Some sites are too badly plowed for any entrance to be identified, or they have wide gaps where portions of the bank have been leveled (these may or may not have included an original entrance). Also, cowpaths have created breaks in the henges that are hard to distinguish from true entrances, or there are narrow cuts that appear to be recent additions, and so on. In addition to the thirteen sites where the entrance orientations cannot be determined (which forced me to leave these sites out of my analysis entirely), there are sixteen entrances in sites with more than one entrance that I have omitted from my analysis because of their doubtful nature. (I would like to emphasize here that in no case did I decide an entrance was "doubtful" because of its orientation; all simply appear to be wholly or partly more recent than the monument itself.)

The entrance orientations could be identified in three ways: (1) by marking the edges of the causeways, (2) by marking the bases of the bank on each side of the entrance gap, or (3) by marking the points where the bank begins to slope down on each side of the gap. Because of variations in the nature and condition of the entrances, in each case I have used whichever method seemed the most appropriate, or a combination of methods. Each has its limitations. By and large, causeways seem to have been least subject to damage through time, but they also are likely to be the least reliable as indicators of an intended orientation, since the break in the bank would have been more evident to an observer and would have been a better pointer to the horizon. A variety of other evidence exists to support the view that the bank, rather than the ditch with its accompanying causeway, was the important element. However, the beginning of the slope at the top of the bank, while a more reliable indicator of entrance orientation in new sites, is also the element most likely to have suffered damage through time. In addition, the base of the bank, thanks to slumping and other damage, is not always easy to identify. At one site--the Giant's Ring in County Down--the measurements used for the base of the bank were in fact simply the lowest spots in the various entrance notches.

When I began studying the Irish henges, I was simply interested in identifying them; I was still highly skeptical of the astronomical interpretations offered for Stonehenge and the other stone circles, and had virtually ignored that part of the literature. However, in analyzing my survey data, it quickly became clear that the builders of the Irish henges had preferences when orienting the entrances to the henges (Fig. 27.1). The east-northeast was the preferred direction, followed by west-southwest, with relatively few entrances

---

[2]A complete list of the sites included, with detailed entrance data, is available from the author on request and will be included in the full report now in preparation.

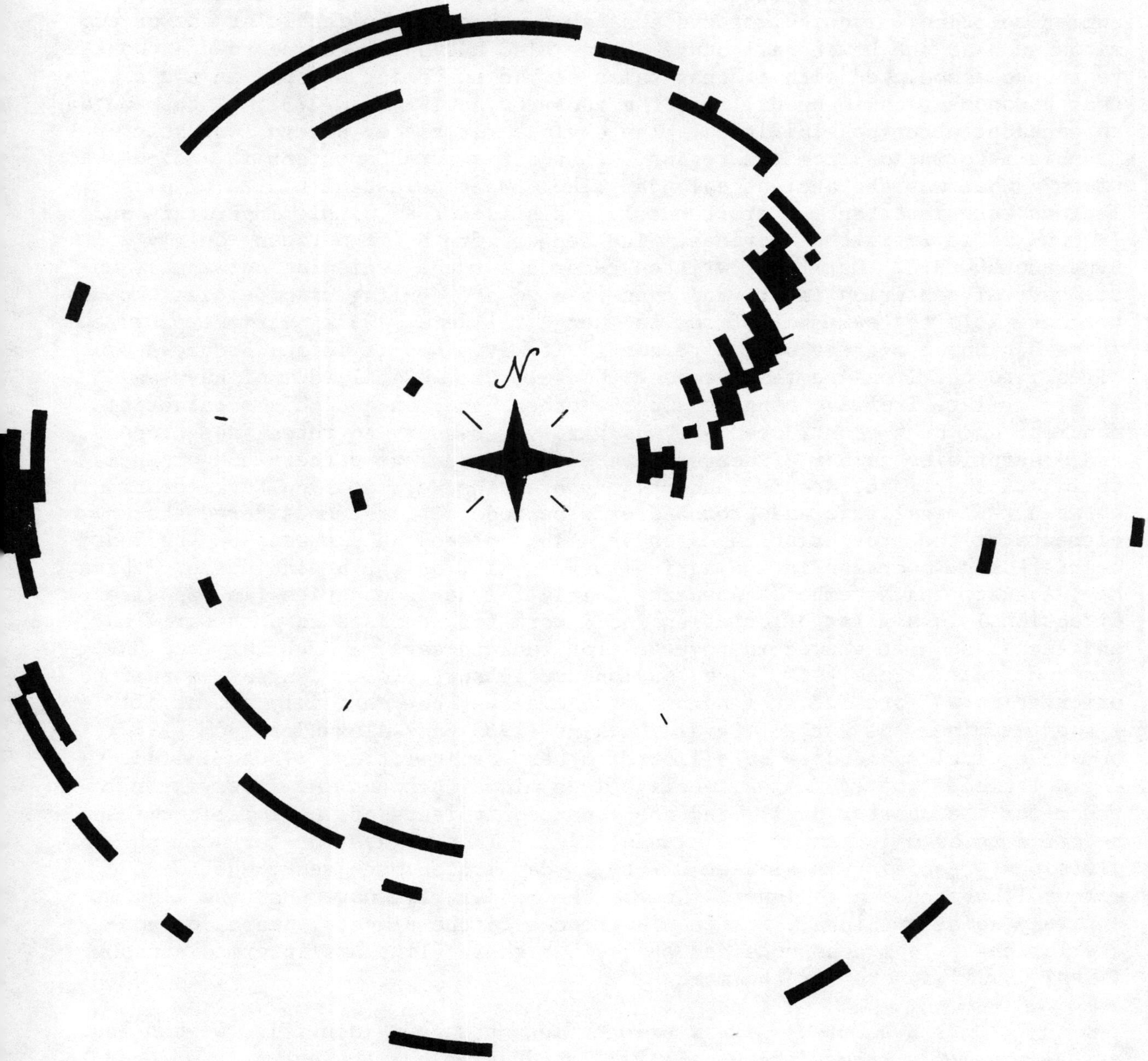

Fig. 27.1 Orientations of henge entrances

elsewhere. This observation forced me to consider astronomical explanations, although I am not yet fully convinced that the orientations of the entrances of henge monuments really have astronomical implications in all (or even most) cases.

If the entrances were indeed astronomical, what orientations were most likely? My first thought, prompted by Stonehenge, was to consider the summer solstice. I quickly found that this was only applicable at one or two sites such as the Dowth Hall henge, which also has a legend recorded a thousand years ago associated with it that talks of the sun being stopped in the sky so that a monument could be built to the gods (Gwynn 1924:270-273). I then began to consider other possibilities. The obvious first step was to look at ethnographic information from the region. What dates were important in ancient times? What was the ancient calendar like? What celestial bodies or constellations were important? Unfortunately, this is not as simple in Britain and Ireland as it is in the Americas. The henges were built between roughly 2500 B.C. and 500 B.C. Our first written records for the region do not appear until several centuries later, and even then we are dealing with peoples who may not have held the same beliefs as the henge builders. Work over the past few years has shown a great deal more continuity from Neolithic and Bronze Age times into the Iron Age than we once thought, but we still do not have any middle or late Iron Age henges. On the other hand, once such a substantial monument had been constructed, it was hardly necessary to repeat the process, and a surprising number of henges have yielded Iron Age pottery and other materials. Therefore, despite uncertainty over the degree of cultural continuity between the megalithic and protohistoric periods, I first considered the known elements of the pre-Christian calendar. This placed an emphasis on the four quarter-days important in the agricultural cycle, at the beginnings of February, May, August, and November. Assuming fairly low horizon angles (in opposite directions), a similar orientation would work for sunrises in both early May and August and (if viewed in reverse) for sunsets early in February and November. My initial analysis, based on inadequate survey data, suggested that this orientation was present at a number of the sites, but certainly not at all. I then examined the work of Gerald Hawkins (1965) and Alexander Thom (1967) on stone circles and drew up a list of other possibilities. These included the solstices, the equinoxes, declinations midway between them (very near to those for the quarter days), and the lunar equivalents of the solstices (usually referred to as major and minor standstills). Thom argues for further subdivisions, resulting in sixteen 22- to 24-day months each year, but for the moment I have chosen to ignore these. The reader will note that the list is entirely solar and lunar, with no references to the planets, stars, or constellations. Thom does consider stars, but these (like his intermediate sun lines) I will ignore for the moment.

Once a group of likely orientations has been identified we can look at the monuments themselves. We have already seen how the entrance orientations cluster (Fig. 27.1). Note that the entrances sometimes occupy quite a wide arc. I perhaps should offer the additional caution that it is by no means certain that the center of the henge was originally the place to stand. It is entirely possible that one was supposed to stand in the entrance, or against the far side, and look somewhere else. However, for the purposes of the present analysis we will assume the center was the proper observation post.

Orientations by themselves tell us nothing. They merely provide the azimuths from which declinations can be calculated. Figure 27.2 shows the *range* of declinations to be seen if one stands in the center of each henge and looks out through the entrance. The heavy line in each bar indicates the declination of the centerline of the causeway and/or bank gap. The graph is divided into four quadrants--northeast, southeast, southwest, and northwest--so that the entrances are shown in something approximating their proper geographical position. From the center of the graph outwards, I have arranged the sites first by the proximity of an entrance to east in the northeast quadrant, and then outward toward north, west, south, and then east again. Some of the bars narrow at one end; this indicates that the entrance in question lies across the quadrant division and its range is shown in two quadrants of the graph. Thus entrances crossing true east or west are shown as narrowing to the vertical axis; the remainder of each can be found an equal distance on the other side of the horizontal axis. As one can see, of the 42 sites for which entrances could be identified in one form or another, eight (or 19%) had entrances that crossed the east/west line; only one crossed the north/south line, and is similarly represented on the outer edges of the graph. I have drawn vertical lines on Figure 27.2 to indicate the solar and lunar declinations I feel likely to be significant. Note that the lunar standstill lines are not the same in the top and bottom halves of the graph; this is to allow for the parallax caused by the moon's proximity to the earth (Thom 1967:118).

As can be easily seen, the declinations indicated by the henge entrance centerlines do *not* fall where predicted, although most fall within the solar range and a few are quite close to one of the expected lines. How does this affect the astronomical hypothesis? There are several possible explanations: (1) the henge orientations are not astronomical at all; (2) there were errors in my surveying; (3) I have chosen the wrong celestial phenomena; or (4) perhaps we should not be looking at the centerlines. Let us examine each of these possibilities.

The possibility that astronomical factors are not involved seems unlikely. As we saw earlier, preferred directions certainly exist. In addition, the work of Gerald Hawkins, Alexander Thom, and others confirms astronomical orientations for closely related monuments--stone circles and passage graves. I therefore feel that it is very unlikely that astronomical considerations of *some* sort are not involved in the henge orientations.

The possibility does exist that errors occurred in the course of my surveying. I carried out the surveys using magnetic north, and then corrected my figures with magnetic variations for the latitude and longitude, date, and time of day of each survey.[3] These corrections obviously would not take into account local magnetic anomalies, and they might be faulted on other grounds. New baselines need to be determined for each site by shooting a sunline. However, I am convinced that in most cases this would not sufficiently account for the discrepancies.

The possibility that the wrong phenomena were selected might have some relevance, although again we can look at the work done by others. I am sure that my list does not contain *all* the relevant phenomena. It is interesting to note that several of the centerlines fall between the line marking

---

[3]This information was kindly provided by James B. Sullivan, Officer-in Charge, Valentia Observatory, Cahirciveen, County Kerry, Ireland.

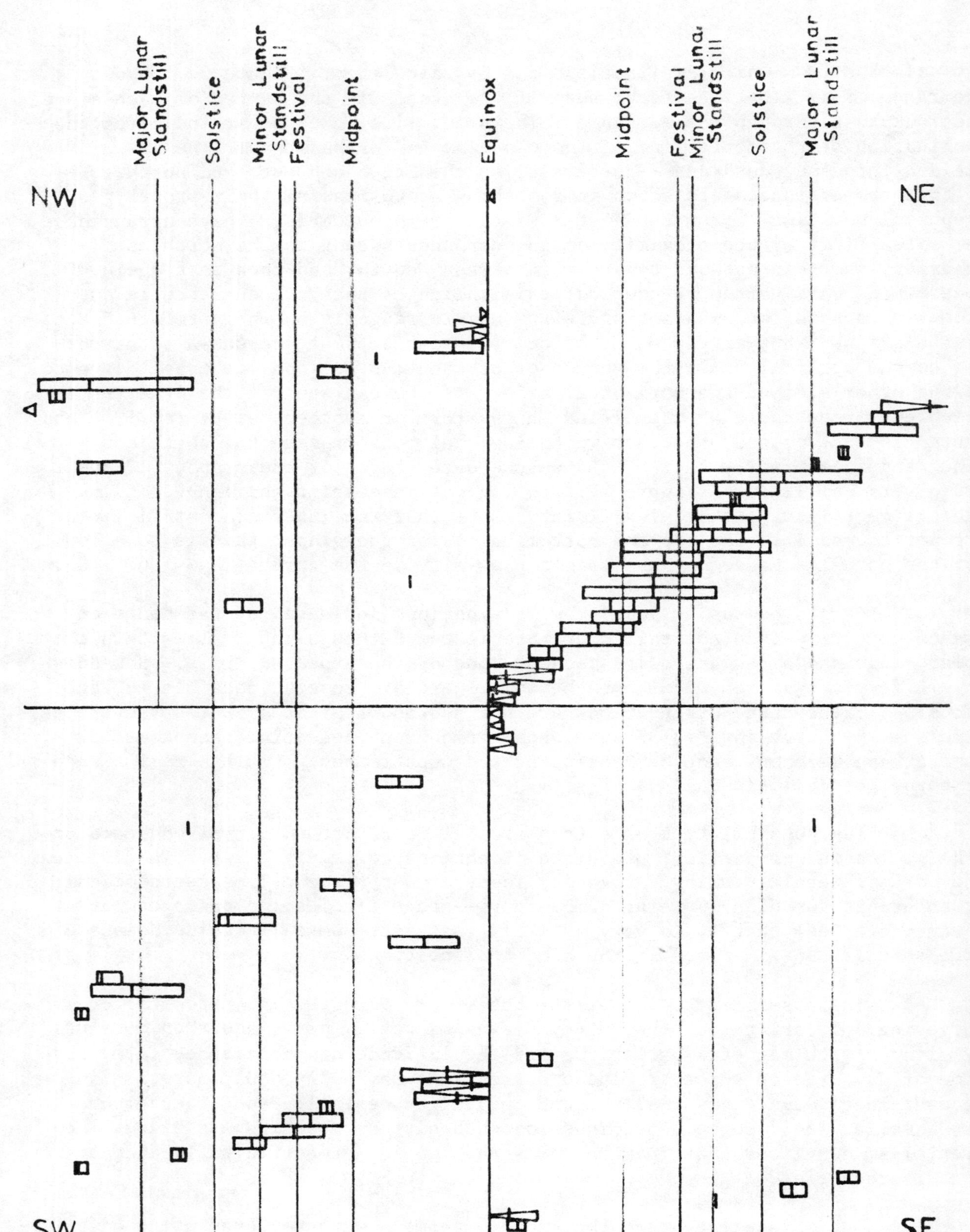

Fig. 27.2 Declinations bracketed from Irish henge entrances

a declination midway between that for the equinoxes and solstices (±11.9°) and the declination line for a *date* midway between the equinoxes and solstices (±16.5°)--i.e., the seasonal festivals. This might indicate a compromise. I have already mentioned two other possibilities--Thom's intermediate sun lines for one, and the planets, stars, and constellations for another. Unfortunately, I have no ethnographic clues to follow here.

My feeling at the moment is that whatever the validity of these suggestions, most of the problem is probably traceable to considering only the entrance centerlines. This is an unsafe procedure, given the eroded nature of many of the sites. If we look beyond the centerline, we find that entrances in most cases cover an arc that allows for at least one (and sometimes two or three) of the predicted observations to be made, given some additional marker such as a post or stone or some horizon feature as a foresight. If we accept this interpretation, we find that the midpoint, festival line, minor lunar standstill, and equinox could all be frequent orientations.

This last explanation seems a little too imprecise, and even if it is true it leaves several orientations unaccounted for; therefore, I will mention a few other factors that should perhaps be taken into consideration as well.

First, I have looked only at the entrance orientation. In a few cases the banks form an artificial horizon surrounding the site, but in most the horizon profile provides a wealth of features that could be used as markers. I am currently beginning an analysis of those data.

Second, in some cases it appears that small henges--as is known to have been the case with passage graves--were oriented toward other, more important, monuments nearby. This would seem to be the case with a line of four henges on the Curragh (in County Kildare), where the two end monuments are pointed toward the center of the line.

Third, there is a long tradition in the British Isles, with clearly pre-Christian roots, of orienting graves and churches roughly east-west. Anthony Aveni (1979:68) has pointed out a seventeenth-century reference that states that at that time the practice was to orient churches toward the sunrise when the foundation was laid. This would serve to explain both the emphasis on east-northeast and the failure of the sites to match exactly the predicted orientations.

Parenthetically, there is proto-historical evidence suggesting that the calendar in use was a combined solar/lunar one based on the 19-year Metonic cycle. If this were the case, the festival dates in any given year would not have been strictly solar, but rather would have been lunar dates linked to solar dates (e.g., the date of the full moon nearest a quarter day). If construction of a henge were started at the time of a festival assembly, the point of sunrise on a date such as this might have been used. As possible support for this, it was noted earlier that several sites have entrance centerlines that appear to mark declinations between those for the dates when the solar *position* is exactly midway between solstice and equinox and for a solar *date* exactly midway between. This gap covers a period of about fourteen days, or approximately half a lunar month.

Finally, we might ask if the purposes to which the sites were put always required an exact orientation. Some may have been primarily intended as observatories, but evidence from archaeology, folklore, and early Irish literature suggests that the larger henges may have been primarily royal or ecclesiastical enclosures. The smallest may have been primarily sepulchral--increasing the sacredness of the burial for high-ranking individuals. The more ordinary-sized henges may have served as places for seasonal assemblies, games, and rituals.

We have, therefore, no simple answer to the question of whether or not Irish henges were oriented astronomically. But we do have a number of alternative hypotheses that can be investigated, and we can hope that further work will eventually enable us to confirm one of them with some confidence.

28 Teaching the History of Astronomy *or* Seven Challenges for Archaeoastronomy

T. Scott Smith

Perhaps in teaching more than in any other activity, one is constantly forced to reassess and rationalize one's basic assumptions. One is thrust into the role of critic simply by the choices involved in syllabus construction and lecture design. The teacher may thus become the happiest sort of critic--one criticizing out of the love for a subject and the desire to see it perfected. Having taught archaeoastronomy in a variety of different contexts, I am led to a number of suggestions for its enrichment.

The first area of concern is that of the standard introductory astronomy course. Despite the fact that the past decade has shown a remarkable explosion of archaeoastronomical knowledge, there has been an opposite trend in introductory astronomy textbooks. Ten years ago when Abell (1969) was the standard introductory textbook, it was customary for texts to include the required picture of solstice at Stonehenge with a few lines admiring the dedication and perseverence of those "howling savages" who had erected it. The situation today is perplexing. The general response to the information explosion in archaeoastronomy has been to drop the entire subject from many standard textbooks. Although there are exceptions, such as Hartmann's fine book (1978) which spends a full chapter on naked eye astronomy and the search for the calendar, trends elsewhere are to present astronomy from an ahistoric or nonhistoric point of view. For example, a recent text (Pasachoff 1978) notable for its completeness (763 pages, 10 appendices, a 16-page glossary, a 3-page list of selected readings, and selections on the scientific enterprise, grants, and publications) makes no reference to Stonehenge, let alone archaeoastronomy. Is this the astronomical application of Brush's thesis (1974) that the history of science should be X-rated for science students? So this is the first challenge to be faced: convincing professional astronomers that the history of astronomy is a worthy focus of effort, and that this effort is reflected in the teaching of introductory astronomy courses.

The second challenge is related--a fairly typical reason given for studying dead-end astronomies is to broaden cosmological perspectives, challenging traditional western notions of the structure and origin of the universe and man's place in it. This point of view superficially glosses over the range and diversity of cosmologies within the western tradition. Are primitive cosmologies merely a trivial example of geocentrism revisited? Why study native American cosmology at all? Are there better (more heuristically valuable) examples of non-western cosmologies? If we wish a more nature-oriented calendar/cosmology, does China represent a greater tradition of thought than does native America? If a rich pantheon of gods is desired, why not Hindu cosmology? By pushing native America, do we run the risk of substituting one ethnocentrism for another? Are we vitally interested in cosmology or merely paying lip service? How many of us have recently taken the time to re-read or even to read Plato's *Timaeus*? So the second challenge can be expressed as follows: what new important visions of the universe are brought to light by a study of archaeoastronomy?

The next area of discussion involves interdisciplinary courses. This should closely parallel the interdisciplinarity of the investigations. Archaeology, art history, folklore, and geography have been listed as disciplines with strong links to archaeoastronomy. By listening carefully during this conference, one could add geology, ethnography, cultural anthropology, and paleomagnetism. If we recall the impact that lithology has had upon the interpretation of megaliths at such sites as Stonehenge and Long Meg, we can speculate that a geologist might be a requisite part of a research team rather than a consulting specialist. Without stretching our imaginations too much we can easily add linguistics, paleoclimatology, paleobotany, and (perhaps most important of all) the history of religions. We can readily imagine all sorts of contexts for interdisciplinary conversations in academic courses. At an absolute minimum, any survey of cultures and civilizations should have the archaeoastronomical perspective expressed in lecture support. If we truly believe in its importance, we must be prepared to educate our academic colleagues. The obvious challenge here is to demonstrate the centrality of astronomical concepts and derivations to the cultures in question.

A less obvious but perhaps deeper challenge is the selection of *where* to do archaeoastronomy. As an anthropologist friend (V. Kavolis, private communication 1979) has said, anthropology is the study of 100,000 societies over 40,000 years--where do you start? How do you decide which of these societies may produce the most enlightening or surprising results? Or (in a different setting) what can we learn from the fact that the American Indians were able to determine the date of the solstice accurately--that is, the length of the year to within a day--when 1500 years earlier the Greeks had the length of the year within minutes (not to mention Chinese accuracies within seconds)?

This is, in fact, related to my earlier evaluation of the importance of a dialogue with the history of religions. As long as we continue to misemphasize the technological necessity of accurate alignments rather than their ritual significance, we will have a culturally-impoverished view. So far as I know, all serious scholars (Heisner 1973) place the origins of agriculture millenia before most of the alignments we are speaking of. Simple stick or tree-shadow measurements of horizon calendars can provide more than sufficiently accurate data for agriculture. This unawareness of prehistory leads us to the fifth challenge: how can archaeoastronomy refine our view of prehistory? Remember, if we place the first modern man at 40,000 years ago, a tremendous length of time exists for the development of primitive sky awareness. Like Marshack (1972), I am disturbed about the "suddenly" model of man's prehistory. SUDDENLY (fill in the blank) formal science, writing, art, mathematics, and the calendar BEGAN. Suddenly astro-alignments began? If we acknowledge the importance of an early interest in time-factored and space-factored activities (indicated by lunar and perhaps seasonal calendars carved into reindeer antlers) we enrich our view of prehistory. We may speculate with Marshack that perhaps the first sky orientation was cave selection--that is, passive solar alignments. Imagine the stimulus to hierophany! We may further speculate as to what astronomical secrets emerged from the ice age as part of man's intellectual baggage. There are alternatives to the "suddenly" model--we may imagine a slow diffusion and modification of ideas. Given the common problem of survival the world around, and the likelihood of memories (both cultural and genetic) of climate and other factors, one could never accept a totally diffusionistic model for archaeoastronomy like that applied to astro-myth in *Hamlet's Mill* (de Santillana and von Dechend 1969)--although that book does contain a number of interesting suggestions which may be tested by a more intensive study of early astronomies. Two examples of such suggestions are (1) the origin of the worldwide earth-deluge myth in astronomical precession and (2) the cultural diffusion of the strong-man figure plus animal jawbone motif evident in ethnographic data on the Orion, Taurus, and Hyades region of the sky. I do not propose diffusionist models as premises for archaeoastronomy but as testable hypotheses. It seems obvious that the final answer to a model of prehistory must be some combination of naturally arising ideas fitting indigenous conditions in conjunction with the local adoption and modification of external ideas. Archaeoastronomy can substantially sharpen such models, since its external reality (positions of sun, moon, planets, and stars) can be almost flawlessly reconstructed; thus it is easier to separate local, idiosyncratic aspects from universally-true observations. Local adaptations or inventions stick out because they fail to work at other global locations (e.g., Stonehenge-type structures "work" only at a certain latitude--north or south of certain limits, different structures are necessary for the same solar and/or lunar observations).

Just as we can look backward beyond megalith construction and building orientation to inquire as to what came before these activities, so we can ask what came after the alignments. The sixth challenge can become: what followed archaeoastronomy? What did it lead to? It seems that the alignments were intimately connected with calendars, either as devices for setting calendars or for ritually celebrating them. And once calendars exist, they not only mark the passage of events but also enable other events to occur. In other words, the calendar serves to correlate everyday life with celestial phenomena. This celestial correlation clearly leads to astrology in a wide variety of cultures. Not only is this an item of theoretical academic interest, but the multiple perspectives obtained by studying the astrological impulse in a variety of cultural settings can lead to interesting practical teaching applications. A perennial problem of astronomers is that of de-bunking astrology. The most recent organized effort culminated in the statement repudiating astrology, signed by 192 leading scientists (including 19 Nobel Prize winners) and many astronomers, which was published in a pamphlet entitled *Objections to Astrology* (Bok and Jerome 1975).

Despite these and other well-motivated efforts, astrology remains a vital force in the world today, even in the educated western world. One is tempted to inquire why. One possible answer may be the failure of astronomers to realize that astrology is an interdisciplinary concern related perhaps more strongly to psychology and religion than to astronomy. (Do mathematicians worry about de-bunking numerology?) I have some success in de-bunking astrology in a course entitled "Comparative Astrology," which draws together readings from a variety of fields including archaeoastronomy, history of religions, sociology, psychology, and anthropology. The major educational thrust is that students are encouraged to see the diversity of ways in which early peoples approached astronomy; examples include the emphasis (in what was to develop into western astronomy/astrology) upon the Zodiac and sidereal periods, the Mesoamerican emphasis upon horizon-based astronomy and synodic periods, and the Chinese emphasis upon polar astronomy and culminations. As astronomies became used as socially-relevant sciences (i.e., astrologies), the interaction of these "sciences" with their societies had profound cultural overtones which conditioned different methods of operating. For example, the Greek obsession with geometry is reflected in the astrological importance accorded by Ptolemy to conjunctions, trines, oppositions, and other aesthetically-pleasing and displeasing geometrical arrangements of the gods around the Zodiac. The Chinese obsession with bureaucracy and dynasty leads to an astronomical emphasis upon the mundane--the state of the state, not necessarily the personal horoscopes of its common people. The symbolic identification of the emperor with the pole star possibly reflects the mechanics of astronomy as practiced by the Chinese. Mesoamerican astrology is profoundly numerological, as reflected in the Sacred Calendar or "The Book of Good and Bad Days," which assigns good things to the number 13--an illuminating contrast with modern hotels and airplanes, which religiously avoid the number. Rapidly leaping to the other side of the world, Hindu astrology (with its emphasis upon Karmic reality and reincarnation) operates with a precessed zodiac and is sidereal as opposed to tropical, thus evading one of the standard textbook criticisms of astrology. If one were an eternal optimist, one could argue that astrology could be a useful tool for cultural analysis, reflecting the major interests and obsessions of the people who practice it--in pyschological terms, a projective experience. More realistically, astrologies seem not unlike our preconceptions of various cultures; perhaps students can be encouraged to see that worldwide the heavens have been the seat of man's projected consciousness and the repository of a number of competing visions of man's destiny. After having spent ten frustrating years with standard de-bunkings of astrologies and ancient cosmonauts, I feel far greater pedagogical success can be gained by introducing students to the range and diversity of early astronomies, so that they may view astrology in a reasonable historical perspective--as a natural outgrowth of megalithic astronomy and the search for the calendar.

The final challenge is that of finding an acceptable methodology which both acknowledges the importance of multiple working hypotheses (the possibility of other than astronomical alignments) and extends the range of conversation to all academic disciplines concerned with our early intellectual development. These are linked. For an alignment to be acceptable it must be ethnographically substantiated. The whole question of uniqueness of solution is approached by including hypotheses from a range of disciplines, hence evading the criticism of always being able to find some astronomical alignment for any site.

To summarize, the seven challenges to archaeoastronomy as a legitimate academic discipline are as follows: (1) to demonstrate the importance of history to astronomy; (2) to define new/important philosophical views of the universe; (3) to show the centrality of astronomy in cultural development; (4) to pay reasonable attention to the study of astronomy in all major cultures (i.e., to have a global perspective); (5) to refine our view of prehistory in significant ways; (6) to show where archaeoastronomy leads intellectually; and (7) to develop secure methodologies. The great challenge remains: can archaeoastronomy be shown to be an indispensible part of our intellectual history and heritage?

BIBLIOGRAPHY

Abell, G.
1969 *Exploration of the Universe*. New York: Holt, Rinehart, Winston.

Alexander, G.
1978 *Nature/Science Annual*, pp. 132-145. Alexandria, VA: Time/Life Books.

Allen, Richard H.
1963 *Star Names*. New York: Dover (reprint of 1899 edition).

Anders, Ferdinand
1963 *Das Pantheon der Maya*. Graz, Austria: Akademische Druck- und Verlagsanstalt.

Anderson, Arthur J. O., and Charles E. Dibble
1953 *The Florentine Codex: General History of the Things of New Spain*. Book 7, No. 14, Part VIII. Santa Fe: School of American Research.

Andrews, G. F.
1975 *Maya Cities. Place-Making and Urbanization*. Norman, OK: University of Oklahoma Press.

Anonymous Chronicler
n.d. Discourso de la Sucesión y Gobierno de los Yngas. In *Judicio de Limites entre el Peru y Bolivia*, Vol. 8, V. Maurtua (editor).

Atkinson, R. J. C.
1951 The Henge Monuments of Great Britain. In *Excavations at Dorchester, Oxon*, R. J. C. Atkinson, C. M. Piggott, and N. K. Sandars, pp. 81-107. Oxford: Ashmolean Museum.

1953 *Field Archaeology*. New York: Methuen and Company.

1966 Moonshine on Stonehenge. *Antiquity*, Vol. 40, pp. 212-216.

Aveni, Anthony F.
1972a Astronomical Tables Intended for Use in Astro-Archaeological Studies. *American Antiquity*, Vol. 37, No. 4, pp. 531-540.

1972b Computer Printout Tables of Rise-Set Azimuths and Heliacal Rise-Set Dates of Prominent Celestial Bodies. Colgate University Computer Center.

Aveni, Anthony F. [cont'd.]

1975 Possible Astronomical Orientations in Ancient Mesoamerica. In *Archaeoastronomy in Pre-Columbian America*, Anthony Aveni (editor), pp. 163-190. Austin: University of Texas Press.

1977a *Native American Astronomy* (editor). Austin: University of Texas Press.

1977b Concepts of Positional Astronomy Employed in Ancient Mesoamerican Architecture. In *Native American Astronomy*, Anthony Aveni (editor), pp. 3-19. Austin: University of Texas Press.

1977c Astronomy in Ancient America. In *In Search of Ancient Astronomies*, E. C. Krupp (editor), pp. 165-202. New York: Doubleday.

1978 Old and New World Naked Eye Astronomy. *Technology Review*, Vol. 81, No. 2, pp. 60-72.

1980 The View from the Tropics. Paper read at the AAAS Symposium on Archaeoastronomy and the Roots of Science, San Francisco.

n.d. Ancient Tropical Astronomies. *Scientific American*. (In press.)

Aveni, Anthony F., Sharon L. Gibbs, and Horst Hartung

1975 The Caracol Tower at Chichen Itzá: An Ancient Astronomical Observatory? *Science*, Vol. 188, pp. 977-985.

Aveni, Anthony F., and Horst Hartung

1976 Investigación Preliminar de las Orientaciones Astronómicas de Copán. *Yaxkin*, Vol. I, No. 3, pp. 8-13. Tegucigalpa (Honduras): Inst. Hondureño de Antropología e Historia.

1979 Some Suggestions about the Arrangement of Buildings at Palenque. *Tercera Mesa Redonda de Palenque*, Vol. IV, Merle Greene Robertson and Donnan Call Jeffers (editors), pp. 173-177. Monterey (California): Pre-Columbian Art Research, Herald Printers.

n.d. The Cross Petroglyph: An Ancient Mesoamerican Astronomical and Calendric Symbol. *Indiana No. 6*. Berlin: Ibero-Amerikanisches Institut. (In press.)

Aveni, Anthony F., H. Hartung, and B. Buckingham

1978 The Pecked Cross Symbol in Ancient Mesoamerica. *Science*, Vol. 202, pp. 267-279.

Aveni, Anthony F., and Robert M. Linsley

1972 Mound J, Monte Alban: Possible Astronomical Orientation. *American Antiquity*, Vol. 37, No. 4, pp. 528-531.

Barber, John

1973 The Orientation of the Recumbent-Stone Circles of the South-west of Ireland. *Journal of the Kerry Archaeological and Historical Society*, Vol. 6, pp. 26-39.

Barrera Vásquez, Alfredo

1965 *El Libro de los Cantares de Dzitbalché*. México, D. F.: Instituto Nacional de Antropología e Historia.

Barrera Vásquez, Alfredo, and Silvia Rendón
1948 *El Libro de los Libros de Chilam Balam*. México, D.F.: Fondo de Cultura Económica.

Barthel, T. S.
1959 Ein Frühlingsfest der Atacameños. *Zeitschrift für Ethnologie*, Band 84 Heft 1, Braunschweig.

Benchley, Elizabeth Carling
1974 *Mississippian Secondary Mound Loci: A Comparative Functional Analysis in Time-Space Perspective*. Unpublished Ph.D. dissertation, University of Wisconsin, Milwaukee.

Benedict, Ruth
1935 *Zuni Mythology*. 2 vols. New York: Columbia University Press.

Berlin, Heinrich
1968 Estudios Epigráficos II. *Antropología e Historia de Guatemala*, Vol. 20, No. 1, pp. 13-24.

Berlin, Heinrich, and David H. Kelley
1961 The 819-Day Count and Color-Direction Symbolism among the Classic Maya. *Middle American Research Institute Publication* 26, pp. 9-20. Tulane University, New Orleans.

Bernal, Ignacio
1969 *The Olmec World*. Berkeley: University of California Press.

Bertonio, Ludovico
1612 *Vocabulario de la lengua Aymara*. La Paz: Don Bosco (1956).

Betanzos, Juan de
1551 *Suma y narración de los Incas*. Madrid: Biblioteca de Autores Españoles, Cronicas de interes indigena.

Bidou, Patrice
1972 Representations de l'Espace dans la Mythologie Tatuyo (Indiens Tucana). *Journal de la Sociéte des Americanistes*, Vol. 61, pp. 45-108. Paris.

Black, Glenn A.
1967 *Angel Site: An Archaeological, Historical, and Ethnographical Study*, Vols. 1 and 2. Indiana Historical Society, Indianapolis.

Bok, B. J., and L. B. Jerome
1975 *Objections to Astrology*. Buffalo: Prometheus Books.

Bordieu, Pierre
1977 Outline of a Theory of Practice. *Cambridge Studies in Social Anthropology*, Vol. 16. Cambridge: Cambridge University Press.

Boscana, Father Geronimo
1933 *Chinigchinich; A Historical Account of the Origin, Customs, and Traditions of the Indians at the Missionary Establishment of San Juan Capistrano, Alta California*. Translated by Alfred Robinson, Annotations by John P. Harrington. Santa Ana: Fine Arts Press.

Bourne, Edward G.
1904 *Narratives of the Career of Hernando DeSoto*, Vol. 1. New York: A. S. Barnes.

Bowditch, Charles P.
1906 *The Temples of the Cross, of the Foliated Cross and of the Sun at Palenque*. Cambridge: Privately printed.

1910 *The Numeration, Calendar Systems and Astronomical Knowledge of the Maya*. Cambridge: Privately printed.

Bradfield, Maitland
1971 The Changing Pattern of Hopi Agriculture. *Royal Anthropological Institute Occasional Paper*, Vol. 20, London.

Brandt, John C., et al.
1975 Possible Rock Art Records of the Crab Nebula Supernova in the Western United States. In *Archaeoastronomy in Pre-Columbian America*, Anthony Aveni (editor), pp. 45-58. Austin: University of Texas Press.

Brandt, John C., and Ray A. Williamson
1979 The 1054 Supernova and Native American Rock Art. *Archaeoastronomy*, Supp. to *Journal for the History of Astronomy*, No. 1, pp. 1-38. Cambridge.

Brasseur de Bourbourg, C. E.
1964 *Relation des choses de Yucatan de Diego de Landa . . . accompagné de documents divers historiques et chronologiques. . . .* Paris.

Brinton, D. G.
1895 A Primer of Mayan Hieroglyphics. *University of Pennsylvania Series in Philosophy, Literature and Archeology*, Vol. 3, No. 2. Philadelphia.

1896 *Myths of the New World: A Treatise on the Symbolism and Mythology of the Red Race of America*. New York: Haskell House (reprinted in 1968).

Britt, Claude, Jr.
1975 Early Navaho Astronomical Pictographs in Canyon de Chelly, Northeastern Arizona, U.S.A. In *Archaeoastronomy in Pre-Columbian America*, Anthony Aveni (editor), pp. 89-108. Austin: University of Texas Press.

Broda De Casas, J.
1969 The Mexican Calendar as Compared to Other Mesoamerican Systems. *Acta Ethnological et Linguistica NR. 15, Series Americana* 4. Wien.

Brotherston, G.
1976 Mesoamerican Description of Space II: Signs for Direction. *Ibero-Amerikanisches Archiv*. N.F. 2, pp. 39-62.

1979 Continuity in Maya Writing: New Readings of Two Passages in the Book of Chilam Balam of Chumayel. In *Maya Archaeology and Ethnohistory*, N. Hammond and G. R. Willey (editors), pp. 241-258. Austin: University of Texas Press.

Brown, Peter Lancaster
1976 *Megaliths, Myths, and Men.* Poole, Dorset: Blandford.

Brownrigg, Leslie A.
1973 A Model of the Andean System of Time: Dispersed and Congregated Ritual Calendars. Presented at the Symposium on Andean Time: Ritual Calendars and Agricultural Cycles, 72d meeting of the American Anthropological Association, New Orleans.

Bruce, R. D., C. Robles U., and E. Ramos Chao
1971 Los Lacandones 2, Cosmovision Maya. *Instituto Nacional de Antropología y Historia, Departamento de Investigaciones Antropológicas*, Pub. 26.

Brundage, B. C.
1979 *The Fifth Sun.* Austin: University of Texas Press.

Brush, Stephen G.
1974 Should the History of Science Be Rated X? *Science*, Vol. 183, pp. 1164-1172.

Bullen, Ripley P.
1966 Stelae at the Crystal River Site, Florida. *American Antiquity*, Vol. 31, pp. 861-865.

Bunzel, Ruth L.
1932 Introduction to Zuni Ceremonialism; Zuni Katchinas; Zuni Origin Myths; Zuni Ritual Poetry. *Bureau of American Ethnology Annual Report*, Vol. 47, pp. 467-1086.

Carlson, John B.
1975 Lodestone Compass: Chinese or Olmec Primacy? *Science*, Vol. 189, pp. 753-760.

1976 Astronomical Investigations and Site Orientation Influences at Palenque. In *The Art, Iconography and Dynastic History of Palenque, Part III*, Merle Greene Robertson (editor), pp. 107-122. Pebble Beach (California): The Robert Louis Stevenson School, Pre-Columbian Art Research.

1978 On Classic Maya Monumental Recorded History. Paper presented at the Tercera Mesa Redonda de Palenque 11-12 June. To be published in the conference proceedings, Vol. V, University of Texas Press, Austin, 1980.

Caso, A.
1971 Calendrical Systems of Central Mexico. In *Handbook of Middle American Indians*, Robert Wauchope and Gordon Willey (editors), Vol. 10, pp. 333-348. Austin: University of Texas Press.

Castetter, E. F., and W. H. Bell
1942 *Pima and Papago Indian Agriculture.* Albuquerque: University of New Mexico Press.

Chamberlain, Von Del
n.d. *When Stars Came Down to Earth: Cosmology of the Skidi Pawnee Indians of North America.* Los Altos: Ballena Press. (In press.)

Charency, Hyacinthe, Comte de
1884 *Vocabulaire Français--Maya*. Alençon, E. Renaut-de-Broise, imprimeur.

Cieza de León, Pedro de
1553 *La Cronica del Peru*. Collección Austral, Espasa Calpe S.A., Buenos Aires--Mexico (1945).

Closs, Michael P.
1977 The Date-Reaching Mechanism in the Venus Table of the Dresden Codex. In *Native American Astronomy*, Anthony Aveni (editor). Austin: University of Texas Press.

1978 The Initial Series on Stela 5 at Pixoy. *American Antiquity*, Vol. 43, No. 4, pp. 690-694. Washington, D.C.

Closs, M. P., A. F. Aveni, and Bruce Crowley
n.d. The Planet Venus and Temple 22 (the Temple of Venus) at Copan. (Ms.)

Cobo, Bernabé
1653 *Historia del Nuevo Mundo*. Madrid: Biblioteca de Autores Españoles (1964).

Codex Bodley 2858
1960 Commentary and Interpretation by Alfonso Caso. Mexico City: Sociedad Mexicana de Antropología.

Codex Peresianus (Codex Paris)
1968 Bibliothèque Nationale Paris. Einleitung und Summary: F. Anders. (Codices Selecti Phototypice Impressi, Vol. 9). Graz: Akademische Druck- und Verlagsanstalt.

Codex Selden 3135 (A. 2)
1964 Commentary and Interpretation by Alfonso Caso. Mexico City: Sociedad Mexicana de Antropología.

Codex Tro-Cortesianus (Codex Madrid)
1967 Museo de América Madrid. Einleitung und Summary: M. Anders. (Codices Selecti Phototypice Impressi, Vol. 8). Graz: Akademische Druck- und Verlagsanstalt.

Codex Vindobonensis Mexicanus 1
1963 History and Description of the Manuscript by Otto Adelhofer (Codices Selecti Phototypice Impressi, Vol. 5). Graz: Akademische Druck- und Verlagsanstalt.

Coe, Michael D.
1964 The Chinampas of Mexico. *Scientific American*, Vol. 211, pp. 90-98.

1965 Archaeological Synthesis of Southern Veracruz and Tabasco. In *Handbook of Middle American Indians*, Gordon R. Willey (editor), Archaeology of Southern Mesoamerica. Vol. 3, pp. 679-715. Austin: University of Texas Press.

1966 *The Maya*. New York: Frederick A. Praeger.

1973 *The Maya Scribe and His World*. New York: Grolier Club.

Coe, Michael D. [cont'd.]
1975 Native Astronomy in Mesoamerica. In *Archaeoastronomy in Pre-Columbian America*, Anthony Aveni (editor), pp. 3-31. Austin: University of Texas Press.

1976 Early Steps in the Evolution of Maya Writing. In *Origins of Religious Art and Iconography in Pre-Classic Mesoamerica*, H. B. Nicholson (editor), pp. 107-122. Los Angeles: UCLA Latin American Center Publications.

1978 *Lords of the Underworld: Masterpieces of Classic Maya Ceramics*. Princeton: Princeton University Press.

Cohodas, M.
1974 The Iconography of the Panels of the Sun, Cross, and Foliated Cross at Palenque, Part II. In *Primera Mesa Redonda de Palenque Part I*, M. G. Robertson (editor), pp. 85-108. Pebble Beach (California): The Robert Louis Stevenson School.

Colton, R., and R. L. Martin
1967 Eclipse Cycles and Eclipses at Stonehenge. *Nature*, Vol. 213, pp. 476-478.

Cook, Warren L. (editor)
1978 *Ancient Vermont*. Castleton: Castleton State College.

Cope, Leona
1919 Calendars of the Indians North of Mexico. *University of California Publications in American Archaeology and Ethnology*, Vol. 16, pp. 119-176.

Covarrubias, Miguel
1942 Origen y desarrollo del estilo artístico "Olmeca." *Sociedad Mexicana de Antropología, Mexico*, pp. 46-49.

Cowan, Thaddeus M.
1975 Effigy Mounds and Stellar Representations. In *Archaeoastronomy in Pre-Columbian America*, Anthony Aveni (editor). Austin: University of Texas Press.

Craine, Eugene R., and Reginald C. Reindorp
1979 *The Codex Pérez and the Chilam Balam of Maní*. Norman: University of Oklahoma Press.

Crawford, O. G. S.
1957 *The Eye Goddess*. London: Phoenix House.

Crow Wing
1925 *A Pueblo Indian Journal, 1920-1921* (E. C. Parsons, editor), *Memoirs of the American Anthropological Association*, Vol. 32.

Curtis, Frank H.
1922 *The North American Indian, Vol. 12: The Hopi*. New York: The Plimpton Press. (Reprinted 1970.)

Cushing, Frank H.
1896 Outlines of Zuni Creation Myths. *Bureau of American Ethnology Annual Report 13*, pp. 321-447.

1901 *Zuni Folk Tales*. New York: G. P. Putnam's Sons.

1941 *My Adventures in Zuni*. Santa Fe: Peripatetic Press.

Deregowski, Jan B.
1972 Pictorial Perception and Culture. *Scientific American*, Vol. 227, No. 5, pp. 82-88.

Diccionario de la Lengua Española
1970 Madrid: Real Academia Española.

DiPeso, Charles
1968 Casas Grandes and the Gran Chichimeca. *El Palacio*, Vol. 75, No. 4, pp. 45-61.

Dix, B. E.
1978a Possible Heliolithic Sites in Vermont. *Archaeoastronomy, Bulletin of the Center for Archaeoastronomy*, Vol. 2, No. 1.

1978b Possible Calendar Sites in Vermont. In *Ancient Vermont*, W. E. Cook (editor). Castleton: Castleton State College.

Dix, B. E., and J. W. Mavor, Jr.
1980 Possible Astronomical Alignments, Date and Origins of the Pearson Stone Chamber. *Bulletin, Early Sites Research Society*, Vol. 7, No. 3.

Dix, B. E., J. W. Mavor, Jr., and J. Potts
1980 Progress Report on the Calendar I Area in Central Vermont. *Archaeoastronomy, Bulletin of the Center for Archaeoastronomy*, Vol. 3, No. 1.

Douglass, William B.
1912 A World-Quarter Shrine of the Tewa Indians. *Records of the Past*, Vol. 2, pp. 159-173.

Dow, J.
1967 Astronomical Orientations at Teotihuacán: A Case Study in Astroarchaeology. *American Antiquity*, Vol. 32, pp. 326-334.

Doyel, D. E.
1977 Rillito and Rincon Period Settlement Systems in the Middle Santa Cruz River Valley: Alternative Models. *The Kiva*, Vol. 43, No. 2, pp. 93-109. Tucson.

Dozier, E. P.
1966 *Hano, A Tewa Indian Community in Arizona*. New York: Holt, Rinehart and Winston, Inc.

DuBois, Constance Goddard
1908 The Religion of the Luiseño Indians of Southern California. *University of California Publications in American Archaeology and Ethnology*, Vol. 8, No. 3, pp. 69-166.

Dumont, Jean-Paul
1976 *Under the Rainbow: Nature and Supernature among the Panare Indians.* Austin: University of Texas Press.

Dutton, Bertha P.
1963 *Sun Father's Way.* Albuquerque: University of New Mexico Press.

Earls, John
1972 *Andean Continuum Cosmology.* Ph.D. dissertation, University of Illinois, Urbana-Champaign.

Eddy, J. A.
1974 Astronomical Alignment of the Big Horn Medicine Wheel. *Science*, Vol. 184, pp. 1035-1043.

1977 Medicine Wheels and Plains Indian Astronomy. In *Native American Astronomy*, Anthony Aveni (editor), pp. 147-169. Austin: University of Texas Press.

Edmonson, M. S.
1976 The Maya Calendar Reform of 11.16.0.0.0. *Current Anthropology*, Vol. 17, No. 4, pp. 713-717.

n.d. *The Ancient Future of the Itzá: The Book of Chilam Balam of Tizimin.* (Ms.)

Eggan, Fred
1950 *Social Organization of the Western Pueblos.* Chicago: University of Chicago Press.

Eidenbach, Peter L.
1979 Wizard's Roost and Wally's Dome, Continuing Investigations of Prehistoric Observatory Sites in the Sacramento Mountains, New Mexico. In *Jornada Mogollon Archaeology*, Beckett and Wiseman (editors). New Mexico State University.

Ekholm, S. M.
1969 Mound 30a and the Early Preclassic Ceramic Sequence of Izapa, Chiapas, Mexico. *Papers of the New World Archaeological Foundation*, No. 25. Provo, Utah.

Ellis, Florence H.
1975 A Thousand Years of the Pueblo Sun-Moon-Star Calendar. In *Archaeoastronomy in Pre-Columbian America*, Anthony Aveni (editor). Austin: University of Texas Press.

1976 Datable Ritual Components Proclaiming Mexican Influence in the Upper Rio Grande of New Mexico. In *Collected Papers in Honor of Marjorie Ferguson Lambert*, pp. 85-107. Albuquerque: Papers of the Archaeological Society of New Mexico.

Escalona Ramos, A.
1940 *Cronologia y Astronomia Maya-Mexica.* Mexico.

Farabee, William Curtis
1918 *The Central Arawaks. University Museum Anthropological Publications*, Vol. 9. Philadelphia.

Farrer, Claire R.
1977 *Play and Inter-ethnic Communication: A Practical Ethnography of the Mescalero Apache*. Ph.D. dissertation, University of Texas, Austin.

1978 Mescalero Ritual Dance: A Four Part Fugue. In *Discovery 1978*, pp. 1-13. School of American Research, Santa Fe, New Mexico.

1980 Singing for Life: The Mescalero Apache Girls' Puberty Ceremony. In *Southwestern Indian Ritual Drama*, Charlotte J. Frisbie (editor), pp. 125-159. Albuquerque and Santa Fe: The University of New Mexico Press and the School of American Research, Advanced Seminar Series.

Fell, Barry
1976 *America, B.C.* New York: New York Times Quadrangle Books.

Ferdon, E. N., Jr.
1955 A Trial Survey of Mexican-Southwestern Architectural Parallels. *School of American Research Monograph* 21, Santa Fe.

Fewkes, Jesse Walter
1892 A Few Summer Ceremonials at the Tusayan Pueblos. *American Ethnology and Archaeology*, Vol. 2, pp. 1-158.

Feyerabend, Paul
1975 *Against Method*. London: Verso.

Fisher, Howard J.
1978 Geometry and Astronomy at Casa Rinconada. *Archaeoastronomy, Bulletin of the Center for Archaeoastronomy*, Vol. 1, No. 3, p. 5.

Fitzhugh, W.
1978 Maritime Archaic Cultures of the Central and Northern Labrador Coast. *Arctic Anthropology*, Vol. 15, No. 2.

Forbes, Jack D.
1971 *Apache, Navaho, and Spaniard*. Norman: University of Oklahoma Press (first published 1960).

Förstemann, Ernst W.
1887 *Zur Entzifferung der Mayahandschriften*. Dresden.

1904 Aids to the Deciphering of the Maya Manuscripts. *Bureau of American Ethnology Bulletin* 28, pp. 397-472. Originally published in 1895 in *Zur Entzifferung der Mayahandschriften*, Vol. 5.

1906 Commentary on the Maya Manuscript in the Royal Public Library of Dresden. *Papers of the Peabody Museum of Harvard University*, Vol. 4, No. 2.

Fowler, Melvin
1977 The Cahokia Site. In *Explorations into Cahokia Archaeology*. Illinois Archaeological Survey, Bulletin 7, pp. 1-30.

1978 Cahokia and the American Bottom: Settlement Archeology. In *Mississippian Settlement Patterns*, Bruce D. Smith (editor), pp. 455-478. New York: Academic Press.

Gadow, H. F.
1908 *Travels Through Southern Mexico*. New York: C. Scribner and Sons.

Garcilaso de la Vega, Gomez S.
1951 *The Florida of the Inca*, John Varner and Jeanette Varner (translators and editors). Austin: University of Texas Press.

1969 *Royal Commentaries of the Incas*, 2 vols, Harold V. Livermore (translator). Austin: University of Texas Press.

Garibay, A. M.
1957 Supervivencias de cultura intelectual Precolumbina entre los Otomies de Huizquilucan. *Instituto Indigenista Interamericano, Ediciones especiales 33*, Mexico.

Gates, William
1931 *An Outline Dictionary of Maya Glyphs*. Baltimore: The Johns Hopkins Press. (Reprinted in 1978; New York: Dover Publications, Inc.)

Gentleman of Elvas
1907 The Narrative of the Expedition of Hernando DeSoto. In *Spanish Explorers in the Southern United States 1528-1543*, Frederick Hodge and Theodore Lewis (editors). New York: Barnes and Noble.

Gibbs, Sharon L.
1977 Mesoamerican Calendrics as Evidence of Astronomical Activity. In *Native American Astronomy*, Anthony Aveni (editor), pp. 21-35. Austin: University of Texas Press.

Gifford, Edward Winslow
1918 Clans and Moieties in Southern California. *University of California Publications in American Archaeology and Ethnology*, Vol. 14, No. 2. Berkeley.

1940 Culture Element Distributions:XII--Apache-Pueblo. *University of California Anthropological Records*, Vol. 4, pp. 1-208. Berkeley.

Girard, R.
1949 *Los Chortis Ante el Problema Maya*, Vol. 2. Antigua Libreria Robredo, Mexico.

1962 *Los Mayas Eternos*. Libro Mex, Mexico.

Gladwin, H. S.
1929 The Red-on-Buff Culture of the Gila Basin. *Medallion Papers*, No. 3. Tucson: University of Arizona Press.

Gladwin, H. S., E. W. Haury, E. B. Sayles, and Nora Gladwin
1938 Excavations at Snaketown, Material Culture. *Medallion Papers*, No. 25. Tucson: University of Arizona Press.

Gleadow, R.
1968 *The Origin of the Zodiac*. New York: Castle Books.

Glendrop, P.
1972 *El México Antiguo*. Mexico: Editorial Trillas.

Gonzalez Holguin, Diego
1608 *Vocabulario de la Lengua General del Peru Llamada Qquichua, o del Inca*. Lima: Imprenta Santa María (1952).

Goodman, J. T.
1887 The Archaic Maya Inscriptions. In Appendix to *Archeology* by A. P. Maudsley. London.

1905 Maya Dates. *American Anthropologist*, Vol. 7, pp. 642-647.

Goodwin, W. B.
1946 *The Ruins of Great Ireland in New England*. Boston: Meador.

Gossen, Gary H.
1974a A Chamula Solar Calendar Board from Chiapas, Mexico. In *Mesoamerican Archaeology: New Approaches*, N. Hammond (editor), pp. 217-253. Austin: University of Texas Press.

1974b *Chamulas in the World of the Sun: Time and Space in a Maya Oral Tradition*. Cambridge: Harvard University Press.

Griffin, J. B.
1971 The Northeast Woodlands Area. In *Prehistoric Man in the New World*, Jesse Jennings and Edward Norbeck (editors), pp. 223-258. Chicago: University of Chicago Press.

Grinsell, Leslie V.
1976 *Folklore of Prehistoric Sites in Britain*. London: David & Charles.

Guamán Poma de Ayala, Felipe
1584-1614 *El Primer Nueva Coronica y Buen Gobierno*. Paris: Musée de l'Homme (1936).

Gwynn, Edward
1924 The Metrical Dindsenchas. Part IV. *Royal Irish Academy Todd Lecture Series* 11. Dublin.

Hadingham, E.
1975 *Circles and Standing Stones*. New York: Walker and Company.

Hanson, Lee H.
1975 The Buffalo Site. *West Virginia Geological and Economic Survey, Report of Archaeological Investigations* No. 5. Morgantown, West Virginia.

Hardman, Clark
1971 The Primitive Solar Observatory at Crystal River and Its Implications. *Florida Anthropologist*, Vol. 24, pp. 135-167.

Harrington, John P.
1916 The Ethnogeography of the Tewa Indians. *Bureau of American Ethnology Annual Report 29*, pp. 37-69.

n.d. Unpublished manuscripts and papers pertaining to field work at Zuni Indian Reservation. National Anthropological Archives, Smithsonian Institution.

Hartmann, W. K.
1978 *Astronomy: The Cosmic Journey*. Belmont: Wadsworth.

Hartung, Horst
1971 Die Zeremonialzentren der Maya. Ein Beitrag zur Untersuchung der Planungsprinzipien. Graz (Austria): Akademische Druck- und Verlagsanstalt.

1974 Monte Alban--Concepto Espacial de un Centro Ceremonial Zapoteco. *Centro de Investigaciones Históricas y Estéticas, Bolletin 19*. Caracas.

1975 A Scheme of Probable Astronomical Projections in Mesoamerican Architecture. In *Archaeoastronomy in Pre-Columbian America*, A. F. Aveni (editor), pp. 191-204. Austin: University of Texas Press.

1976 El Espacio Exterior en el Centro Ceremonial de Palenque. In *The Art, Iconography and Dynastic History of Palenque*, Merle Greene Robertson (editor), pp. 123-135. Pebble Beach (California): The Robert Louis Stevenson School, Pre-Columbian Art Research.

1977 Ancient Maya Architecture and Planning: Possibilities and Limitations for Astronomical Studies. In *Native American Astronomy*, A. F. Aveni (editor), pp. 111-129. Austin: University of Texas Press.

Hatch, M. P.
1975 An Astronomical Calendar in a Portion of the Madrid Codex. In *Archaeoastronomy in Pre-Columbian America*, A. F. Aveni (editor), pp. 283-240. Austin: University of Texas Press.

Haury, E. W.
1945 The Excavation of Los Muertos and Neighboring Ruins in the Salt River Valley, Southern Arizona. *Papers of the Peabody Museum of American Archaeology and Ethnology*, Vol. 24, No. 1. Cambridge: Harvard University Press.

1976 *The Hohokam: Desert Farmers and Craftsmen*. Tucson: University of Arizona Press.

Hawkes, J.
1967 God in the Machine. *Antiquity*, Vol. 41, pp. 174-180.

Hawkins, G. S., and S. K. Rosenthal
1967 5,000 and 10,000-Year Star Catalogs. *Smithsonian Contributions to Astrophysics*, Vol. 10, No. 2, Washington, D.C.

Hawkins, G. S., and John B. White
1965 *Stonehenge Decoded*. New York: Doubleday.

Hedges, Ken
1977 Preliminary Report on a Winter Solstice Rock Painting at La Rumorosa, Baja California. Paper presented at the Fourth Annual Rock Art Symposium, May 1977, Tempe, Arizona.

Heilman, James M.
1975 Ceramic and Spatial Patterning at Incinerator Site (33 My 57). Paper presented at the 40th Annual Meeting of the Society for American Archaeology, Dallas, Texas.

Heisner, C. B.
1973 *Seed to Civilization: The Story of Man's Food.* San Francisco: W. H. Freeman.

Hewitt, J. M.
1977 The Sedentary-Classic Transition in the Tucson Basin: Indications from Ceramics. Tucson: Arizona State Museum. (Ms.)

Heyden, D.
1975 An Interpretation of the Cave Underneath the Pyramid of the Sun at Teotihuacán, Mexico. *American Antiquity*, Vol. 40, pp. 131-147.

Heyden, D., and P. Gendrop
1975 *Pre-Columbian Architecture of Mesoamerica.* New York: Harry N. Abrams.

Hibben, Frank
1975 *Kiva Art of the Anasazi at Pottery Mound.* Las Vegas: K. C. Publications.

Hicks, Ronald
1977 Thom's Megalithic Yard and Traditional Measurements. *Irish Archaeological Research Forum*, Vol. 4, pp. 1-7.

Hopi Indians Collection
1921-1922 Hopi Indians Collection (NAU 31), Special Collections Division, Northern Arizona University Library, Flagstaff.

Howard, John H.
1968 The Southeastern Ceremonial Complex and Its Interpretation. *Memoir of the Missouri Archaeological Society*, No. 6.

Hoyle, Fred
1966a Stonehenge--An Eclipse Predictor. *Nature*, Vol. 211, pp. 454-456.

1966b Speculations on Stonehenge. *Antiquity*, Vol. 40, pp. 262-276.

1977 *On Stonehenge.* San Francisco: W. H. Freeman.

Hudson, Travis, and Thomas Blackburn
1978 The Integration of Myth and Ritual in South-Central California: The "Northern Complex." *Journal of California Anthropology*, Vol. 5, pp. 225-250.

Hudson, Travis, Georgia Lee, and Ken Hedges
1979 Solstice Observers and Observatories in Native California. *Journal of California and Great Basin Anthropology*, Vol. 1, pp. 39-63.

Hudson, Travis, and Earnest Underhay
1978 *Crystals in the Sky.* Socorro: Ballena Press.

Hunt, Eva
1977 *The Transformation of the Hummingbird: Cultural Roots of a Zinacantecan Mythical Poem.* Ithaca: Cornell University Press.

Icazbalceta, Joaquín García
1891 Historia de los Mexicanos por sus Pinturas. In *Nueva Colección de Documentos para la Historia*, Vol. 3, pp. 207-240. México, D.F.: Editorial Salvador Chávez Hayhoe.

Isbell, B. J.
1978 *To Defend Ourselves*. Austin: Foundation for Latin American Studies.

Isbell, William H.
1978 Cosmological Order Expressed in Prehistoric Ceremonial Centers. *Actes du XLII*[e] *Congrès International des Américanistes*, Vol. 4, pp. 269-297.

Ives, Joseph C.
1861 *Report upon the Colorado River of the West, Explored in 1857 and 1858 . . .*, Washington.

Izikowitz, K. G.
1955 Rhythmical Aspects of Canella Life. Proceedings of the 31st International Congress of Americanists, pp. 195-209. *31 Int. Cong. of Americanists*. São Paulo.

James, Harry C.
1974 *Pages from Hopi History*. Tucson: University of Arizona Press.

Johnson, Mary Elizabeth
1914 *Indian Legends of the Cuyamaca Mountains*. San Diego: Privately printed.

Jones, C. C.
1873 *Antiquities of the Southern Indians*. New York.

Joralemon, Peter David
1971 A Study of Olmec Iconography. *Studies in Pre-Columbian Art and Archaeology*, No. 7. Washington, D.C.: Dumbarton Oaks.

1976 The Olmec Dragon: A Study in Pre-Columbian Iconography. In *Origins of Religious Art and Iconography in Pre-Classic Mesoamerica*, H. B. Nicholson (editor), pp. 27-71. Los Angeles: UCLA Latin American Center Publications.

Kehoe, Thomas F., and Alice B. Kehoe
1979 Solstice-Aligned Boulder Configurations in Saskatchewan. *National Museum of Man Mercury Series, Canadian Ethnology Service Paper* No. 48.

Kelley, David B.
1962 A History of the Decipherment of Maya Script. *Anthropological Linguistics*, Vol. 4, No. 8, pp. 1-48.

1965 The Birth of the Gods at Palenque. In *Estudios de Cultura Maya*, Vol. 5, pp. 93-134.

1975 Planetary Data on Caracol Stela 3. In *Archaeoastronomy in Pre-Columbian America*, Anthony F. Aveni (editor), pp. 257-262. Austin: University of Texas Press.

Kelley, David H. [cont'd.]
1976 *Deciphering the Maya Script.* Austin: University of Texas Press.

1977 Maya Astronomical Tables and Inscriptions. In *Native American Astronomy*, Anthony Aveni (editor), pp. 57-73. Austin: University of Texas Press.

1978 Comment. *Current Anthropology*, Vol. 19, No. 3, pp. 573-574.

Kelley, David H., and K. Ann Kerr
1973 Maya Astronomy and Astronomical Glyphs. In *Mesoamerican Writing Systems*, Elizabeth P. Benson (editor), pp. 179-215. Washington, D.C.: Dumbarton Oaks Research Library and Collections.

Kelly, J. Charles, and Ellen A. Kelly
1974 An Alternative Hypothesis for the Explanation of Anasazi Culture History. In *Collected Papers in Honor of Florence Hawley Ellis*, pp. 178-223. Albuquerque: Papers of the Archaeological Society of New Mexico 2.

Kingsborough, E. K.
1831-1838 *Antiquities of Mexico.* 9 vols. London.

Kissam, P.
1971 *Surveying Practice.* New York: McGraw-Hill Book Company.

Knick, Stanley G.
1977 Paleopathology of North American Indians at Incinerator Site. Paper presented at the American Association of Physical Anthropologists Annual Meeting, St. Louis, Missouri.

Koenig, Seymour H.
1979 Stars, Crescents, and Supernovae in Southwestern Indian Art. *Archaeoastronomy*, Supp. to *Journal for the History of Astronomy*, No. 1, pp. 39-50. Cambridge.

Kroeber, Alfred L.
1925 Handbook of the Indians of California. *Bureau of American Ethnology Bulletin* 78. Washington, D.C.

Krupp, Edwin C.
1977 Cahokia Corn, Commerce and the Cosmos. *Griffith Observer*, Vol. 41, No. 5, pp. 10-20.

Kubler, George
1967 The Iconography of the Art of Teotihuacán. *Studies in Pre-Columbian Art and Archaeology*, No. 4. Washington, D.C.: Dumbarton Oaks.

Kubler, G., and C. Gibson
1951 The Tovar Calendar. *Memoirs of the Connecticut Academy of Arts and Sciences*, Vol. 11. New Haven.

Kuhn, Thomas S.
1962 *The Structure of Scientific Revolutions.* 2d ed. Chicago: The University of Chicago.

Lacombe, C. H., and S. S. Block
1975 The Maya Calendar Decoded. *Museum* (November). Miami Museum of Science.

LaFarge, O., and D. Byers
1931 The Year Bearer's People. *Middle American Research Series*, Pub. 3. New Orleans: Tulane University.

Lamb, Weldon
1979 Términos olvidados de la astronomía maya. *Boletín de la Escuela de Ciencias Antropológicas de la Universidad de Yucatán*, Vol. 6, No. 36 (Mayo-Junio), pp. 19-32.

n.d. Yucatec Maya Astronomy since the Conquest. Las Cruces, New Mexico. (Ms.)

Laughlin, R. M.
1969 The Tzotzil. In *Handbook of the Middle American Indians*, Vol. 7, R. Wauchope (editor), pp. 152-194. Austin: University of Texas Press.

Leach, Edmund
1976 *Culture and Communication: The Logic by Which Symbols Are Connected.* Cambridge: Cambridge University Press.

Leal, Marcia Castro
1973 *El Juego de Pelota--The Ball Game.* Translated by Doris Heyden. Mexico, D.F.: Ediciones Lara, S.A.

León, Juan de
1954 *Diccionario Quiché-Español.* Guatemala: Guatemala Editorial Landivar.

Leopold, A. S.
1959 *Wildlife of Mexico.* Berkeley: University of California Press.

Lévi-Strauss, Claude
1948 The Nambicuara. *Handbook of South American Indians*, Vol. 3, Bureau of American Ethnology, pp. 361-370.

1964 *Tristes Tropiques.* New York: Atheneum.

1969 *The Raw and the Cooked.* New York: Harper & Row.

1973 *From Honey to Ashes.* New York: Harper & Row.

Lipkind, William
1940 Carajá Cosmography. *Journal of American Folklore*, Vol. 53, pp. 248-251.

Lira, Jorge A.
1944 *Diccionairo Kkechwa-Español.* Tucuman: Universidad de Tucuman.

1946 *Farmacopea tradicional indigena y practicas rituales.* Lima.

Lockyer, J. N.
1894 *The Dawn of Astronomy.* London: Cassell. (Reprinted by MIT Press, Cambridge, MA, 1964.)

Lockyer, J. N. [cont'd.]

1909 *Stonehenge and Other British Stone Monuments Astronomically Considered.* London: Macmillan.

Lothrop, S. K.

1930 A Modern Survival of the Ancient Maya Calendar. *Proceedings of the XXIII International Congress of Americanists*, pp. 652-655. New York.

Lounsbury, Floyd G.

1973 On the Derivation and Reading of the "Ben-Ich" Prefix. In *Meso-American Writing Systems*, E. Benson (editor). Washington, D.C.: Dumbarton Oaks.

1976 A Rational for the Initial Date of the Temple of the Cross at Palenque. In *The Art, Iconography and Dynastic History of Palenque, Part III*, Merle G. Robertson (editor), pp. 211-224. Pebble Beach, CA: The Robert Louis Stevenson School, Pre-Columbian Art Research.

1978 Maya Numeration, Computation, and Calendrical Astronomy. In *Dictionary of Scientific Biography*, Vol. 15, pp. 759-818.

MacGowan, K.

1945 The Orientation of Middle American Sites. *American Antiquity*, Vol. 11, p. 118.

MacKie, Euan

1977 *Science and Society in Prehistoric Britain.* New York: St. Martin's Press.

MacNeill, Eóin

1926-1928 On the Calendar of Coligny. *Ériu*, Vol. 10, pp. 1-67.

Magin, P. Juan

1740 *Revista de Indias*, Vol. 1, No. 1, pp. 151-185. Madrid: Instituto Fernando de Oriedo, 1940.

Makemson, Maud

1946 The Maya Correlation Problem. *Publications of the Vassar College Observatory*, No. 5. Poughkeepsie.

Malmström, V. H.

1973 Origin of the Mesoamerican 260-Day Calendar. *Science*, Vol. 181, pp. 939-941.

1978 A Reconstruction of the Chronology of Mesoamerican Calendrical Systems. *Journal for the History of Astronomy*, Vol. 9, pp. 105-116.

Marcus, Joyce

1976a *Emblem and State in the Classic Maya Lowlands.* Washington, D.C.: Dumbarton Oaks.

1976b The Origins of Mesoamerican Writing. *Annual Review of Anthropology*, Vol. 5, pp. 35-67.

Markham, C. R.

1910 *The Incas of Peru.* London and New York: E. P. Dutton and Company.

Marquina, I., and L. Ruiz
1934 La Orientación de las Pirámides. *Proceedings of the 25th International Congress of Americanists*, pp. 101-106. Buenos Aires: Universidad de la Plata.

Marshack, Alexander
1972 *The Roots of Civilization.* New York and London: McGraw-Hill and Weidenfeld and Nicholson.

Marshall, James A.
1978 American Indian Geometry. *Ohio Archaeologist*, Vol. 28, No. 1, pp. 29-33.

1979 Geometry of the Hopewell Earthworks. *Early Man* (Spring), pp. 1-5.

Matheny, R. T.
1976 Maya Lowland Hydraulic Systems. *Science*, Vol. 193, pp. 639-646.

Maudslay, A. P.
1889-1902 *Archeology.* 5 vols. London.

Maurtua, V. (editor)
1906 Discurso de la Sucesión y gobierno de los Yngas (Manuscrito anonimo, sin fecha). *Juicio de Limites entre el Peru y Bolivia; Prueba Peruana*, Vol. 8, *Chunchos*, pp. 149-165.

Maybury-Lewis, David
1971 *Akwẽ-Shavante Society.* Oxford: Clarendon Press.

Mayer, Dorothy
1975 Star Patterns in Great Basin Petroglyphs. In *Archaeoastronomy in Pre-Columbian America*, Anthony Aveni (editor), pp. 89-130. Austin: University of Texas Press.

1976 Some Comments on an Astronomical Petroglyph Panel in Capitol Reef National Park, Reported and Described by Klaus F. Wellmann. *Southwestern Lore*, Vol. 52, pp. 14-20.

1977 An Examination of Miller's Hypothesis. In *Native American Astronomy*, Anthony Aveni (editor), pp. 59-87. Austin: University of Texas Press.

1979 Miller's Hypothesis: Some California and Nevada Evidence. *Archaeoastronomy*, Supp. to *Journal for the History of Astronomy*, No. 1, pp. 551-574.

McCluskey, Stephen C.
1977 The Astronomy of the Hopi Indians. *Journal for the History of Astronomy*, Vol. 8, Part 2, pp. 174-195.

1980 Evangelists, Educators, Ethnographers and the Establishment of the Hopi Reservation. *Journal of Arizona History*, Vol. 21, pp. 363-390.

McNitt, Frank
1962 *The Indian Traders.* Norman: University of Oklahoma Press.

Means, P.
1931 *Ancient Civilizations of the Andes*. New York: Scribner's.

Meeker, Ralph L., W. H. Brug, F. Matanzo, R. L. Googins, and G. M. Schaffer
1973 *Soil Survey of Fayette County, Ohio*. Washington, D.C.: U.S. Department of Agriculture Soil Conservation Service.

Meinshausen, Martin
1913 Über Sonnen-und Mondfinsternisse in der Dresdener Mayahandschrift. *Zeitschrift für Ethnologie*, Vol. 45, pp. 221-227.

Mejía, Medardo
1969 *Historia de Honduras*, Vol. 1. Honduras, C.A.: Editorial Andrade.

Melody, Michael E.
1977 *The Apaches: A Critical Bibliography*. Bloomington: Indiana University Press.

Merrill, R. H.
1945 Maya Sun Calendar Dictum Disproved. *American Antiquity*, Vol. 10, pp. 307-311.

Milbrath, S.
n.d.a Star Gods and Astronomy of the Aztecs. In *Religion and History in Actec Art*, C. F. Klein (editor). (In press.)

n.d.b The Star Calendar of the Codex Madrid. In *La Antropología Americanista en la Actualidad*, homenaje a Raphäel Girard. (In press.)

Miles, S. W.
1957 The Sixteenth-Century Pokom-Maya: A Documentary Analysis of Social Structure and Archaeological Setting. *Transactions of the American Philosophical Society*, Vol. 47, Pt. 4. Philadelphia.

1965 Summary of Preconquest Ethnology of the Guatemala-Chiapas Highlands and Pacific Slopes. In *Handbook of Middle American Indians*, Robert Wauchope and G. Willey (editors), Vol. 2, pp. 276-287. Austin: University of Texas Press.

Miller, William C.
1955 Two Possible Astronomical Pictographs Found in Northern Arizona. *Plateau*, Vol. 27, pp. 6-12.

Millon, René
1967 Teotihuacán. *Scientific American*, Vol. 216, No. 6, pp. 38-63.

1973 *Urbanization at Teotihuacán* (editor). Vol. 1, Parts 1 and 2. Austin: University of Texas Press.

Mills, William C.
1904 Exploration of the Gartner Mound and Village Site. *Ohio Archaeological and Historical Quarterly*, Vol. 13. Columbus, Ohio.

Molina, Cristobal de
1573 *Relación de las fabulas y ritos de los Incas*. Anotaciones y concordancias por Horacio H. Urteaga. Lima: Sanmarti (1916).

Molloy, J. P., R. E. White, T. P. Culhert, and D. H. Kayser
1973 The Casa Grande Archaeological Zone Pre-Columbian Astronomical Observatory. (Preprint.)

Morey, Robert V.
1971 Guahibo Time - Reckoning. *Anthropological Quarterly*, Vol. 44, No. 1, pp. 22-36.

Morley, Sylvanus G.
1946 *The Ancient Maya*. Stanford: Stanford University.

Motul Dictionary
1929 *Diccionario de Motul, maya-español, español-maya, atribuído a Fray Antonio de Cuidad Real; y Arte de la Lenga Maya, por Juan Coronel*, J. Martínez Hernández (editor). Merida: Talleres de la Compañía Tipográfica Yucateca, S.A.

n.d. Seventeenth century manuscript copy of lost sixteenth century original, in John Carter Brown Library, Providence.

Mulholland, D.
1978 Archaeoastronomy. *Science*, Vol. 199, pp. 1197-1198.

Nebrija, Elio Antonio de
1951 *Vocabulario Español-Latino*. Salamanca, ca. 1495. Facsimile reprint by Real Academia Española.

Neuenswander, Helen
1978 Vestiges of Early Maya Time Concepts in a Contemporary Maya (Cubulco Achi) Community: Implications for Epigraphy. Paper presented at the 77th American Anthropological Association Meetings, Los Angeles.

Newham, C. A.
1966 Stonehenge: A Neolithic "Observatory." *Nature*, Vol. 211, pp. 456-458.

Newman, Stanley A.
1958 Zuni Dictionary. *International Journal of American Linguistics*, Vol. 24, No. 1, pp. 1-22.

New York Times
1978 A Little Stonehenge Found in Africa (May).

Nimuendajú, Curt
1942 *The Sherénte*. Publications of the Frederick Webb Hodge Anniversary Publication Fund, Los Angeles.

1946 The Eastern Timbira. *University of California Publications in American Archaeology and Ethnology*, Vol. 41. Berkeley.

1967 *The Apinayé*. Oosterhout N.B.-The Netherlands: Anthropological Publications.

Norrish, Richard
1978 Woodhenge--Work of a Genius. *Cahokian* (February), pp. 21-4. Collinsville, IL: Cahokia Mounds Museum Society.

Nuttall, Zelia

1906 Astronomical Methods of the Ancient Mexicans. *Boas Anniversary Volume*, pp. 290-298.

1928 Nouvelles Lumieres sur les Civilisations Americaines et le Systeme du Calendrier. *Proceedings of the 22d International Congress of Americanists*, pp. 119-148. Rome.

1975 *The Codex Nuttall: A Picture Manuscript from Ancient Mexico.* New York: Dover Publications. (First published in 1902 by the Peabody Museum of American Archaeology and Ethnology, Cambridge.)

O'Kelly, Claire

1973 Passage-Grave Art in the Boyne Valley, Ireland. *Proceedings of the Prehistoric Society*, Vol. 39, pp. 354-382.

Opler, Morris E.

1965 *An Apache Life-Way.* New York: Cooper Square Publishers. (First published in 1941.)

1969 *Apache Odyssey: A Journey Between Two Worlds.* New York: Holt, Rinehart, and Winston.

Opperman, Renee

1979 Pima and Papago Cosmology. Paper presented at the conference, Archaeoastronomy in the Americas, Santa Fe, New Mexico.

Ortiz, A.

1969 *The Tewa World.* Chicago: University of Chicago Press.

1972 Ritual Drama and the Pueblo World View. In *New Perspectives on the Pueblos*, Alfonso Ortiz (editor), pp. 135-162. Albuquerque: University of New Mexico Press.

Ovenden, M., and D. A. Rodger

1979 Megaliths and Medicine Wheels. *Bulletin of the American Astronomical Society*, Vol. 10, No. 4, pp. 611-612.

Owen, Nancy Kelly

1975 The Use of Eclipse Data to Determine the Maya Correlation Number. In *Archaeoastronomy in Pre-Columbian America*, Anthony Aveni (editor), pp. 237-246. Austin: University of Texas Press.

Packard, Gar, and Maggy Packard

1974 *Suns and Serpents.* Santa Fe: Packard Publications.

Paddock, John

1966 *Ancient Oaxaca.* Stanford: Stanford University.

Parker, R. A.

1974 Ancient Egyptian Astronomy. *Philosophical Transactions of the Royal Society of London, A*, Vol. 276, pp. 51-65.

Parsons, Elsie C.

1923 The Origin Myth of Zuni. *Journal of American Folklore*, Vol. 33, pp. 135-162.

Parsons, Elsie C. [cont'd.]
1930a Isleta, New Mexico. *Bureau of American Ethnology Annual Reports*, Vol. 47, pp. 193-466.

1930b Zuni Tales. *Journal of American Folklore*, Vol. 43, pp. 1-58.

1939 *Pueblo Indian Religion*. 2 vols. Chicago: University of Chicago Press.

Pasachoff, J. J., and M. L. Kutner
1978 *University Astronomy*. Philadelphia: W. B. Saunders.

Peebles, Christopher S.
1975 *Moundville. The Organization of a Prehistoric Community and Culture.* University of Windsor Press.

1978 Determinants of Settlement Size and Location in the Moundville Phase. In *Mississippian Settlement Patterns*, Bruce D. Smith (editor), pp. 369-416. New York: Academic Press.

Perez Toro, A.
1942 *La Milpa*. Mérida, Yucatan.

Pío Pérez, Juan
1842 Ancient Chronology of Yucatan; or a True Exposition of the Method Used by the Indians for Computing Time. In *Incidents of Travel in Yucatan*, J. L. Stephens (editor), Vol. I, pp. 287-303. Dover, NY (1963 unabridged republication).

1866-1877 *Diccionario de la Lengua Maya*. Mérida: Imprenta Literaria de Juan F. Molina Solis.

1898 Coordinación alfabética de las voces del idioma maya. Mérida: Imprenta de la Ermita.

n.d. Apuntes al *Diccionario de la Lengua Maya*. (Ms.)

Polo de Ondegardo, Juan
1559-1560 Los errores y supersticiones de los indios, sacadas del tratado y averiguación que hizo el Licenciado Polo, H. H. Urteaga and C. A. Romero (editors), *Colección de Libros y Documentos referentes a la historia del Perú*, Series 1, Vol. 3, pp. 1-43 (1916).

1561 Informe del Licenciado Polo al Licenciado Briviesca de Muñatones sobre la perpetuidad de las encomiendas en el Perú, C. A. Romero (editor), *Revista Histórica*, Vol. 13, pp. 125-196. Lima (1940).

1571a Relación de los fundamentos acerca del notable daño que resulta de no guardar a los indios sus fueros, H. H. Urteaga and C. A. Romero (editors), *Colección de Libros y Documentos referentes a la historia del Perú*, Series 1, Vol. 3, pp. 45-203 (1916).

1571b Del Linage de los Ingas y como conquistaron, H. H. Urteaga and C. A. Romero (editors), *Colección de Libros y Documentos referentes a la historia del Perú*, Series 1, Vol. 4, pp. 45-95 (1917).

Proskouriakoff, Tatiana
1960 Historical Implications of a Pattern of Dates at Piedras Negras, Guatemala. *American Antiquity*, Vol. 25, No. 4, pp. 454-475.

Proskouriakoff, Tatiana [cont'd.]
1961 The Lords of the Maya Realm. *Expedition*, Vol. 4, No. 1, pp. 14-21.

1963 Historical Data in the Inscriptions of Yaxchilan, Part I: The Reign of Shield Jaguar. *Estudios de Cultura Maya*, Vol. 3, pp. 149-167. Universidad Nacional Autonoma de Mexico, Mexico.

1964 Historical Data in the Inscriptions of Yaxchilan, Part II: The Reigns of Bird-Jaguar and His Successors. *Estudios de Cultura Maya*, Vol. 4, pp. 177-201. Universidad Nacional Autonoma de Mexico, Mexico.

Prufer, Olaf H., and Orrin C. Shane III
1970 *The Blain Village and the Fort Ancient Tradition in Ohio*. Kent, OH: Kent State University Press.

Quine, Willard Van Orman
1953 *From a Logical Point of View*. New York: Harper & Row.

Redfield, Robert
1941 *The Folk Culture of Yucatan*. Chicago: University of Chicago Press.

Redfield, Robert, and Alfonso Villa Rojas
1934 *Chan Kom*. Carnegie Institute of Washington, Publication No. 448. Washington, D.C.

Reed, Nelson
1977 Monks and Other Mississippian Mounds. In *Explorations into Cahokia Archaeology*. Illinois Archaeological Survey Bulletin 7, pp. 31-42.

Reichel-Dolmatoff, Gerardo
1950 Los Kogi: Un Tribu Indígena de la Sierra Nevada de Santa Marta, Colombia: Vol. 1, *Revista del Instituto Etnológico Nacional*, Vol. 4, Nos. 1-2. Bogotá.

1951 *Los Kogi: Una Tribu Indígena de la Sierra Nevada de Santa Marta, Colombia*, Vol. 2. Bogotá: Editorial Iqueima.

1967 Notas sobre el simbolismo religioso de los indios de la Sierra Nevada de Santa Marta. *Razón y Fabula*, No. 1, pp. 55-72. Bototá.

1975 Templos Kogi: Introduccion al simbolismo y a la astronomía del espacio sagrado. *Revista Colombia de Antropología*, Vol. 19, pp. 201-245.

1978 The Loom of Life: A Kogi Principle of Integration. *Journal of Latin American Lore*, Vol. 4, No. 1, pp. 5-27.

Remington, J. A.
1977 Current Astronomical Practices among the Maya. In *Native American Astronomy*, Anthony F. Aveni (editor), pp. 75-88. Austin: University of Texas Press.

Renaud, E. B.
1942 Recent Work in the Upper Rio Grande Valley of Colorado and New Mexico. *University of Denver Archaeological Series*, Third Paper.

Renfrew, Colin
1973 *Before Civilization*. New York: Knopf.

Reyman, Jonathan
1971 *Mexican Influence on Southwestern Ceremonialism*. Unpublished Ph.D. dissertation, Southern Illinois University.

1975a The Nature and Nurture of Archaeoastronomical Studies. In *Archaeoastronomy in Pre-Columbian America*, Anthony F. Aveni (editor), pp. 205-215. Austin: University of Texas Press.

1975b Sun Temple, Mesa Verde National Park: A Re-evaluation. Paper presented at a symposium held at Colgate University, September 23-26.

1976 Astronomy, Architecture, and Adaptation at Pueblo Bonito. *Science*, Vol. 193, pp. 957-962.

1977 Archaeoastronomy at Pueblo Bonito. *Science*, Vol. 79, pp. 12-13.

1978 Wupatki: Rejecting False Profits. *American Antiquity*, Vol. 43, pp. 729-733.

1979 Some Observations on Archaeology and Archaeoastronomy. *Archaeoastronomy*, Vol. 2, No. 2, pp. 11-13.

Reyman, Jonathan, and J. Eddy
1975 Big Horn Medicine Wheel: Why Was It Built? *Science*, Vol. 188, pp. 278-279.

Risser, Anna
1941 Seven Zuni Folk Tales. *El Palacio*, Vol. 58, pp. 215-226. Santa Fe.

Rivard, Jean-Jacques
1970 A Hierophany at Chichen Itzá. *Katunob*, Vol. 7, No. 3, pp. 51-55.

Robbins, Louise M.
1975 The Investigation of Infanticide in an Ohio Fort Ancient Site. Paper presented at the 40th annual meeting of the Society for American Archaeology, Dallas, Texas, May.

Robelo, Cecilio A.
1951 *Diccionario de Mitología Nahuatl*, 2d ed. México, D.F.: Librería Navarro.

Robicsek, Francis
1975 *A Study in Maya Art and History: The Mat Symbol*. New York: Museum of the American Indian, Heye Foundation.

Rostworowski de Diez Canseco, María
1978 *Mediciones y Computos en el Antiguo Peru*. Cuadernos Prehispánicos. Seminário de Historia de América, Universidad de Valladolid.

Roth, Walter E.
1908/1909 An Inquiry into the Animism and Folklore of the Guiana Indians. *Bureau of American Ethnology Annual Report* 30, pp. 103-387.

Rowe, J. H.
1946 Inca Culture at the Time of the Spanish Conquest. *Handbook of the South American Indians*, Vol. I, No. 2, pp. 183-330. Washington, D.C.

1979 An Account of the Shrines of Ancient Cuzco. *Nawpa Pacha*, Vol. 17, pp. 1-80.

Roys, Ralph Loveland
1965 *Ritual of the Bacabs.* Norman: University of Oklahoma Press.

1967 *The Book of Chilam Balam of Chumayel*, 2d ed. Norman: University of Oklahoma Press.

1972 *The Indian Background of Colonial Yucatan.* Norman: University of Oklahoma Press.

Sadler, D. H.
1966 Prediction of Eclipses. *Nature*, Vol. 211, p. 1120.

Sahagún, Bernadino de
1950-1969 *Florentine Codex.* A. J. O. Anderson and C. E. Dibble (editors). Santa Fe: The School of American Research.

San Francisco Dictionary
1870 *Diccionario Maya-Español del Convento de San Francisco en Mérida*, C. H. Berendt (compiler). Mérida.

Santacruz Pachacuti Yamqui Salcamaygua, Juan de
1613 *Relación de Antigüedades de Este Reyno del Peru.* Asunción de Paraguay: Ed. Guaranía (1950).

de Santillana, G., and H. von Dechend
1969 *Hamlet's Mill.* Boston: Gambit.

Sarmiento de Gamboa, Pedro
1561 Informe del Licenciado Polo al Licenciado Briviesca de Muñatones sobre la perpetuidad de las encomiendas en el Pefũ, C. A. Romero (editor), *Revista Histórica*, Vol. 13, pp. 125-96. Lima (1940).

1571a Relación de los fundamentos acerca del notable daño que resulta de no guardar a los indios sus fueros, H. H. Urteaga and C. A. Romero (editors), *Colección de Libros y Documentos referentes a la historia del Perú*, Series 1, Vol. 3, pp. 45-203 (1916).

1571b Del Linage de los Ingas y como conquistaron, H. H. Urteaga and C. A. Romero (editors), *Colección de Libros y Documentos referentes a la historia del Perú*, Series 1, Vol. 4, pp. 45-95 (1917).

1572 *Historia de los Incas.* Buenos Aires: Biblioteca Emece (1947).

Schaafsma, Polly
1980 *Indian Rock Art of the Southwest.* Albuquerque: University of New Mexico Press.

Schele, Linda

1976 Accession Iconography of Chan-Bahlum in the Group of the Cross at Palenque. In *The Art, Iconography and Dynastic History of Palenque, Part III*, Merle Greene Robertson (editor), pp. 9-34. Pebble Beach, CA: Robert Louis Stevenson School.

1977 Palenque: The House of the Dying Sun. In *Native American Astronomy*, Anthony F. Aveni (editor). Austin: University of Texas Press.

Schmidt, E. F.

1928 Time Relations of Prehistoric Pottery Types in Southern Arizona. *Anthropological Papers*, Vol. 30, No. 5. American Museum of Natural History, New York.

Schwartz, H. F.

1925 Eclipses as Interpreted by the American Aborigines. *Natural History*, Vol. 25, pp. 162-165.

Scott, O.

1942 *The Stars in Myth and Fact*. Caxton: Caldwell.

Shane III, Orrin C., and Michael Barber

1975 Analysis of the Vertebrate Fauna from the Incinerator Site and Its Implication for Fort Ancient Settlement Patterning. Paper presented at the 40th annual meeting of the Society for American Archaeology, Dallas, Texas, May.

Silverberg, Robert

1970 *The Mound Builders*. New York: Ballentine Books.

Sims, S. C.

1903 A Wheel-Shaped Stone Monument in Wyoming. *American Anthropologist*, N.S., Vol. 5, pp. 107-110.

Smiley, Charles H.

1973 The Thix and the Fox, Mayan Solar Eclipse Intervals. *Journal of the Royal Astronomical Society of Canada*, Vol. 67, No. 4, pp. 175-182.

1975 The Solar Eclipse Warning Table in the Dresden Codex. In *Archaeoastronomy in Pre-Columbian America*, Anthony F. Aveni (editor), pp. 247-256. Austin: University of Texas Press.

Smith, Mary Elizabeth

1973 *Picture Writing from Ancient Southern Mexico: Mixtec Place Signs and Maps*. Norman: University of Oklahoma Press.

Smith, Watson

1952 Kiva Mural Decorations at Awatovi and Kawaika-a. *Papers of the Peabody Museum*, Vol. 37. Cambridge: Harvard University Press.

Snow, C. P.

1959 *Two Cultures and the Scientific Revolution*. Cambridge: Cambridge University Press.

Sofaer, Anna, Volker Zinser, and Rolf Sinclair
1979a A Unique Solar Marking Construct of the Ancient Pueblo Indians. *American Indian Rock Art, Vol. V.* El Toro: American Rock Art Research Association.

1979b A Unique Solar Marking Construct. *Science*, Vol. 206, pp. 283-291.

Solís Alcalá, Ermilio
1949 *Códice Pérez.* Mérida: Imprenta Oriente.

Somerville, Boyle
1922 Instances of Orientation in Prehistoric Monuments of the British Isles. *Archaeology*, Vol. 73, pp. 193-224.

Spier, Leslie
1923 Southern Diegueño Customs. *University of California Publications in American Archaeology and Ethnology*, Vol. 20, No. 16, pp. 297-358. Berkeley.

Spinden, Herbert J.
1977 Diffusion of Maya Astronomy. In *The Maya and Their Neighbors*, Hay, Linton, Lothrop, Shapiro, and Vaillant (editors). New York: Dover. (Republication of 1940 Appleton-Century edition.)

Stephen, Alexander M.
c.1883 Hopi Notebooks, Vol. 1, Elsie Clews Parsons Collection, Rare Book and Manuscript Library, Columbia University.

1936 *Hopi Journal of Alexander M. Stephen*, E. C. Parsons (editor), *Columbia University Contributions to Anthropology*, Vol. 23, 2 vols. New York.

Stephens, John L.
1963 *Incidents of Travel in the Yucatan.* 2 vols. New York: Dover. (Republication of 1863 edition.)

Stevenson, Matilda Cox
1904 The Zuni Indians. *Bureau of American Ethnology Annual Report* 23, pp. 13-608. Smithsonian Institution.

Stirling, M. W.
1942 The Origin Myth of Acoma and Other Records. *Bureau of American Ethnology Bulletin* 135. Smithsonian Institution.

Swadesh, Morris, Ma. Christina Alvarez, and Juan R. Bastarrachea
1970 *Diccionario de Elementos del Maya Yucateco Colonial*, 1st ed. Universidad Nacional Autónoma de México. Centro de Estudios Mayas, Cuaderno 3. México.

Swanton, John R.
1928 Sun Worship in the Southeast. *American Anthropologist*, Vol. 30, pp. 206-213.

1946 The Indians of the Southeastern United States. *Bureau of American Ethnology Bulletin* 137. Smithsonian Institution.

Talayesva, D.
1942 *Sun Chief*, Leo W. Simmons (editor). New Haven: Yale University Press.

Tedlock, Dennis
1972 *Finding the Center*. New York: The Dial Press.

1975 The Girl and the Protector: A Zuni Story. Performed by Walter Sanchez. *Alcheringa* (n.s.), Vol. 1, No. 1, pp. 110-150.

Teeple, John E.
1930 Maya Astronomy. *Carnegie Institute, Publication 403*, Contribution 2. Washington.

Thom, Alexander
1955 A Statistical Analysis of the Megalithic Sites in Britain. *Journal of the Royal Statistics Society, A*, Vol. 188, pp. 275-295.

1962 The Megalithic Unit of Length. *Journal of the Royal Statistics Society, A*, Vol. 125, pp. 243-251.

1967 *Megalithic Sites in Britain*. Oxford: Clarendon.

1968 The Metrology and Geometry of Cup-and-Ring Marks. *Systematics*, Vol. 6, pp. 173-189.

1971 *Megalithic Lunar Observations*. Oxford: Clarendon.

Thom, Alexander, and A. S. Thom
1972 The Carnac Alignments. *Journal for the History of Astronomy*, Vol. 3, pp. 11-26.

1978 A Reconsideration of the Lunar Sites in Britain. *Journal for the History of Astronomy*, Vol. 9, pp. 170-179.

Thompson, J. E. S.
1960 *Maya Hieroglyphic Writing: An Introduction*. 2d ed. Norman: University of Oklahoma Press.

1970 *Maya History and Religion*. Norman: University of Oklahoma Press.

1971 *Maya Hieroglyphic Writing*. 3d ed. Norman: University of Oklahoma Press.

1972 *A Commentary on the Dresden Codex*. Philadelphia: American Philosophical Society.

1974 Maya Astronomy. In *The Place of Astronomy in the Ancient World, Philosophical Transactions of the Royal Society*, Vol. 276, pp. 83-98, London.

Thompson, Laura
1950 *Culture in Crisis: A Study of the Hopi Indians*. New York: Harper & Brothers.

Thwaites, R. G. (editor)
1896 *The Jesuit Relations and Allied Documents*. 73 vols. Cleveland: Barrows Bros.

Titiev, Mischa
1949 Old Oraibi: A Study of the Hopi Indians of Third Mesa. *Papers of the Peabody Museum of American Archaeology and Ethnology*, Vol. 22, No. 1. Cambridge: Harvard University Press.

Tozzer, A. M.
1907 *A Comparative Study of the Mayas and Lacandones*. New York: MacMillan.

1941 Landa's Relación de las Cosas de Yucatan. *Papers of the Peabody Museum of Archaeology and Ethnology*, Vol. 18. Cambridge: Harvard University Press.

Trento, S.
1978 *The Search for Lost America*. New York: Contemporary Books.

Trotter, R. J.
1972 Tracing the Roots of Civilization. *Science News*, Vol. 101, pp. 124-126.

Turton, David, and Clive Ruggles
1978 Agreeing to Disagree: The Measurement of Duration in a Southwestern Ethiopian Community. *Current Anthropology*, Vol. 19, pp. 585-593.

Udall, Louise [and Helen Sekaquaptewa]
1969 *Me and Mine*. Tucson: University of Arizona Press.

Urton, Gary
1978a Beasts and Geometry: Some Constellations of the Peruvian Quechuas. *Anthropos*, Vol. 73, pp. 32-40.

1978b Orientation in Quechua and Incaic Astronomy. *Ethnology*, Vol. 17, No. 2, pp. 157-167.

1979 *The Astronomical System of a Community in the Peruvian Andes*. Ph.D. dissertation, University of Illinois, Urbana-Champaign.

1980a Celestial Crosses: The Cruciform in Quechua Astronomy. *Journal of Latin American Lore*, Vol. 6, No. 1.

1980b *At the Crossroads of the Earth and Sky*. Austin: University of Texas Press.

n.d. Yana Phuyu--The Quechua Dark Cloud Constellations. (Ms.)

Uspensky, J. V.
1937 *Introduction to Mathematical Probability*. New York: McGraw Hill.

Vallée, Lionel
1972 Cycle Écologique et Cycle Rituel: Le Cas d'un Village Andin. *Canadian Review of Sociology and Anthropology*, Vol. 9, No. 3, pp. 238-254.

Vienna Dictionary
1972 *Vocabulario de Mayathan*, Vienna Codex N.S. 3833. Graz: Akademische Druck- und Verlagsanstalt. Dated to ca. 1670.

1978 *Vocabulario de Mayathan*, Dorothy Andrews and Heath de Zapata (compilers). Mérida: Area Maya.

Villacorta, C. J., and C. A. Villacorta
1930 *Codices Mayas: Dresdensis, Peresianus, Tro-Cortesianus*. Reproduction and commentary. Guatemala City: Tipografia Nacional.

Villa Rojas, A.
1945 The Maya of East Central Quintana Roo. *Carnegie Institution of Washington, Publication 559*. Washington, D.C.

1973 The Concepts of Space and Time among the Contemporary Maya. In *Time and Reality in the Thought of the Maya*, M. Leon-Portilla (editor), pp. 113-171. Boston: Beacon Press.

Vivian, G., and P. Reiter
1960 The Great Kivas of Chaco Canyon and Their Relationships. *School of American Research and the Museum of New Mexico, Monograph 22*. Santa Fe.

Wagley, Charles
1940 World View of the Tapirape Indians. *Journal of American Folklore*, Vol. 53, pp. 252-260.

Wagner, Gail
1977 *The Dayton Museum of Natural History Flotation Procedure Manual*. Dayton, OH.

Wainwright, Geoffrey J.
1970 Woodhenges. *Scientific American*, Vol. 223, pp. 30-38.

Wallace, Anthony F. C.
1970 *Culture and Personality*, 2d ed. New York: Random House.

Waters, Frank [and Oswald White Bear Fredericks]
1969 *Book of the Hopi*. New York: Ballantine Books.

Wedel, Waldo R.
1977 Native Astronomy and the Plains Caddoans. In *Native American Astronomy*, Anthony Aveni (editor), pp. 131-145. Austin: University of Texas Press.

Weigand, Phil C., Gorman Harbottle, and Edward V. Sayre
1977 Turquoise Sources and Source Analysis: Mesoamerica and the Southwestern U.S. In *Exchange Systems in Prehistory*, T. K. Earle and J. E. Ericson (editors). New York: Academic Press.

Wilbert, J.
1975 Eschatology in a Participatory Universe: Destinies of the Soul among the Warao Indians of Venezuela. In *Death and the Afterlife in Pre-Columbian America*, E. Benson (editor), pp. 163-189. Washington, D.C.: Dumbarton Oaks.

Wilcox, D. R., and L. O. Shenk
1977 The Architecture of the Casa Grande and Its Interpretation. *Arizona State Museum Archaeological Series*, No. 115. Tucson: Arizona State Museum.

Williamson, Ray A.
1977 Archaeoastronomy at Pueblo Bonito. *Science*, Vol. 79, p. 12.

1978 Sir Norman Lockyer Archaeoastronomy Symposium (proceedings). *Archaeoastronomy*, the *Bulletin of the Center for Archaeoastronomy*, Vol. 1, No. 3, pp. 4-12.

1979 Hovenweep National Monument--Field Report. *Archaeoastronomy*, the *Bulletin of the Center for Archaeoastronomy*, Vol. 2, No. 3, pp. 11-12.

Williamson, Ray A., H. J. Fisher, and D. O'Flynn
1977 Anasazi Solar Observatories. In *Native American Astronomy*, Anthony Aveni (editor), pp. 203-217. Austin: University of Texas Press.

Williamson, Ray A., H. J. Fisher, A. F. Williamson, and C. Cochran
1975 The Astronomical Record in Chaco Canyon, New Mexico. In *Archaeoastronomy in Pre-Columbian America*, Anthony Aveni (editor), pp. 33-43. Austin: University of Texas Press.

Williamson, Ray A., and M. J. Young
1979 An Equinox Sun Petroglyph Panel at Hovenweep National Monument. *American Indian Rock Art*, Vol. V. El Toro: American Rock Art Research Association.

Willson, Robert
1924 Astronomical Notes on the Maya Codices. *Papers of the Peabody Museum of American Archaeology and Ethnology*, Vol. 6, No. 3. Cambridge.

Wilson, Gilbert
1917 Agriculture of the Hidatsa Indians--An Indian Interpretation. *Studies in the Social Sciences*, No. 9. Minneapolis.

Wimberly, Mark, and Peter Eidenbach
1977 *Inventory of the Cultural Resources: Sierra Blanca Ski Area Land Transfer (BIA Contract No. MOOC14202478)*. Tularosa, NM: Human Systems Research.

Winkler, L.
1972 Astronomically Determined Dates and Alignments. *American Journal of Physics*, Vol. 40, No. 1.

Winship, George Parker
1892 The Coronado Expedition, 1540-1542. *Bureau of American Ethnology Annual Report* 14, pp. 339-613.

Winslow Mail
1938 *Winslow* [Arizona] *Mail*, Friday, August 19, p. 1, col. 7.

Wisdom, C.
1940 *The Chorti Indians of Guatemala*. Chicago: University of Chicago Press.

Witthoft, John

1949 Green Corn Ceremonialism in the Eastern Woodlands, *Occasional Contributions from the Museum of Anthropology of the University of Michigan*, No. 13. Ann Arbor: University of Michigan Press.

Wittry, Warren

1961 Report of Phase 3 Archaeological Salvage Project FAI 255 Section 60-6-1, Tract 15-A Project I-70-1 (40)4. The Cahokia Site 3000 Feet West of Monks Mound, Madison County, Illinois. *Preliminary Reports Series*, No. 2. Illinois State Museum.

1964 An American Woodhenge. *Cranbrook Institute of Science Newsletter*, Vol. 33, No. 9, pp. 102-107.

1973 An American Woodhenge. In *Explorations into Cahokia Archaeology*, Bulletin 7. Illinois Archaeological Survey.

Wood, J.

1978 *Sun, Moon and Standing Stones*. London: Oxford.

Young, M. Jane

1979a In Search of Archaeoastronomy: What Is It and Where Is It Going? *Archaeoastronomy, the Bulletin of the Center for Archaeoastronomy*, Vol. 2, No. 2, pp. 14-15.

1979b Seminar on the Role of Archaeoastronomy. *Archaeoastronomy, the Bulletin of the Center for Archaeoastronomy*, Vol. 2, No. 3, pp. 5-6.

Zuidema, R. Tom

1964 *The Ceque System of Cuzco: The Social Organization of the Capital of the Inca*. Leiden: E. J. Brill.

1977a The Inca Calendar. In *Native American Astronomy*, Anthony Aveni (editor), pp. 219-259. Austin: University of Texas Press.

1977b Mito y Historia en el Antiguo Peru. *Allpanchis Phuturinqa*, Vol. 10, pp. 15-52.

1978a Lieux sacres et irrigation Tradition historique, mythes et rituals au Cuzco. *Les Annales*. Paris.

1978b Shafttombs and the Inca Empire. *Journal of the Steward Anthropological Society*, Vol. 9, Nos. 1 and 2 (Fall 1977), (Spring 1978).

1980 El Ushnu. *Revista de la Universidad Complutense*, Editorial de la Universidad Complutense de Madrid, Vol. 28, No. 117, pp. 317-362.

n.d.a El Sunturhuasi. (Ms.)

n.d.b Llamas and the Inca Calendar. (Ms.)

Zuidema, R. Tom, and G. Urton

1976 La Constellation de la Llama en los Andes Peruana. *Allpanchis Phuturinga*, Vol. 9, pp. 59-119.

INDEX

CONTRIBUTORS

NEV E. AUTREY attended the American Institute of Technology and Pima Community College. He has a diversified background in surveying, data processing, and archaeological fieldwork, and is presently employed by Tucson Electric Power as a Computer Programmer.

WANDA R. AUTREY attended Pima Community College. Her background is in archaeological fieldwork and cultural anthropology; she is currently employed by Science Applications, Inc. as a Technical Illustrator.

ANTHONY F. AVENI, Professor of Astronomy and Anthropology at Colgate University, organized the first two major conferences on American archaeoastronomy at Mexico City and Colgate University. His latest book, *Skywatchers of Ancient Mexico*, is published by University of Texas Press.

JOHN CARLSON, an astronomer, is Director of the Center for Archaeoastronomy at the University of Maryland and Editor of *Archaeoastronomy*. His research interests include the astronomical methods of the ancient Mesoamericans.

BETH A. COLLEA studied at Colgate University and the State University of New York at Binghamton. She is currently pursuing graduate studies in archaeology at Harvard University.

ANN DANIEL-HARTUNG is a former engineer who worked at NASA on the Apollo Project. A computer specialist for government and industry as well as an archaeoastronomer who specializes in prehistoric southeast cultures, she now works for Sperry Corporation in Salt Lake City.

BYRON E. DIX is an active member of Associated Scientists at Woods Hole and has been conducting archaeoastronomical research in central Vermont for the past seven years. He is on the technical staff of the Air Force Geophysics Laboratory in Cambridge, Massachusetts.

JOHN A. EDDY, an astronomer and Senior Scientist at the High Altitude Observatory, Boulder, Colorado, has published a number of articles on the physical history of the sun and solar-terrestrial relationships. His interests in archaeoastronomy have focused primarily on the possible astronomical meaning of the Medicine Wheels of the United States and Canada.

JOHN H. EVANS, an astronomer in the Department of Physics and Astronomy of the University of Wisconsin at Oshkosh, has also worked at NASA, Lockheed Aircraft, and Los Alamos Scientific Laboratory.

CLAIRE R. FARRER is Assistant Professor of Anthropology at the University of Illinois, Urbana. Her research and writing interests focus on expressive culture, especially that of the Mescalero Apache, with whom she lived in 1974-75 and with whom she maintains contact.

HORST HARTUNG, a practicing architect and urban planner, is Professor of City Planning at the University of Guadalajara. He has been involved in archaeoastronomical studies since 1970 and has published numerous articles on Mesoamerican archaeoastronomy.

KEN HEDGES is Curator of Ethnology and Archaeology at the San Diego Museum of Man, San Diego, California. His current research interests are southern California archaeology and ethnography, rock art in western North America, shamanism and rock art, and archaeoastronomy.

J. M. HEILMAN is Curator of Anthropology at the Dayton Museum of Natural History. His major research has focused on the excavation and analysis of the Incinerator Site, a twelfth-century Fort Ancient village which is scheduled for reconstruction.

RONALD HICKS, Associate Professor of Anthropology and Director of the Archaeological Resources Management Service at Ball State University, has done extensive field research on the folklore and archaeoastronomy of Irish henges. He has recently studied the Indiana mounds.

HARRY HILLMAN is a civil engineer and professional land surveyor who has owned his own engineering firm and taught at Central Arizona College. He is now Assistant Professor at the University of Arizona and a private engineering consultant.

ROGER R. HOEFER was Curator of Astronomy and Director of the planetarium at the Dayton Museum of Natural History. He now manages the family business in Oklahoma.

TRAVIS HUDSON is Curator of Anthropology at the Santa Barbara Museum of Natural History. He is the author or co-author of several books on the Chumash Indians of Southern California, including *Crystals in the Sky*, a study of Chumash astronomy (Ballena Press).

E. C. KRUPP, an astronomer and Director of Griffith Observatory, has edited and co-authored *In Search of Ancient Astronomies*. He has visited and photographed more than 300 ancient and prehistoric sites throughout the world.

WELDON LAMB is a graduate student in anthropology at Tulane University in New Orleans. His publications include studies of Mayan hieroglyphics and Mayan astronomy.

VINCENT H. MALMSTRÖM is a specialist in cultural and historical geography and a Professor at Dartmouth College in Hanover, New Hampshire. His interest in archaeoastronomy grew out of fieldwork he conducted into the origins of early civilizations in the Near East, Europe, and Latin America.

JAMES W. MAVOR, JR. is an engineer who has served on the faculties of MIT, Woods Hole Oceanographic Institution, Northeastern University, and the U.S.

Naval Academy. He was responsible for the design and construction of the manned, deep submersible ALVIN, and he also initiated the Minoan excavations on Thera, Greece. He has practiced archaeoastronomy for the past eleven years in Europe, North Africa, and New England, and is presently a founder-director of Associated Scientists at Woods Hole, Inc.

STEPHEN C. McCLUSKEY, an historian of science at West Virginia University, has done extensive fieldwork among the Hopi Indians of Arizona, and has made major contributions to the understanding of their traditional solar-lunar calendar.

SUSAN MILBRATH, an art historian and archaeologist, has taught at the University of Florida in Gainesville, and was a Tinker Fellow at Yale (1979-1980). She is now a Research Associate and guest curator at the Museum of the American Indian, preparing a major traveling exhibition and catalogue of Pre-Columbian astronomical art.

JUDITH REMINGTON is an anthropologist who has worked among both the Maya of Guatemala and the Navajo. She is currently teaching at Navajo Community College, Shiprock, New Mexico. She has a strong interest in Maya calendrics.

BERNARD SECOND is a Mescalero Apache Singer of the girls' puberty ceremony. He is accomplished in traditional plastic and verbal arts. When not pursuing his primary vocation, he works for the Tribe's programs in recreation and leisure.

T. SCOTT SMITH is Associate Professor of Physics and Astronomy, Dickinson College. An astronomer by training, he is interested in science/society interactions of the past (megalithic cultures), present (environmental/energy crisis), and future (science fiction).

GARY URTON is Assistant Professor of Anthropology at Colgate University. His book, *At the Crossroads of the Earth and the Sky: An Andean Cosmology* (University of Texas Press) is now in press.

M. JANE YOUNG, a folklorist and linguist at the University of Pennsylvania, has authored several articles on Southwestern ethno- and archaeoastronomy. Her current research interests are Pueblo rock art and its relationship to Pueblo society. She has worked extensively at Zuni Pueblo.

RAY A. WILLIAMSON is a Project Director and science policy analyst in energy and space for the Congressional Office of Technology Assessment (OTA). Prior to that he was Assistant Dean and Tutor at St. John's College, Annapolis, Maryland. His research interests in archaeoastronomy include the historic and prehistoric astronomical practices of the Pueblo Indians. His photographs of calendrical phenomena at archaeoastronomical sites have been published in a number of articles and books.

R. THOMAS ZUIDEMA is Professor of Anthropology at the University of Illinois, Champaign-Urbana. He has published a number of articles on the Incas and their descendents and comes to ethno- and archaeoastronomy by way of a deep interest in the Inca calendar and its effects on social organization.